AF581648

Potential Health Benefits of Fruits and Vegetables

Potential Health Benefits of Fruits and Vegetables

Editors

Luca Mazzoni
María Teresa Ariza Fernández
Franco Capocasa

MDPI • Basel • Beijing • Wuhan • Barcelona • Belgrade • Manchester • Tokyo • Cluj • Tianjin

Editors

Luca Mazzoni
Agricultural, Food and Environmental Sciences
Università Politecnica delle Marche
Ancona
Italy

Maria Teresa Ariza Fernández
Breeding and Biotechnology
Instituto Andaluz de Investigación y Formación Agraria y Pesquera (IFAPA)
Malaga
Spain

Franco Capocasa
Agricultural, Food and Environmental Sciences
Università Politecnica delle Marche
Ancona
Italy

Editorial Office
MDPI
St. Alban-Anlage 66
4052 Basel, Switzerland

This is a reprint of articles from the Special Issue published online in the open access journal *Applied Sciences* (ISSN 2076-3417) (available at: www.mdpi.com/journal/applsci/special_issues/Benefits_Fruits_Vegetables).

For citation purposes, cite each article independently as indicated on the article page online and as indicated below:

LastName, A.A.; LastName, B.B.; LastName, C.C. Article Title. *Journal Name* **Year**, *Volume Number*, Page Range.

ISBN 978-3-0365-2421-4 (Hbk)
ISBN 978-3-0365-2420-7 (PDF)

Contents

MDPI

Editorial

Potential Health Benefits of Fruits and Vegetables

Luca Mazzoni [1,*], Maria Teresa Ariza Fernández [2,*] and Franco Capocasa [1,*]

1 Agricultural, Food and Environmental Sciences, Università Politecnica delle Marche, Via Brecce Bianche 10, 60131 Ancona, Italy
2 Breeding and Biotechnology, Instituto Andaluz de Investigación y Formación Agraria y Pesquera (IFAPA), IFAPA-Centro de Churriana, Cortijo de la Cruz, s/n, 29140 Málaga, Spain
* Correspondence: l.mazzoni@staff.univpm.it (L.M.); mariat.ariza@juntadeandalucia.es (M.T.A.F.); f.capocasa@staff.univpm.it (F.C.)

In recent decades, the consciousness of consumers regarding the importance of a balanced diet to prevent the occurrence of chronic diseases has significantly increased. In particular, the consumption of plant-based foods, both vegetables and fruits, have been demonstrated to have a central role in the prevention of many chronic diseases, due to the high amount of bioactive compounds they contain. To date, many researchers and scientists in different fields of research have contributed with great efforts to the characterization of the phytochemical pattern of hundreds of fruits and vegetables, and have elucidated several mechanisms of actions and metabolic pathways through which fruits and vegetables exert their health-promoting and/or disease-preventing activities.

The aim of this special issue was to compile the most recent research on fruit and vegetable phytochemical composition, on the health-promoting effects and mechanisms of action of their application/assumptions in different models, such as in vitro cellular models, in vivo animal trials and in vivo human trials.

The first step of the evaluation of the potential health benefits of fruit and vegetables is the evaluation of the phytochemical composition. In this special issue, six studies were focused on the evaluation of the phytochemical composition and the antioxidant capacity of different fruit/vegetables species.

Li et al. characterized the phenolic composition of five different varieties of apples grown in Australia ('Royal Gala', 'Pink Lady', 'Red Delicious', 'Fuji' and 'Smitten') through liquid chromatography. The results underlined that different genotypes showed different amounts of total phenolic and total flavonoid content. Furthermore, a total of 97 different phenolic compounds were detected in the five apple varieties, highlighting the interest of Australian apple varieties as a rich source of bioactive compounds [1].

Among fruits, berries are raising interest for their proven high nutritional quality, due to the high amount of several phytochemicals. Two berries were analyzed in this special issue: in the first study, Kruger et al. evaluated the effect of cultivar, environmental variations and their interaction on anthocyanin composition of six strawberry cultivars grown in five locations from the North to South of Europe, for two different years. As a general trend, fruits grown in southern locations were richer in total anthocyanins and pelargonidin-3-glucoside content. Principal component analysis revealed that anthocyanin content of cultivars is influenced by environmental factors; in particular, the minor anthocyanins (cyanidin-3-glucoside, cyanidin-3-(6-*O*-malonyl)-glucoside, pelargonidin-3-rutinoside and pelargonidin-3-(6-*O*-malonoyl)-glucoside) were sensitive to the maximum temperature value. However, different cultivars changed their anthocyanin pattern in relation to the environmental conditions to varying extent, with 'Gariguette' and 'Clery' cultivars remaining unaffected [2].

The second study on a berry species involved the phytochemical characterization of 30 haskap berry genotypes. In this manuscript, different spectrophotometric and spectrofluorimetric methods were used to evaluate the antioxidant capacity and the amount of

Citation: Mazzoni, L.; Ariza Fernández, M.T.; Capocasa, F. Potential Health Benefits of Fruits and Vegetables. *Appl. Sci.* **2021**, *11*, 8951. https://doi.org/10.3390/app11198951

Received: 3 September 2021
Accepted: 13 September 2021
Published: 26 September 2021

different phenolic compounds and vitamin C. This study allowed for the identification of those genotypes, which were more interesting for a high content of the studied phytochemicals, and that could be recommended to consumers for a healthy diet. Furthermore, the heritability and genetic process analyses allowed for the indication of the effectiveness of breeding for the transmission of the analyzed traits to the progeny, suggesting the most suitable genotypes for the implementation of future breeding programs aimed at obtaining healthier fruits [3].

Together with fruits, fresh vegetables are also interesting for their content of specific bioactive compounds. In particular, vegetables belonging to the *Brassicaceae* family have attracted increasing attention in recent years for the quantity and quality of their bioactive compounds, and this attention has also been confirmed in this special issue, where two studies (one review and one research) were published on this argument.

The research study presented an investigation on the amount of nine inorganic elements (Cd, Co, Cr, Cu, Fe, Ni, Mn, Pb and Zn) in genotypes from three different species belonging to the *Brassicaceae* family (*Brassica rapa*, *Eruca vesicaria* and *Sinapis alba*), grown according to both conventional and organic cultivation techniques, during two agricultural seasons on two different experimental farms. The results underlined that the inorganic elements amount is influenced mainly by many factors other than the cultivation technique, comprising the soil characteristics. The organic cultivation technique did not decrease the heavy metal content or increase the nutritional quality of *Brassicaceae*, as commonly believed. As a final result, it was predicted that the consumption of 150–200 g of these vegetables, both from organic and conventional agriculture, fulfill the same percentage of Dietary Reference Intakes for Co, Cr, Cu, Fe, Mn and Zn. Regarding the heavy metal (Cd, Ni and Pb) tolerable intakes, only slight differences (mainly for Pb) have been found between both cropping systems [4].

The second manuscript regarding the quality of *Brassicaceae* is a review article, which aimed to highlight the main phytochemical compounds present in brassicas used as a food vegetable that confer nutritional and sensorial quality to the final product, and to investigate the main factors that affect the phytochemical concentration and the overall quality of Brassica vegetables. In summary, the bioactive molecules responsible for the nutritional quality of *Brassicaceae* can be divided in antioxidant compounds (e.g., phenols, vitamin C) and non-antioxidant compounds (e.g., minerals, glucosinolates). The amount of these compounds in Brassica vegetables could be influenced by many factors, including the genetic source, the environmental conditions and the cultivation techniques adopted for the vegetable production [5].

As mentioned above, the nutritional quality of fresh fruits and vegetables depends on many factors, which could affect the final quality of the products. However, if the product is not stored properly after its harvest, the loss or degradation of phytochemical compounds is a tangible possibility. In this regard, a study in this special issue evaluated different options for prolonging the apricot fruit quality during cold storage and shelf life, decreasing the postharvest losses of apricots. The quality parameters (quality losses, antioxidant properties and enzyme activities) were evaluated at different time periods (from 7 to 21 days) at cold storage (1 °C) and shelf life (25 °C), comparing post-harvest treatments with methyl jasmonate and salicylic acid. As a general trend, both post-harvest treatments significantly decreased the quality loss of chilling injury and fruit decay on all dates. The antioxidant capacity and the phenolic patterns increased for both treatments at all dates, and almost all the antioxidant enzyme activities increased significantly on all dates for both treatments (except catalase activities, which decreased with the methyl jasmonate treatment). In conclusion, both methyl jasmonate and salicylic acid are useful and inexpensive techniques to maintain the apricot fruit quality in both cold storage and shelf life conditions [6].

The second step for the evaluation of the health potential of fruit vegetables is testing the product in an in vitro model. Usually, in this step the capacity of the fruit/vegetable extract is evaluated to limit the viability of pathological cells, or to protect the healthy cells from an induced external stress. In this special issue, a study evaluating the effect of olive (*Olea europaea* L.) vegetation water on human cells regarding its antioxidant properties and radical scavenger bioactivities was published. The study involved the treatment of two cell lines, human hepatocellular carcinoma and human keratinocytes, with two food supplements containing concentrated olive water in combination with 6% lemon juice or 70% grape juice, respectively. The first analysis of the extracts revealed that hydroxytyrosol was the most abundant polyphenol in both formulations, followed by tyrosol and oleuropein (for the olive-derived concentrate with lemon juice), and by proanthocyanidins and tyrosol (for the olive concentrate with grape juice). Both extracts were demonstrated to be effective antioxidants, also preventing the advanced glycation end product formation. In addition, preliminary data indicate that the administration of hydroxytyrosol through these hydrophilic matrices is better absorbed into the human body [7].

After the demonstration of the beneficial effects of fruit and vegetable consumption in in vitro models, the following step is the evaluation of the positive effects also in in vivo animal models. A trial on morbidity and mortality in the context of sepsis and septic shock on male Sprague Dawley® rats was presented in this special issue. In this study, a thiosulfinate-enriched *Allium sativum* extract was used as adjuvant in the management of sepsis induced by intraperitoneal *Escherichia coli* ATCC 25922 inoculation. To evaluate the efficacy in the sepsis-induced management, clinical, analytical, microbiological and histopathological parameters were evaluated in the control group, in the group treated with antibiotic, and in the group treated with antibiotic plus *Allium sativum* extract. The results confirmed that the utilization of *Allium sativum* extract as an adjuvant to antibiotic treatment in the management of sepsis could improve the sepsis attenuation, ameliorating clinical parameters of rats as weight, ocular secretions, whiskers separation and physical activity level, inhibiting *Escherichia coli* proliferation and thus, reducing overall mortality after an animal peritonitis model [8].

The final step for the valorization of the potential health benefits of fruit and vegetable consumption is the introgression of these products in the human diet. With this aim, two studies were published in this special issue. The first study involved a particular group of people affected by prediabetes mellitus, whose glucose levels did not meet the criteria for diabetes but were higher than those considered normal. These people were fed with two servings per day of *Gynura bicolor*, a red purple-colored vegetable, and the effect on glycemic control and antioxidant ability was evaluated. People were divided into control group and *Gynura bicolor*-fed group, and data on anthropometry and biochemical analysis were collected at 0, 8 and 12 weeks. The results clearly showed that *Gynura bicolor* consumption improved both the glycemic control and the antioxidant activity, mainly because of its high content of polyphenols [9].

The second interventional study was focused on the vulnerable category of elderly people, in particular, regarding the respiratory tract infections. The objective of this study was to evaluate the efficacy of the consumption of a combination of elderberry and reishi extracts on the incidence, severity and duration of respiratory tract infections in a group of healthy elderly volunteers. A group of 60 nursing home residents $\geq$65 years of age randomly received a combination of 1.5 g of elderberry + 0.5 g of reishi or a placebo daily for 14 weeks. If the incidence of respiratory infections was similar in both groups, the berry-fed group presented a significant reduction of common cold event duration and of high severity influenza-like illness events. Moreover, the sleep disturbances were significantly reduced in the berry-fed group. Thus, the suitability of the elderberry + reishi extract in reducing the respiratory tract infections was confirmed [10].

To summarize, in this special issue we have published several works demonstrating that fruit and vegetables contain several bioactive compounds, which give high potentiality to these foods in the prevention of many chronic human diseases. Furthermore, we have published some studies showing that fruit and vegetables demonstrated their positive activities both in in vitro and in vivo models. The last study of the special issue that we want to present is the correct conclusion of this editorial because it translates all the previous suggested findings for a healthy life; in fact, the broad recognition of the positive effects of the Mediterranean Diet, the dietary patterns that were followed in specific regions of the area in the 1950s and 1960s on the longevity of Mediterranean populations led to the adoption of this diet in other regions of the world. This study reviewed the scientific knowledge regarding the beneficial health effects of adherence to this diet, underlying that it is not only linked to the consumption of specific food products but also to social, religious, environmental and cultural aspects. Therefore, the Mediterranean Diet represents a healthy lifestyle in general that can allow to optimize the positive effect of fruit and vegetable consumption [11].

Funding: This research received no external funding.

Conflicts of Interest: The authors declare no conflict of interest.

References

1. Li, H.; Subbiah, V.; Barrow, C.J.; Dunshea, F.R.; Suleria, H.A. Phenolic Profiling of Five Different Australian Grown Apples. *Appl. Sci.* **2021**, *11*, 2421. [CrossRef]
2. Krüger, E.; Will, F.; Kumar, K.; Celejewska, K.; Chartier, P.; Masny, A.; Mott, D.; Petit, A.; Savini, G.; Sønsteby, A. Influence of PostFlowering Climate Conditions on Anthocyanin Profile of Strawberry Cultivars Grown from North to South Europe. *Appl. Sci.* **2021**, *11*, 1326. [CrossRef]
3. Gawroński, J.; Żebrowska, J.; Pabich, M.; Jackowska, I.; Kowalczyk, K.; Dyduch-Siemińska, M. Phytochemical Characterization of Blue Honeysuckle in Relation to the Genotypic Diversity of *Lonicera* sp. *Appl. Sci.* **2020**, *10*, 6545. [CrossRef]
4. Cámara-Martos, F.; Sevillano-Morales, J.; Rubio-Pedraza, L.; Bonilla-Herrera, J.; de Haro-Bailón, A. Comparative Effects of Organic and Conventional Cropping Systems on Trace Elements Contents in Vegetable *Brassicaceae*: Risk Assessment. *Appl. Sci.* **2021**, *11*, 707. [CrossRef]
5. Biondi, F.; Balducci, F.; Capocasa, F.; Visciglio, M.; Mei, E.; Vagnoni, M.; Mezzetti, B.; Mazzoni, L. Environmental Conditions and Agronomical Factors Influencing the Levels of Phytochemicals in Brassica Vegetables Responsible for Nutritional and Sensorial Properties. *Appl. Sci.* **2021**, *11*, 1927. [CrossRef]
6. Ezzat, A.; Hegedűs, A.; Szabó, S.; Ammar, A.; Szabó, Z.; Nyéki, J.; Molnár, B.; Holb, I.J. Temporal Changes and Correlations between Quality Loss Parameters, Antioxidant Properties and Enzyme Activities in Apricot Fruit Treated with Methyl Jasmonate and Salicylic Acid during Cold Storage and Shelf-Life. *Appl. Sci.* **2020**, *10*, 8071. [CrossRef]
7. Bender, C.; Straßmann, S.; Heidrich, P. Cellular Antioxidant Effects and Bioavailability of Food Supplements Rich in Hydroxytyrosol. *Appl. Sci.* **2021**, *11*, 4763. [CrossRef]
8. Redondo-Calvo, F.J.; Montenegro, O.; Padilla-Valverde, D.; Villarejo, P.; Baladrón, V.; Bejarano-Ramírez, N.; Galán, R.; Gómez, L.A.; Villasanti, N.; Illescas, S.; et al. Thiosulfinate-Enriched Allium sativum Extract as an Adjunct to Antibiotic Treatment of Sepsis in a Rat Peritonitis Model. *Appl. Sci.* **2021**, *11*, 4760. [CrossRef]
9. Hsia, C.-H.; Tung, Y.-T.; Yeh, Y.-S.; Chien, Y.-W. Effects of Gynura bicolor on Glycemic Control and Antioxidant Ability in Prediabetes. *Appl. Sci.* **2021**, *11*, 5066. [CrossRef]
10. Gracián-Alcaide, C.; Maldonado-Lobón, J.; Ortiz-Tikkakoski, E.; Gómez-Vilchez, A.; Fonollá, J.; López-Larramendi, J.; Olivares, M.; Blanco-Rojo, R. Effects of a Combination of Elderberry and Reishi Extracts on the Duration and Severity of Respiratory Tract Infections in Elderly Subjects: A Randomized Controlled Trial. *Appl. Sci.* **2020**, *10*, 8259. [CrossRef]
11. Chatzopoulou, E.; Carocho, M.; Di Gioia, F.; Petropoulos, S. The Beneficial Health Effects of Vegetables and Wild Edible Greens: The Case of the Mediterranean Diet and Its Sustainability. *Appl. Sci.* **2020**, *10*, 9144. [CrossRef]

Article

Effects of *Gynura bicolor* on Glycemic Control and Antioxidant Ability in Prediabetes

Chu-Hsuan Hsia [1], Yu-Tang Tung [2], Yu-Sheng Yeh [3] and Yi-Wen Chien [1,4,5,6,*]

1. School of Nutrition and Health Sciences, Taipei Medical University, Taipei 11031, Taiwan; da07108003@tmu.edu.tw
2. Graduate Institute of Biotechnology, National Chung Hsing University, Taichung 402, Taiwan; peggytung@nchu.edu.tw
3. Cardiovascular Division, Department of Medicine, Washington University School of Medicine, St. Louis, MO 63112, USA; yu-sheng@wustl.edu
4. Graduate Institute of Metabolism and Obesity Sciences, Taipei Medical University, Taipei 11031, Taiwan
5. Nutrition Research Center, Taipei Medical University Hospital, Taipei 11031, Taiwan
6. Research Center of Geriatric Nutrition, College of Nutrition, Taipei Medical University, Taipei 11031, Taiwan

* Correspondence: ychien@tmu.edu.tw; Tel.: +886-2-2736-1661 (ext. 6556); Fax: +886-2-2737-3112

Abstract: There exists an intermediate group of individuals whose glucose levels do not meet the criteria for diabetes yet are higher than those considered normal (prediabetes mellitus (preDM)). Those people have a higher risk of developing diabetes in the future. *Gynura bicolor* (GB) is a red-purple-colored vegetable, which is common in Taiwan. GB has shown antioxidant, anti-inflammatory and anti-hyperglycemic effects in previous studies. The aim of this study was to assess the effects of serving two serving sizes of GB every day on the glycemic control and antioxidant ability of preDM subjects. According to the age and anthropometry data of the participates, we assigned them into a control or GB group for the 8-week intervention and 4-week washout period. Data of anthropometry and biochemical analysis were collected at 0, 8 and 12 weeks. Oral glucose tolerance tests were performed, and we collected dietary records on the baseline and Week 8. Both groups received nutrition education and a diet plan individually. After intervention, the fasting glucose and malondialdehyde (MDA) values were significantly decreased in the GB group. HOMA-IR and QUICKI values were improved, and antioxidant activity was increased in the GB group. GB could improve glycemic control and decrease oxidative stress because of its large amounts of polyphenols.

Keywords: *Gynura bicolor*; prediabetes; phytochemical; blood glucose; oxidative stress

Citation: Hsia, C.-H.; Tung, Y.-T.; Yeh, Y.-S.; Chien, Y.-W. Effects of *Gynura bicolor* on Glycemic Control and Antioxidant Ability in Prediabetes. *Appl. Sci.* **2021**, *11*, 5066. https://doi.org/10.3390/app11115066

Academic Editors: Luca Mazzoni, Maria Teresa Ariza Fernández and Franco Capocasa

Received: 21 April 2021
Accepted: 27 May 2021
Published: 30 May 2021

1. Introduction

According to The American Diabetes Association (ADA), people whose fasting plasma glucose (FPG) level is 100 to 125 mg/dL were defined as impaired fasting glucose (IFG), or whose 2 h value in the oral glucose tolerance test (OGTT) is 140 to 199 mg/dL were defined as impaired glucose tolerance (IGT), or individuals with glycated hemoglobin (HbA1c) of 5.7–6.4%. People who meet one of three diagnosis criteria were called "prediabetes (preDM)" [1]. Individuals with both IFG and IGT have more severe dysglycemic condition and are especially at high-risk for type 2 diabetes [2]. According to the International Diabetes Federation (IDF), in 2019, there were 463 million people with preDM, and this number is expected to reach 700 million by 2045 [3]. In total, 5–10% of individuals with preDM develop diabetes annually, with up to 70% eventually developing diabetes. Several trials have reported on the risk of diabetes development in preDM individuals after lifestyle and drug-based interventions. PreDM can convert back to normoglycaemia [4,5].

Under hyperglycemia or IR condition, oxidative stress will be increased and endothelial dysfunction will occur [6]. The endothelium is a key regulator of vascular function [7], when dysfunction will cause inflammation, vasoconstriction, thrombosis and platelet activation, which are linked with atherosclerosis [8]. Evidence from prospective studies suggest that cardiovascular disease (CVD) may be associated with preDM [9]. Lifestyle intervention includes diet modification, weight reduction or moderate physical activity. Some studies reported that intensive lifestyle interventions such as achieving and maintaining ideal body weight may prevent the progression to type 2 diabetes in IGT or IFG subjects [10]. In the Da Qing trial in China, 577 individuals with IGT were randomized to dietary counseling, increased exercise, diet plus exercise or control (general recommendations). The cumulative 6-year incidence of diabetes was significantly lower in the diet group (43.8%), the exercise group (41.1%) and diet plus exercise group (46.0%) than the control group (67.7%). In a proportional hazards analysis adjusted for differences in baseline BMI and fasting glucose, the diet, exercise, and diet-plus-exercise interventions were associated with 31% ($p < 0.03$), 46% ($p < 0.0005$), and 42% ($p < 0.005$) reductions in risk of developing diabetes, respectively [11]. In a meta-analysis, it was reported that consuming large amounts of vegetables and fruits may reduce the risk of diabetes and associate to inflammation and oxidative stress [12]. Giugliano reported that diets containing vegetables and fruits, n-3 fatty acids and fiber attenuate the inflammation [13]. Phytochemicals such as vitamin C, vitamin E, carotenoids, phytosterols, anthocyanins and alkaloids showed anti-inflammation and antioxidant effects [14].

Anthocyanins belong to the widespread class of phenolic compounds collectively named flavonoids. Mechanic studies support the beneficial effects of flavonoids, including anthocyanins, on the biomarkers of CVD risk such as NO, inflammation and endothelial dysfunction. The role of anthocyanins in CVD prevention is strongly related to against oxidative damage [15]. Red-purple vegetables are rich in anthocyanins, the common vegetables in Taiwan such as purple-leaved sweet potato, *Gynura bicolor* (GB) and purple-leaved celosia. GB is widely distributed in Asia and common cuisine in Taiwan. The leaves of GB distinctively show a reddish purple color on the abxial side and a green color on the adaxial side. *Gynura* is used as traditional Chinese medicine for treatment of inflammation, fever, hypertension and diabetes [16,17]. GB has been shown to have antioxidant, anti-inflammatory and anti-hyperglycemic effects [18–20]. However, limited clinical human studies have examined the effects on glycemic control and antioxidant ability in preDM subjects.

The aim of this study was to assess the effects of serving two exchanges of GB per day on glycemic control and antioxidant ability on preDM subjects.

2. Materials and Methods

2.1. GB Extract Preparation

Fresh GB was obtained from Taiwan Seed Improvement and Propagation Station, COA, Taiwan. Leaves of GB were removed, washed, dried and ground to fine powder. The extract was produced from 100 mg of the lyophilized vegetable powder with a 10-fold volume of 70% methanol (0.1% of HCl) in the shaker for 30 min at room temperature, then centrifuged at 1400× *g* for 10 min at 4 °C, adjusted the volume of supernant and repeated extraction for 3 times. The collected extracts were kept at −20 °C until used for total polyphenol and total anthocyanin analysis.

2.2. Total Polyphenol and Total Anthocyanin Analysis of GB Extracts

Total polyphenol contents were determined by the colorimetric method using Folin–Ciocalteau reagent [21]. Briefly, the extracts, blank or standard (galic acid), were added to 2 mL of 2% Na_2CO_3. After 2 min, 100 μL of 50% Folin–Ciocalteau reagent was added. The mixture was left at room temperature in the absence of light for 30 min. The absorbance of the colored product was measured at 750 nm. The standard curve of gallic acid was used for calculating the polyphenol contents.

Total anthocyanins were determined by using the pH differential method [22]. Before analysis, two dilutions of the sample were prepared, one for pH 1.0 using potassium hydroxide buffer (0.025 M, KOH) and the other for pH 4.5 using sodium acetate buffer (0.4 M, 54.4 g CH_3COONa). Samples were diluted 100 times and waited 15 min. The absorbance was measured at 520 and 700 nm. The concentration (mg/L) of each anthocyanin was calculated according to the following formula and expressed as cyanidin-3-glucoside equivalent (CGE):

$$\text{Total anthocyanins} = \frac{(A \times MW \times DF \times 100)}{\varepsilon \times L}$$

where A is the absorbance $=(A_{520nm} - A_{700nm})_{pH1.0} - (A_{520nm} - A_{700nm})_{pH4.5}$, MW is the molecular weight (g/mol) = 449.2 g/mol for cyanidin-3-glucoside, DF is the dilution factor = 100, and ε is the extinction coefficient = 26,900, where L (path length in cm) was 1.

2.3. Subjects

Subjects were recruited from the neighborhood near by Taipei Medical University. Inclusion criteria include whose blood glucose data meet the preDM diagnosis: FPG 100–125 mg/dL or 2 h values in the OGTT 140–190 mg/dL or HbA1c 5.7–6.4% and 20–70 years old. Exclusion criteria were individuals with a history of diabetes or use hypoglycemic agents/insulin in past 3 months, hepatic or renal disease, history of cardiovascular disease or cancer, pregnancy, breast feeding, or intending to become pregnant during the study period, thyroid or pituitary disease, gastrointestinal disease, hematological disorders or neurological disease, drinking alcohol or smoker. Written informed consent was obtained from all subjects, and the study was approved by the Joint Institutional Review Board at Taipei Medical University in Taiwan (JIRB number: 201307030).

2.4. Study Design

The 12-week study consisted of an 8-week intervention period (Week 0 to Week 8) and a 4-week washout period (Week 8 to Week 12). Subjects were assigned to the control group (prohibit eating red-purple food) or GB group (diet including 2 serving sizes of GB, equivalent to 200 g of edible portion (E.P)). The experimenter weighed GB, and asked the participants to take it home every week and cook it it by themselves. Both groups received dietary counseling based on the guideline of the Taiwanese Association of Diabetes Educators. Subjects were asked to do a 3-day dietary record (2 weekdays and 1 weekend) at Week 0 and 8, and the compliance was checked by dietary record. Every 4 weeks of the intervention period (Week 0, 4 and 8) and 4 weeks washout period (Week 12), anthropometric data and blood pressure were measured. Anthropometric data included height, weight, body composition, body mass index (BMI), waist and hip circumference and waist to hip ratio. Body composition was measured by Inbody 3.0 (Biospace, Seoul, Korea) according to the principle of biochemical impedance analysis (BIA). At Week 0 and Week 8, blood samples were collected after they had fasted overnight.

2.5. Blood Sample Analysis

Plasma collected by centrifugation at 1500× *g* for 10 min at 4 °C was stored at −80 °C until further analysis. Fasting glucose, fasting insulin and glycated hemoglobin (HbA1c), total cholesterol (TC), high-density lipoprotein cholesterol (HDL-C), low-density lipoprotein cholesterol (LDL-C), aspartate transaminase (AST) and alanine transaminase (ALT) levels were analyzed at the laboratory of Taipei Medical University Hospital. Homeostatic model assessment-insulin resistance (HOMA-IR) was calculated according to the formula insulin (μU/mL) × glucose (mmol/L)/22.5, and the quantitative insulin sensitivity check index (QUICKI) was calculated according to the formula 1/[log (fasting insulin) + log (fasting glucose)]. Total polyphenol levels in plasma were measured by the method described by Serafini et al. [23] using Folin–Ciocalteau reagent. Absorption at 750 nm was measured spectrophotometrically. The other blood measurements included the ferric reducing ability of plasma (FRAP), vitamin C and malondialdehyde (MDA), which involved standardized methods. Glutahione peroxidase (GSH-Px) and IL-6 were measured using Cayman standard glutathione assay kit and R&D human IL-6 high sensitivity ELISA kit, respectively. The total polyphenol content was expressed as gallic acid equivalent (GAE).

2.6. Dietary Nutrient Intake Analysis

Daily nutrient intake was calculated as the mean daily intake from a 3-day dietary record using the nutrient analysis software (E-kitchen, Taichung, Taiwan). Nutrient assessment included total caloric intake, dietary macronutrient intake (protein, fat and carbohydrate), dietary fiber, vitamin C and vitamin E.

2.7. Statistical Analysis

All data were expressed as the mean ± SEM. The data were analyzed using one-way ANOVA followed by Duncan's multiple range test, Mann–Whitney U test and paired *t*-test. A value of $p < 0.05$ was used to indicate statistical significance.

3. Results

3.1. Total Polyphenols and Total Anthocyanins of GB

The total polyphenol of GB extract was 25.54 ± 1.77 mg GAE/g DW. The total anthocyanin of GB extract was 1.53 ± 0.15 mg CGE/g DW.

3.2. Baseline Characteristics and Dietary Intake of Subjects

Nine subjects (two male and seven female) assigned to the control group and nine subjects (two male and seven female) assigned to the GB group completed the whole study period. General baseline characteristics were shown in Table 1. There were no differences in age or height. Although the dietary fiber intake of both groups did not show significant differences at baseline, the intake in the GB group was significantly increased (36%) after intervention (Table 2).

Table 1. Baseline subject characteristics.

	Control	GB
Sex		
Male (%)	22.2	22.2
Female (%)	77.8	77.8
Age	57.6 ± 2.2	61.8 ± 2.6
Height (cm)	159.9 ± 2.8	157.9 ± 1.5
Body weight (kg)	62.6 ± 3.0	64.0 ± 3.1
BMI (kg/m^2)	24.4 ± 0.6	26.0 ± 1.4
Body fat (%)	31.8 ± 1.5	32.4 ± 2.5
Fasting plasma glucose (mg/dL)	111.2 ± 6.4	125. 0 ± 3.8
Fasting insulin (mU/L)	10.6 ± 0.9	12.9 ± 1.3
HbA1c (%)	5.79 ± 0.12	6.41 ± 0.17 §

Values are mean ± SEM (n = 9), § $p < 0.05$ between groups by Student's *t*-test, GB, *Gynura bicolor*; BMI, Body mass index; HbA1c, Glycated hemoglobin.

Table 2. Dietary intake during study period.

	Control		GB	
	Week 0	Week 8	Week 0	Week 8
Energy (kcal)	1370.3 ± 144.5	1382.3 ± 88.3	1440.2 ± 143.7	1391.6 ± 123.7
Protein (%)	15.3 ± 1.0	16.2 ± 1.2	15.6 ± 1.0	17.2 ± 1.1
Fat (%)	28.9 ± 3.2	29.9 ± 1.8	28.7 ± 3.1	28.0 ± 1.7
Carbohydrate (%)	60.0 ± 3.6	54.1 ± 2.3	56.0 ± 3.3	55.7 ± 1.7
Dietary fiber (g)	14.0 ± 1.6	13.0 ± 1.2	15.0 ± 2.3	20.4 ± 2.0 §
Vitamin C (mg)	114.38 ± 40.89	104.52 ± 23.17	120.44 ± 34.40	164.72 ± 44.74
Vitamin E (α-TE)	3.38 ± 0.52	3.46 ± 0.51	3.66 ± 0.52	3.74 ± 0.66

Values are mean ± SEM (n = 9), § $p < 0.05$ between groups by Mann–Whitney U test, GB, *Gynura bicolor*; E, energy.

3.3. Glycemic-Control-Related Markers

Glycemic-control-related markers were shown in Table 3. At baseline, there was no difference in FPG, fasting insulin, HOMA-IR and QUICKI between groups. After an 8-week intervention, FPG and HOMA-IR were significantly decreased in the GB group when compared with baseline. In addition, the results remain maintain after the 4-week washout period (Week 12). The changes of fasting insulin, HOMA-IR and QUICKI in the GB group were significant when compared with the control group (Δ0–8 week).

Table 3. Effects of *Gynura bicolor* on glycemic control profile.

	Control	GB
Fasting plasma glucose (mg/dL)		
Week 0	111.22 ± 6.43	125.00 ± 3.83 §
Week 8	100.11 ± 2.30	108.22 ± 2.90 §*
Week 12	101.78 ± 3.60	106.00 ± 3.51 *
△0–8 week	−11.11 ± 6.04	−16.78 ± 2.72
△0−12 week	−9.1 ± 18.8	−19.8 ± 9.3 §
△8–12 week	2.1 ± 9.6	−1.4 ± 8.1
Fasting insulin (mU/L)		
Week 0	10.60 ± 0.91	12.87 ± 1.32
Week 8	12.10 ± 1.61	9.89 ± 0.93
Week 12	8.99 ± 0.87	9.00 ± 1.03 *
△0–8 week	1.50 ± 1.30	−2.98 ± 1.37 §
△0–12 week	−1.30 ± 2.7	−4.20 ± 4.8
△8–12 week	−2.30 ± 3.4	−1.20 ± 3.8
HbA1C (%)		
Week 0	5.79 ± 0.12	6.41 ± 0.17 §
Week 8	5.76 ± 0.11	6.30 ± 0.13 §
Week 12	5.69 ± 0.11	6.14 ± 0.14 §
△0–8 week	−0.03 ± 0.05	−0.11 ± 0.08
△0–12 week	−0.10 ± 0.11	−0.26 ± 0.28
△8–12 week	−0.07 ± 0.10	−0.13 ± 0.18
HOMA-IR		
Week 0	2.97 ± 0.39	3.95 ± 0.40
Week 8	3.02 ± 0.44	2.63 ± 0.25 *
Week 12	2.30 ± 0.27	2.34 ± 0.26 *
△0–8 week	0.05 ± 0.31	−1.32 ± 0.42 §
△0–12 week	−0.52 ± 0.85	−1.52 ± 1.52
△8–12 week	−1.00 ± 1.59	−0.36 ± 0.96
QUICKI		
Week 0	0.33 ± 0.01	0.31 ± 0.00
Week 8	0.33 ± 0.01	0.33 ± 0.00
Week 12	0.34 ± 0.01	0.34 ± 0.00
△0–8 week	−0.00 ± 0.01	0.02 ± 0.01 §
△0–12 week	−0.02 ± 0.11	−0.01 ± 0.11
△8–12 week	−0.02 ± 0.11	0.00 ± 0.11

Values are mean ± SEM (n = 9), § $p < 0.05$ between groups by Mann–Whitney U test; * $p < 0.05$ compared with Week 0 by Mann–Whitney U test, HOMA-IR, homeostasis model assessment-insulin resistance; QUICKI, quantitative insulin sensitivity check index.

3.4. Anthropometric Measures, Blood Pressure, Lipid Profile and Liver Function

Anthropometric measures included weight, BMI, body fat, fat-free mass, waist circumference, hip circumference and waist:hip ratio. There were no differences at baseline, Week 4, Week 8 and Week 12 between groups or within groups (Table 4). Blood pressure, TC, HDL-C, LDL-C, AST, and ALT levels did not differ between groups or within groups at baseline, Week 4, Week 8 and Week 12 (Tables 5 and 6).

Table 4. Effects of *Gynura bicolor* on anthropometric and body composition measures.

	Control	GB
Body weight (kg)		
Week 0	62.56 ± 3.10	64.03 ± 3.10
Week 4	62.38 ± 2.67	64.00 ± 2.94
Week 8	62.70 ± 2.84	63.66 ± 3.11
Week 12	59.51 ± 1.92	63.31 ± 3.11
BMI (kg/m^2)		
Week 0	24.36 ± 0.62	25.96 ± 1.35
Week 4	24.32 ± 0.41	25.94 ± 1.25
Week 8	24.41 ± 0.42	25.54 ± 1.17
Week 12	24.00 ± 0.61	25.64 ± 1.24
Body fat (%)		
Week 0	31.83 ± 1.46	32.41 ± 2.47
Week 4	30.84 ± 1.34	32.47 ± 2.46
Week 8	30.33 ± 1.22	31.38 ± 2.34
Week 12	30.88 ± 1.62	31.19 ± 2.34
Fat-free mass (kg)		
Week 0	42.82 ± 2.75	42.98 ± 1.81
Week 4	43.29 ± 2.51	42.98 ± 1.89
Week 8	43.88 ± 2.66	43.40 ± 1.82
Week 12	41.11 ± 1.60	43.29 ± 1.88
Waist circumference (cm)		
Week 0	85.61 ± 1.29	87.33 ± 3.00
Week 4	84.33 ± 1.39	87.56 ± 3.22
Week 8	83.00 ± 1.90	87.39 ± 2.48
Week 12	82.38 ± 1.29	84.89 ± 2.85
Hip circumference (cm)		
Week 0	99.61 ± 1.50	99.56 ± 2.57
Week 4	100.11 ± 1.56	101.00 ± 2.77
Week 8	98.89 ± 1.30	98.39 ± 2.57
Week 12	95.50 ± 1.88	98.44 ± 2.19
Waist to hip ratio		
Week 0	0.92 ± 0.01	0.93 ± 0.02
Week 4	0.91 ± 0.01	0.94 ± 0.02
Week 8	0.91 ± 0.01	0.93 ± 0.01
Week 12	0.91 ± 0.01	0.93 ± 0.01

Values are mean ± SEM (n = 9).

Table 5. Effects of *Gynura bicolor* on blood pressures.

	Control	GB
Systolic (mmHg)		
Week 0	122.89 ± 2.80	135.33 ± 5.06
Week 4	120.89 ± 5.78	136.78 ± 6.28
Week 8	115.22 ± 6.09	133.44 ± 4.19
Week 12	110.88 ± 7.18	120.56 ± 5.23
Diastolic (mmHg)		
Week 0	71.22 ± 5.17	78.22 ± 3.05
Week 4	71.78 ± 5.25	80.22 ± 3.03
Week 8	67.67 ± 4.66	77.11 ± 3.12
Week 12	63.25 ± 4.63	72.00 ± 3.54

Values are mean ± SEM (n = 9).

Table 6. Effects of GB on blood lipid profile and liver function [1–3].

	Control			GB		
	Week 0	Week 8	Week 12	Week 0	Week 8	Week 12
Blood lipid profile						
TC (mg/dL)	204.2 ± 11.8	197.0 ± 11.0	199.4 ± 11.6	197.7 ± 6.5	199.9 ± 8.9	189.6 ± 10.1
HDL-C (mg/dL)	58.7 ± 4.6	58.9 ± 5.0	61.4 ± 6.2	58.9 ± 4.2	56.2 ± 3.6	56.9 ± 3.6
LDL-C (mg/dL)	136.6 ± 11.0	128.2 ± 9.0	123.1 ± 10.4 *	131.4 ± 7.3	135.8 ± 7.3	123.7 ± 10.6
Liver function						
AST (IU/L)	32.7 ± 7.3	22.8 ± 2.5 *	21.9 ± 1.3	24.8 ± 3.6	23.4 ± 2.7	21.8 ± 3.1
ALT (IU/L)	36.7 ± 11.4	23.6 ± 3.7	21.2 ± 2.4	32.1 ± 8.9	29.9 ± 6.8	26.2 ± 7.1

[1] Values are mean ± SEM, [2,*] $p < 0.05$ compared with Week 0 by Mann–Whitney U test, [3] GB, *Gynura bicolor*; TC, total cholesterol; HDL-C, high-density lipoprotein cholesterol; LDL-C, low-density lipoprotein cholesterol; AST, aspartate transaminase; ALT, alanine transaminase.

3.5. Oxidative-Stress-Related Markers

Oxidative-stress-related markers were shown in Figure 1. At baseline, vitamin C (Figure 1a), total polyphenol (Figure 1b) and MDA (Figure 1e) in the GB group were significantly different when compared with the control group. The baseline total polyphenol level in the GB group was significantly lower when compared to the control group (0.39 ± 0.01 vs. 0.43 ± 0.01). After asking clients to avoid red-purple food for 8 weeks, the total polyphenol level in the control group was significantly decreased when comparing Week 8 to Week 0, though neither group showed any differences at Week 8. Although FRAP was significantly increased in both control and GB groups after 8 weeks, the delta value from Week 0 to Week 8 in the GB group was significantly higher than the control group (3.92 ± 0.56 vs. 1.80 ± 0.19) (Figure 1c). At Week 8, GSH-Px activity in the GB group was significantly higher than baseline (Figure 1d). At baseline, the levels of MDA in the GB group were significantly higher than in the control group (9.32 ± 0.62 vs. 5.88 ± 0.65). After the intervention, the MDA levels of the GB group were significantly decreased when comparing Week 8 to Week 0, and there were no significant differences between Week 0 and Week 8 of the control group (Figure 1e). Moreover, the differences of MDA levels between Week 8 and Week 0 were significantly higher in the GB group when compared to the control group (−5.23 ± 0.66 vs. −0.81 ± 0.43). There were no differences in IL-6 between groups or within groups (Figure 1f).

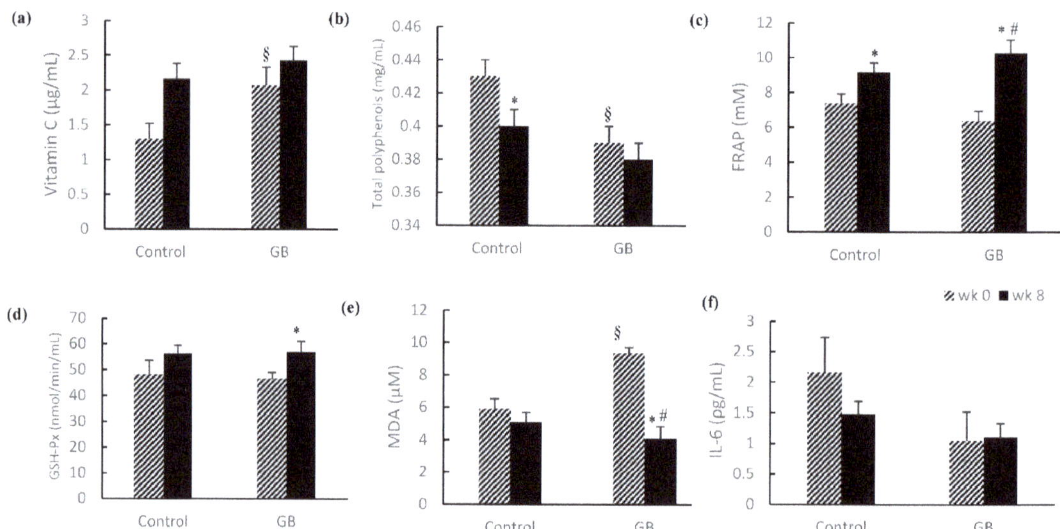

Figure 1. Effects of *Gynura bicolor* on antioxidants, antioxidant capacity and inflammation marker. (**a**) Vitamin C (μg/mL), (**b**) total polyphenols (mg/mL), (**c**) FRAP (mM), (**d**) GSH-Px (nmol/min/mL), (**e**) MDA (μM), and (**f**) IL-6 (ρg/mL). Values are mean ± SEM (n = 9). § $p < 0.05$ between groups by Mann–Whitney U test; * $p < 0.05$ compared with Week 0 by Mann–Whitney U test; # the difference value of Week 0 and Week 8 with a significant difference between groups by Mann–Whitney U test ($p < 0.05$) FRAP, ferric reducing ability of plasma; GSH-Px, glutathione peroxidase; MDA, malondialdehyde; IL-6, interlukin-6.

4. Discussion

In 1988, Reaven described a cluster of risk factors for DM and CVD as "syndrome X", the former name of "metabolic syndrome (Mets)" [24]. Mets have several definitions. In Taiwan, Mets is defined as any three of the following five features: waist circumference >90 cm in men or >80 cm in women, HDL-C < 40 mg/dL in men or <50 mg/dL in women, FPG $\geq$ 100 mg/dL, TG $\geq$ 150 mg/dL and BP $\geq$ 130/85 mmHg. Insulin resistance (IR) is considered one of the main pathophysiology that caused Mets. IR causes abnormal nutrient metabolism such as hepatic gluconeogenesis, lipid peroxidation and TG synthesis. Moreover, vasoconstriction, oxidative stress and inflammation happened. Under this condition, it increases the risk of developing CVD [25]. In previous studies, it showed that IR and β-cell dysfunction were already happening before developing to diabetes [26]. Ford et al. performed a meta-analysis; it concluded that IFG and IGT are associated with modest increases in the risk of cardiovascular disease. Therefore, preDM is related to Mets [27]. In our study, all subjects' FPG levels were over 100 mg/dL and have a large waist circumference. We considered the subjects in this study to be at a higher risk of developing Mets and CVD than others.

HOMA-IR was proposed as a simple and inexpensive alternative to more sophisticated techniques by Matthew et al., and had been evaluated to be reliably used in large-scale or epidemiological studies [28,29]. QUICKI was first performed by Katz et al. QUICKI is a novel, simple, accurate and reproductive method for determining insulin sensitivity in humans [30]. HOMA-IR and QUICKI can only use FPG and fasting insulin to assess the levels of insulin resistance (IR) and insulin sensitivity. The higher the HOMA-IR, the higher the levels of IR. QUICKI is the reverse to HOMA-IR. Several studies showed the different cutoff point of HOMA-IR and QUICKI. Ascaso et al. observed 65 subjects aged 30–60 years in Spain, showing that subjects had IR when HOMA-IR > 2.6 and QUICKI < 0.33 [31]. Keskin et al. described that the cutoff point for diagnosis of IR is >2.5 for adults and >3.15 for adolescents [32]. In China, a total of 2217 subjects were observed. It increases the risks of developing Mets and diabetes when the quartile1 (Q1) of HOMA-IR > 2.8. No matter which cutoff point they are based on, HOMA-IR and QUICKI in both groups were abnormal before the intervention period. After an 8-week intervention of GB and a 4-week washout period, HOMA-IR in the GB group were significantly decreased when compared with baseline. The difference of QUICKI between baseline and the 8-week intervention was significantly decreased when compared to the control group. Therefore, we speculated that an intake of GB may improve the insulin resistance and elevate the insulin sensitivity. Furthermore, the efforts of improving insulin resistance can be maintained.

Increased consumption of vegetables, whole grains, and soluble and insoluble fiber is recommended to treat preDM and diabetes individuals. Wolfram and Ismail-Beigi collected 14 randomized clinical trials on diabetes from the past decade. They concluded that improving insulin sensitivity and glucose homeostasis on a plant-based diet is more effective than other commonly used diets, because plant-based diets contain fiber, micronutrients (potassium, folate and magnesium) and phytochemicals such as anthocyanins, flavonoids and chlorophyll [33–35]. The PREDIMED study (PREvencio'n con DIetaMEDiterra'nea) conducted a 3-month clinical trial. It showed increasing dietary fiber intake with natural food (22 g per day) on risk factors for CVD in subjects at high risk decreased FPG, TC and increased HDL-C, which is associated with CVD risk factors [36]. Another randomized clinical trial suggested that a moderate amount of dietary fiber intake (7 g per day) may be beneficial for managing the FPG but no effects on weight and BMI in Japanese men with hyperglycemia and visceral fat obesity [37]. According to "2005–2008 Nutrition and Health Survey in Taiwan (NAHSIT)", the dietary fiber intake in men was 13.7 g and 14 g in women, both of which were less than 25–35 g, the recommended intake from the Ministry of Health and Welfare (MHW). In our study, the dietary intake (13–14 g) in the control and GB groups was no different at baseline. After an 8-week intervention of GB, the dietary intake in the GB group was increased 6 g/day and FPG was decreased when compared with baseline. One exchange of GB contained dietary fiber (3.1 g). Although not reaching the recommended intake from the MHW, it did indeed raise the dietary intake. We speculated that decreased FPG was associated with an increasing intake of dietary fiber, which is from GB. In addition, total polyphenols and total anthocyanins of GB extract were 25.54 ± 1.77 mg GAE/g DW and 1.53 ± 0.15 mg CGE/g DW, respectively. A previous study showed that the main chemical components of GB are phenolic acids, flavonoids, carotenoids and anthocyanins [38]. Using UPLC-MS/MS analysis, 53 phenolics were identified [39]. Most polyphenols exist in the form of glycosylated derivatives in GB. These polyphenols must undergo digestive enzymes and intestinal microbiota metabolism to become bioactive in the human body [40]. Clinical study in obese healthy subjects did not support the use of dietary supplementation with dried purple carrot (259.2 mg/day of phenolic acids) to achieve weight loss, improvements in body composition, LDL-C and blood pressure [41]. The same results were also observed in our study, because the total polyphenol was much lower.

Several antioxidants exist in plasma, such as ascorbic acid, vitamin E, carotenoid and polyphenols. The ability of dietary polyphenols to reduce inflammation is related to acting as antioxidants, interfering with oxidative stress signaling, suppressing the pro-inflammatory signaling transductions. The biological significance of phenolic compounds is not only in direct reaction with ROS, but also in activation of cell signaling pathways [42]. FRAP is presented as a method for assessing total antioxidant power. Ferric ferrous ion (Fe^{3+}) reduction at low pH causes a navy blue ferrous-tripyridyltriazine (Fe^{2+}-TPTZ) complex form. The deeper the color, the better the antioxidant power [43]. We observed that FRAP both in the GB group and control group were increased after an 8-week experimental period. It meant that total antioxidant power was both enhanced. Moreover, the change between baseline and Week 8 in the GB group was more significant than in the control group. MDA is an intermediate product of lipid peroxidation, which usually uses thiobarbituric acid-reactive substances (TBARS) to quantify. One study showed that smoking and alcohol consumption were confounding factors of MDA [44]. GSH-Px is the important enzyme of the cell defense system. GSH-Px could breakdown peroxide to oxygen and water. The activity of GSH-Px would decrease when peroxidant increases [45]. A previous study observed that Aronia extract results from the influence of anthocyanins and possibly other flavonoids on decreasing MDA and catalase and increasing GSH-Px and superoxide dismutase activities [46]. In our study, we excluded smoker and alcohol consumption subjects. MDA and GSH-Px in the GB group were decreased/increased in Week 8 when compared with baseline, respectively. It showed that, after an 8-week intervention, the oxidative stress in subjects was decreased.

Because of increasing oxidative stress, the pathophysiology of preDM and T2D were considered as a chronic inflammation. IL-6 is an important mediator of inflammatory response [47]. Lucas et al. observed that TNF-α and IL-6 were significantly higher in overweight/obese young subjects with preDM (BMI > 33 kg/m^2) than healthy overweight/obese ones [48]. It is not consistent with our study. We speculated that this is because the study population consisted of all overweight individuals (BMI > 25 kg/m^2). The inflammation generated from adipocytes was fewer than obesity.

The bioavailability of polyphenols was influenced by the molecular weight, conjugated form (methyl, glucuronide or sulphate) and cooking method. For most flavonoids absorbed in the small intestine, the plasma concentration then rapidly decreases (elimination half-life period of 1–2 h). The maintenance of a high concentration in plasma thus requires a repeated ingestion of the polyphenols over time. Anthocyanins are quite rapidly absorbed, but their bioavailability seems to be the lowest of all flavonoids [40]. Nielsen et al. concluded that urinary flavonoids may be useful as a new biomarker for vegetables and fruits [49]. In the GB group, there were no differences in total polyphenols of plasma. It could be associated with the fact that the blood sample collection of subjects was not conducted all at the same time. When all were finished to collect, it already passed the half-life time of polyphenols. In the control group, total polyphenols of plasma were significantly decreased. It might be associated with prohibiting eating red-purple food for 8 weeks. We suggested that future studies must use urine tests to assess total polyphenol intake.

Our study was the first interventional trial using GB for preDM, by using natural ingredients and at least eating two exchanges of vegetables every day. GB is not only rich in fiber; it also contains higher polyphenols than other green-leafy vegetables. This might have resulted in improvement of glycemic control and reducing oxidative stress in preDM subjects. Although the beneficial effects of GB could be observed, some limitations needed be addressed. (1) PreDM without medication subjects are difficult to find, because people pay less attention to their own blood report. (2) The intervention period was 8 weeks. We did not consider that the half-life of HbA1c is about 3 months. (3) The baseline characteristics had biases between the control group and GB group. This is because we did not assign randomly, but based on subjects' dietary habit. Further studies, we suggest, should consider expanding the intervention time and the allocations of groups.

In conclusion, the subjects in our study were overweight and preDM subjects. After intervention of two serving sizes of GB for 8 weeks, FPG and MDA were decreased, FRAP and GSH-Px were increased in the intervention group and HOMA-IR and QUICKI were improved. We concluded that GB could improve glycemic control and decrease oxidative stress because of its large amounts of polyphenols. In addition, GB not only has health-care functions, but its pigments also show potential uses as natural food colorings. Therefore, GB extract can be used to make functional foods, such as adding to fruit mousse or yogurt.

Author Contributions: C.-H.H. and Y.-W.C. were involved in the conceptualization and the design of this study. C.-H.H. conducted this study and biochemical analyses. C.-H.H. wrote the original draft preparation. C.-H.H. and Y.-T.T. reviewed and edited this study. Y.-S.Y. supervised this study. Y.-W.C. visualized and supervised this study. All authors provided critical inputs to data analyses and the interpretation of the data. All authors have read and agreed to the published version of the manuscript.

Funding: This research received no external funding.

Institutional Review Board Statement: The study was conducted according to the guidelines of the Declaration of Helsinki, and approved by the Institutional Review Board of Taipei Medical University in Taiwan.

Informed Consent Statement: Informed consent was obtained from all subjects involved in the study.

Data Availability Statement: No new data were created or analyzed in this study. Data sharing is not applicable to this article.

Acknowledgments: We thank all subjects for participating in this study.

Conflicts of Interest: The authors declare no conflict of interest.

References

1. American Diabetes Association. Diagnosis and classification of diabetes mellitus. *Diabetes Care* **2021**, *44*, S15–S33.
2. DeFronzo, R.A.; Abdul-Ghani, M.A. Preservation of beta-cell function: The key to diabetes prevention. *J. Clin. Endocrinol. Metab.* **2011**, *96*, 2354–2366. [CrossRef]
3. International Diabetes Federation. *IDF Diabetes Atlas*, 9th ed.; International Diabetes Federation: Brussels, Belgium, 2019.
4. Tabák, A.G.; Herder, C.; Rathmann, W.; Brunner, E.J.; Kivimäki, M. Prediabetes: A high-risk state for diabetes development. *Lancet* **2012**, *379*, 2279–2290. [CrossRef]
5. Shaw, J.; Sicree, R.; Zimmet, P. Global estimates of the prevalence of diabetes for 2010 and 2030. *Diabetes Res. Clin. Pract.* **2010**, *87*, 4–14. [CrossRef]
6. Su, Y.; Liu, X.-M.; Sun, Y.-M.; Jin, H.-B.; Fu, R.; Wang, Y.-Y.; Wu, Y.; Luan, Y. The relationship between endothelial dysfunction and oxidative stress in diabetes and prediabetes. *Int. J. Clin. Pract.* **2008**, *62*, 877–882. [CrossRef]
7. Antoniades, C.; Shirodaria, C.; Crabtree, M.; Rinze, R.; Alp, N.; Cunnington, C.; Diesch, J.; Tousoulis, D.; Stefanadis, C.; Leeson, P.; et al. Altered plasma versus vascular biopterins in human atherosclerosis reveal relationships between endothelial nitric oxide synthase coupling, endothelial function, and inflammation. *Circulation* **2007**, *116*, 2851–2859. [CrossRef] [PubMed]
8. Schafer, A.; Bauersachs, J. Endothelial dysfunction, impaired endogenous platelet inhibition and platelet activation in diabetes and atherosclerosis. *Curr. Vasc. Pharmacol.* **2008**, *6*, 52–60. [CrossRef] [PubMed]
9. Milman, S.; Crandall, J.P. Mechanisms of Vascular Complications in Prediabetes. *Med. Clin. N. Am.* **2011**, *95*, 309–325. [CrossRef]
10. Glechner, A.; Keuchel, L.; Affengruber, L.; Titscher, V.; Sommer, I.; Matyas, N.; Wagner, G.; Kien, C.; Klerings, I.; Gartlehner, G. Effects of lifestyle changes on adults with prediabetes: A systematic review and meta-analysis. *Prim. Care Diabetes* **2018**, *12*, 393–408. [CrossRef] [PubMed]
11. Pan, X.-R.; Li, G.-W.; Hu, Y.-H.; Wang, J.-X.; Yang, W.-Y.; An, Z.-X.; Hu, Z.-X.; Lin, J.-; Xiao, J.-Z.; Cao, H.-B.; et al. Effects of Diet and Exercise in Preventing NIDDM in People with Impaired Glucose Tolerance: The Da Qing IGT and Diabetes Study. *Diabetes Care* **1997**, *20*, 537–544. [CrossRef] [PubMed]
12. Salas-Salvadó, J.; Martinez-González, M.; Bulló, M.; Ros, E. The role of diet in the prevention of type 2 diabetes. *Nutr. Metab. Cardiovasc. Dis.* **2011**, *21*, B32–B48. [CrossRef] [PubMed]
13. Giugliano, D.; Ceriello, A.; Esposito, K. The effects of diet on inflammation: Emphasis on the metabolic syndrome. *J. Am. Coll. Cardiol.* **2006**, *48*, 677–685. [CrossRef]
14. Zhao, C.; Yang, C.; Wai, S.T.C.; Zhang, Y.; Portillo, M.P.; Paoli, P.; Wu, Y.; Cheang, W.S.; Liu, B.; Carpéné, C.; et al. Regulation of glucose metabolism by bioactive phytochemicals for the management of type 2 diabetes mellitus. *Crit. Rev. Food Sci. Nutr.* **2018**, *59*, 830–847. [CrossRef]
15. Kimble, R.; Keane, K.M.; Lodge, J.K.; Howatson, G. Dietary intake of anthocyanins and risk of cardiovascular disease: A systematic review and meta-analysis of prospective cohort studies. *Crit. Rev. Food Sci. Nutr.* **2018**, *59*, 3032–3043. [CrossRef] [PubMed]
16. Shimizu, Y.; Maeda, K.; Kato, M.; Shimomura, K. Isolation of anthocyanin-related MYB gene, GbMYB2, from Gynura bicolor leaves. *Plant Biotechnol.* **2010**, *27*, 481–487. [CrossRef]
17. Lu, C.-H.; Yang, H.-C.; Chang, W.-L.; Chang, Y.-P.; Wu, C.-C.; Hsieh, S.-L. Development of beverage product from Gynura bicolor and evaluation of its antioxidant activity. *Genom. Med. Biomark. Health Sci.* **2012**, *4*, 131–135. [CrossRef]
18. Lin, J.-Y.; Liu, Y.-L.; Lin, B.-F.; Hsiao, C.-C.; Tsai, S.-J. Antioxidant, Antimutagenic and Immunomodulatory Potentials of Gynura Bicolar and Amaranthus Gangetcaue. *Taiwan J. Agric. Chem. Food Sci.* **2004**, *42*, 231–241.
19. Wu, C.-C.; Lii, C.-K.; Liu, K.-L.; Chen, P.-Y.; Hsieh, S.-L. Antiinflammatory Activity of Gynura bicolor (紅鳳菜 Hóng Fèng Cài) Ether Extract Through Inhibits Nuclear Factor Kappa B Activation. *J. Tradit. Complement. Med.* **2013**, *3*, 48–52. [CrossRef]
20. Li, W.-L.; Ren, B.-R.; Zhuo, M.-; Hu, Y.; Lu, C.-G.; Wu, J.-L.; Chen, J.; Sun, S. The Anti-Hyperglycemic Effect of Plants in Genus Gynura Cass. *Am. J. Chin. Med.* **2009**, *37*, 961–966. [CrossRef]
21. Singleton, V.L.; Rossi, J.A. Colorimetry of Total Phenolics with Phosphomolybdic-Phosphotungstic Acid Reagents. *Am. J. Enol. Vitic.* **1965**, *16*, 144–158.

22. Hosseinian, F.; Li, W.; Beta, T. Measurement of anthocyanins and other phytochemicals in purple wheat. *Food Chem.* **2008**, *109*, 916–924. [CrossRef]
23. Serafini, M.; Maiani, G.; Ferro-Luzzi, A. Alcohol-free red wine enhances plasma antioxidant capacity in humans. *J. Nutr.* **1998**, *128*, 1003–1007. [CrossRef]
24. Reaven, G.M. Banting lecture 1988. Role of insulin resistance in human disease. *Diabetes* **1988**, *37*, 1595–1607. [CrossRef]
25. Mendrick, D.L.; Diehl, A.M.; Topor, L.S.; Dietert, R.R.; Will, Y.; A La Merrill, M.; Bouret, S.; Varma, V.; Hastings, K.L.; Schug, T.T.; et al. Metabolic Syndrome and Associated Diseases: From the Bench to the Clinic. *Toxicol. Sci.* **2018**, *162*, 36–42. [CrossRef] [PubMed]
26. Tabák, A.G.; Jokela, M.; Akbaraly, T.N.; Brunner, E.J.; Kivimäki, M.; Witte, D.R. Trajectories of glycaemia, insulin sensitivity, and insulin secretion before diagnosis of type 2 diabetes: An analysis from the Whitehall II study. *Lancet* **2009**, *373*, 2215–2221. [CrossRef]
27. Ford, E.S.; Li, C.; Sniderman, A. Temporal changes in concentrations of lipids and apolipoprotein B among adults with diagnosed and undiagnosed diabetes, prediabetes, and normoglycemia: Findings from the National Health and Nutrition Examination Survey 1988–1991 to 2005–2008. *Cardiovasc. Diabetol.* **2013**, *12*, 26. [CrossRef]
28. Bonora, E.; Targher, G.; Alberiche, M.; Bonadonna, R.C.; Saggiani, F.; Zenere, M.B.; Monauni, T.; Muggeo, M. Homeostasis model assessment closely mirrors the glucose clamp technique in the assessment of insulin sensitivity: Studies in subjects with various degrees of glucose tolerance and insulin sensitivity. *Diabetes Care* **2000**, *23*, 57–63. [CrossRef] [PubMed]
29. Matthews, D.R.; Hosker, J.P.; Rudenski, A.S.; Naylor, B.A.; Treacher, D.F.; Turner, R.C. Homeostasis model assessment: Insulin resistance and beta-cell function from fasting plasma glucose and insulin concentrations in man. *Diabetologia* **1985**, *28*, 412–419. [CrossRef]
30. Katz, A.; Nambi, S.S.; Mather, K.; Baron, A.D.; Follmann, D.A.; Sullivan, G.; Quon, M.J. Quantitative insulin sensitivity check index: A simple, accurate method for assessing insulin sensitivity in humans. *J. Clin. Endocrinol. Metab.* **2000**, *85*, 2402–2410. [CrossRef]
31. Ascaso, J.F.; Pardo, S.; Real, J.T.; Lorente, R.I.; Priego, A.; Carmena, R. Diagnosing insulin resistance by simple quantitative methods in subjects with normal glucose metabolism. *Diabetes Care* **2003**, *26*, 3320–3325. [CrossRef]
32. Keskin, M.; Kurtoglu, S.; Kendirci, M.; Atabek, M.E.; Yazici, C. Homeostasis Model Assessment Is More Reliable Than the Fasting Glucose/Insulin Ratio and Quantitative Insulin Sensitivity Check Index for Assessing Insulin Resistance Among Obese Children and Adolescents. *Pediatrics* **2005**, *115*, e500–e503. [CrossRef] [PubMed]
33. Wolfram, T.; Ismail-Beigi, F. Efficacy of High-Fiber Diets in the Management of Type 2 Diabetes Mellitus. *Endocr. Pract.* **2011**, *17*, 132–142. [CrossRef] [PubMed]
34. Kaur, J. A Comprehensive Review on Metabolic Syndrome. *Cardiol. Res. Pract.* **2014**, *2014*, 943162. [CrossRef]
35. Castro-Barquero, S.; Ruiz-León, A.M.; Sierra-Pérez, M.; Estruch, R.; Casas, R. Dietary Strategies for Metabolic Syndrome: A Comprehensive Review. *Nutrients* **2020**, *12*, 2983. [CrossRef] [PubMed]
36. Estruch, R.; A Martinez-Gonzalez, M.; Corella, D.; Basora, J.; Ruiz-Gutierrez, V.; I Covas, M.; Fiol, M.; Gomez-Gracia, E.; López-Sabater, M.C.; Escoda, R.; et al. Effects of dietary fibre intake on risk factors for cardiovascular disease in subjects at high risk. *J. Epidemiol. Community Health* **2009**, *63*, 582–588. [CrossRef]
37. Kobayakawa, A.; Suzuki, T.; Ikami, T.; Saito, M.; Yabe, D.; Seino, Y. Improvement of Fasting Plasma Glucose Level After Ingesting Moderate Amount of Dietary Fiber in Japanese Men With Mild Hyperglycemia and Visceral Fat Obesity. *J. Diet. Suppl.* **2013**, *10*, 129–141. [CrossRef] [PubMed]
38. Do, T.V.T.; Suhartini, W.; Mutabazi, F.; Mutukumira, T.N. Gynura bicolor DC. (Okinawa spinach): A comprehensive review on nutritional constituents, phytochemical compounds, utilization, health benefits, and toxicological evaluation. *Food Res. Int.* **2020**, *134*, 109222. [CrossRef]
39. Chen, J.; Mangelinckx, S.; Lü, H.; Wang, Z.T.; Li, W.L.; De Kimpe, N. Profiling and elucidation of the phenolic compounds in the aerial parts of Gynura bicolor and G. divaricata collected from different Chinese origins. *Chem. Biodivers.* **2015**, *12*, 96–115. [CrossRef] [PubMed]
40. Marín, L.; Miguélez, E.M.; Villar, C.J.; Lombó, F. Bioavailability of Dietary Polyphenols and Gut Microbiota Metabolism: Antimicrobial Properties. *BioMed Res. Int.* **2015**, *2015*, 905215. [CrossRef]
41. Wright, O.R.; Netzel, G.A.; Sakzewski, A.R. A randomized, double-blind, placebo-controlled trial of the effect of dried purple carrot on body mass, lipids, blood pressure, body composition, and inflammatory markers in overweight and obese adults: The QUENCH Trial. *Can. J. Physiol. Pharmacol.* **2013**, *91*, 480–488. [CrossRef]
42. Zhang, H.; Tsao, R. Dietary polyphenols, oxidative stress and antioxidant and anti-inflammatory effects. *Curr. Opin. Food Sci.* **2016**, *8*, 33–42. [CrossRef]
43. Benzie, I.F.; Strain, J.J. The ferric reducing ability of plasma (FRAP) as a measure of "antioxidant power": The FRAP assay. *Anal. Biochem.* **1996**, *239*, 70–76. [CrossRef] [PubMed]
44. Nielsen, F.; Mikkelsen, B.B.; Nielsen, J.B.; Andersen, H.R.; Grandjean, P. Plasma malondialdehyde as biomarker for oxidative stress: Reference interval and effects of life-style factors. *Clin. Chem.* **1997**, *43*, 1209–1214. [CrossRef] [PubMed]
45. Ursini, F.; Maiorino, M.; Gregolin, C. The selenoenzyme phospholipid hydroperoxide glutathione peroxidase. *Biochim. Biophys. Acta* **1985**, *839*, 62–70. [CrossRef]

46. Broncel, M.; Kozirog, M.; Duchnowicz, P.; Koter-Michalak, M.; Sikora, J.; Chojnowska-Jezierska, J. Aronia melanocarpa extract reduces blood pressure, serum endothelin, lipid, and oxidative stress marker levels in patients with metabolic syndrome. *Med. Sci. Monit.* **2010**, *16*, CR28–CR34.
47. McCracken, E.; Monaghan, M.; Sreenivasan, S. Pathophysiology of the metabolic syndrome. *Clin. Dermatol.* **2018**, *36*, 14–20. [CrossRef] [PubMed]
48. Lucas, R.; Parikh, S.J.; Sridhar, S.; Guo, D.-H.; Bhagatwala, J.; Dong, Y.; Caldwell, R.; Mellor, A.; Caldwell, W.; Zhu, H.; et al. Cytokine profiling of young overweight and obese female African American adults with prediabetes. *Cytokine* **2013**, *64*, 310–315. [CrossRef] [PubMed]
49. Nielsen, S.E.; Freese, R.; Kleemola, P.; Mutanen, M. Flavonoids in human urine as biomarkers for intake of fruits and vegetables. *Cancer Epidemiol. Biomark. Prev.* **2002**, *11*, 459–466.

Article

Thiosulfinate-Enriched *Allium sativum* Extract as an Adjunct to Antibiotic Treatment of Sepsis in a Rat Peritonitis Model

Francisco Javier Redondo-Calvo [1,2,3,†], Omar Montenegro [1,†], David Padilla-Valverde [3,4], Pedro Villarejo [4], Víctor Baladrón [1], Natalia Bejarano-Ramírez [3,5], Rocío Galán [1], Luis Antonio Gómez [6], Natalia Villasanti [3,7], Soledad Illescas [3,8], Vicente Morales [9], Lucía Medina-Prado [2], José Ramón Muñoz-Rodríguez [2,3] and José Manuel Pérez-Ortiz [2,3,*]

1 Department of Anesthesiology and Critical Care Medicine, University General Hospital, 13004 Ciudad Real, Spain; fjredondo@sescam.jccm.es (F.J.R.-C.); ommontenegro@sescam.jccm.es (O.M.); vbaladron@sescam.jccm.es (V.B.); rgalanm@sescam.jccm.es (R.G.)
2 Translational Research Unit, University General Hospital, 13004 Ciudad Real, Spain; lucia.medinaprado@gmail.com (L.M.-P.); jmunozrodriguez@sescam.jccm.es (J.R.M.-R.)
3 Faculty of Medicine, Universidad de Castilla-La Mancha, 13071 Ciudad Real, Spain; davidp@sescam.jccm.es (D.P.-V.); nbejarano@sescam.jccm.es (N.B.-R.); nvillasanti@sescam.jccm.es (N.V.); msillescasf@sescam.jccm.es (S.I.)
4 Department of Surgery, University General Hospital, 13004 Ciudad Real, Spain; pvillarejo@sescam.jccm.es
5 Department of Pediatrics, University General Hospital, 13004 Ciudad Real, Spain
6 Department of Chemical Engineering, Institute of Chemical and Environmental Technology, Universidad de Castilla-La Mancha, 13071 Ciudad Real, Spain; luisantonio.gomez@alvenpesalud.com
7 Department of Pathological Anatomy, University General Hospital, 13004 Ciudad Real, Spain
8 Department of Microbiology, University General Hospital, 13004 Ciudad Real, Spain
9 Department of Clinical Analysis, University General Hospital, 13004 Ciudad Real, Spain; vmelipe@sescam.jccm.es
* Correspondence: jmperezo@sescam.jccm.es; Tel.: +34-9262-78000 (ext. 79111)
† Both authors contributed equally and should be considered co-first authors.

Citation: Redondo-Calvo, F.J.; Montenegro, O.; Padilla-Valverde, D.; Villarejo, P.; Baladrón, V.; Bejarano-Ramírez, N.; Galán, R.; Gómez, L.A.; Villasanti, N.; Illescas, S.; et al. Thiosulfinate-Enriched *Allium sativum* Extract as an Adjunct to Antibiotic Treatment of Sepsis in a Rat Peritonitis Model. *Appl. Sci.* **2021**, *11*, 4760. https://doi.org/10.3390/app11114760

Academic Editor: Luca Mazzoni

Received: 22 April 2021
Accepted: 19 May 2021
Published: 22 May 2021

Featured Application: Thiosulfinate-enriched *Allium sativum* extract used as an adjuvant to antibiotic treatment and to sepsis management could improve the response profile and attenuate the outcome of the sepsis shock.

Abstract: Up to now, there are no studies that have shown a decrease in morbidity and mortality in the context of sepsis and septic shock, except for antibiotic therapy and the objective-guided resuscitation strategy. The goal was to evaluate the use of thiosulfinate-enriched *Allium sativum* extract (TASE) as an adjuvant in the management of sepsis. An experimental in vivo study was carried out with male Sprague Dawley® rats. Animals were randomized in three treatment groups: the control group (I), antibiotic (ceftriaxone) treatment group (II) and ceftriaxone plus TASE treatment group (III). All animals were housed and inoculated with 1×10^{10} CFU/15 mL of intraperitoneal *Escherichia coli* ATCC 25922. Subsequently, they received a daily treatment according to each group for 7 days. Clinical, analytical, microbiological, and histopathological parameters were evaluated. Statistically significant clinical improvement was observed in the ceftriaxone plus TASE vs. ceftriaxone group in weight, ocular secretions, whiskers separation and physical activity level ($p \leq 0.05$). When comparing interleukins on the third day of treatment between II and III, we found statistically significant differences in IL-1 levels ($p < 0.05$). Blood and peritoneal liquid cultures of group I were positive for multisensitive *E. coli*. Group II and III cultures were negative for *E. coli*, although an overgrowth of *Enterococcus faecalis* was found. In conclusion, TASE used as an adjuvant to antibiotic treatment in the management of sepsis could improve response profiles with sepsis attenuation, thus reducing overall mortality after an animal peritonitis model.

Keywords: garlic; *Allium sativum*; thiosulfinate; allicin; sepsis; immunomodulation; interleukins; rats

1. Introduction

There are few areas in critical medicine that generate as much interest and research as sepsis. Despite diagnostic and therapeutic advances, sepsis morbidity, mortality, and incidence remain very high. Sepsis is an altered host response to an infectious pathogen, causing potentially life-threatening organ dysfunction, and septic shock is a subset of sepsis in which the underlying circulatory and metabolic abnormalities are deep enough to substantially increase mortality [1]. Despite critical care progress in recent years in critical care, sepsis and septic shock account for more than 50% of deaths in critical care units.

Sepsis is now recognized as a multifactorial host response to an infectious pathogen that can be significantly amplified by endogenous factors involving the early activation of both pro- and anti-inflammatory responses along with major modifications in non-immunological pathways such as cardiovascular, neuronal, autonomic, hormonal, bioenergetic, metabolic and coagulative [2,3], all of which are of prognostic importance. Additionally, the biological and clinical heterogeneity of affected individuals is important, as well as age, underlying comorbidities, concurrent injuries (including surgery), medications, and source of infection [1].

What differentiates sepsis from infection is an aberrant or poorly regulated host response with the presence of organ dysfunction. The severity of organ dysfunction has been evaluated with various scoring systems that quantify abnormalities based on clinical findings, laboratory data, or therapeutic interventions. Differences in these scoring systems have also resulted in inconsistent information. The predominant score currently in use is the Sequential Organ Failure Assessment Score (SOFA), which has been simplified in quick SOFA [1,3].

Garlic (*Allium sativum*) has long been a medicinal ingredient used as an antineoplastic and antimicrobial agent. Sulfur compounds (i.e., thiosulfinates) appear to be the active components in the root bulb of the garlic plant [4,5]. Allicin is the main thiosulfinate of *Allium sativum* and could act on four points of the inflammatory cascade. The ability of allicin to inhibit the activation of the nuclear factor NF-κβ [6], prevent the adhesion of T cells to endothelial cells and reduce transendothelial migration [7] have been described. Allicin can also reduce the activity of induced nitric oxide synthase [8] and decrease the amount of nitric oxide and the vasodilatation that may lead to shock. Additionally, it could act by preventing the activation of the coagulation cascade by acting as an antiplatelet [9].

Up to now, there are no studies that have shown a decrease in morbidity and mortality in the context of sepsis and septic shock, except for antibiotic therapy and the objective-guided resuscitation strategy proposed in the 2016 sepsis campaign. This strategy was made by a group of international experts who established a series of based-on-evidence recommendations for the management of acute sepsis and septic shock. It is also the basis for the better outcome of high-mortality critically ill patients [10]. However, many drugs have been used unsuccessfully in both animal models and clinical trials [11–15], so there is still a need of new therapeutics that can overcome the antibiotic resistance.

As allicin is not stable [16], here, we decided to explore whether intraperitoneal applications of thiosulfinate-enriched *Allium sativum* extract (TASE) could be an adjuvant to specific antibiotic treatment in sepsis and septic shock and to evaluate its possible immunomodulatory role.

2. Materials and Methods

2.1. Animals and Sepsis Model

Male 5-week-old Sprague Dawley® rats (Harlan Laboratories Models SL) were used. The study was conducted at the Translational Research Unit of the University General Hospital, Ciudad Real. The procedures were carried out at the same time of day to avoid the possible influence of the circadian cycle on the results of the work.

Rats were kept with food and water ad libitum, in a cycle of 12 h of light and 12 h of darkness, and a room temperature of 22 ± 2 °C with a relative humidity of 50–70% and 15–20 air renewals per hour without recirculation. They were housed according to RD

53/2013 and no rat was caged alone to favor their group behavior. In addition, they were maintained in these environmental conditions to allow acclimatization for a week before the study started. Animals were randomized in three groups. Group I: physiological saline ($n = 6$); group II: ceftriaxone ($n = 9$); and group III: ceftriaxone + TASE ($n = 9$). A model of peritonitis was generated in all groups. Rats from different groups were never housed in the same cage.

To create the peritonitis model, each rat was administered with an intraperitoneal injection of bacteria after anesthesia with ketamine/xylacin (75/10 mg/kg), also directly into the abdominal cavity. Prior to this experiment, we conducted an experimental study to determine the most optimal inoculum dose to generate the sepsis and septic shock model. We determined that it was necessary to use a concentration of *Escherichia coli* ATCC 25922 of 1×10^{10} colony forming units (CFU) in 15 mL of distilled water [17].

2.2. Thiosulfinate-Enriched Allium sativum Extract

Lyophilized *Allium sativum* extract was obtained from the purple garlic ecotype from Las Pedroñeras (Ciudad Real, Spain), the only European region with protected geographical status for garlic (ES/PGI/005/0228/12.03.2002). A patented protocol (WO 2008/102036 A1. Method for obtaining a freeze-dried, stable extract from plants of the *Allium* genus) was employed for extraction to guarantee the stability and concentration of allicin and other thiosulfinates. The standardized composition and concentration of lyophilized *Allium sativum* extract are stable for over 10 months at 4 °C (Table 1, WO 2008/102036 A1). We employed diallyl thiosulfinate (allicin) concentration as the reference for the elaboration of the experimental treatment.

Table 1. Composition of organic and inorganic compounds of lyophilized *Allium sativum* extract from Las Pedroñeras (Ciudad Real, Spain) under optimized conditions (WO 2008/102036 A1).

Compound	Concentration (μg/mg)	Compound	Concentration (μg/mg)
Dimethyl thiosulfinate	18.30	**Se**	9.37
Allylmethyl+Methyl-allyl	4.58	**B**	89.45
Propyl-methyl+Methyl-propylthiosulfinate	3.39	**Zn**	10.60
Diallyl thiosulfinate (**allicin**)	5.62	**Cd**	9.48
Allyl-1-propenyl thiosulfinate	31.02	**P**	1188.87
1-propenyl-allyl+allyl-propyl thiosulfinate	1.76	**Ca**	159.11
Propyl-allyl thiosulfinate	1.59	**K**	3974.85
Di-propyl thiosulfinate	1.65	**Mg**	188.41
Methyl allyl sulfide	3.58	**Cu**	298.16
Methyl allyl disulfide	4.73	**Fe**	95.84
Dimethyl tetrasulfide	6.62	**Cr**	26.37
Di-allyl trisulfide	0.74	**Si**	3665.76
Di-methyl pentasulfide	1.63	**Mn**	1.18
Prostaglandin E1	4.83	**Na**	102.41
(E,Z)-Ajoene	0.07	**Co**	Non-detected
Inulin	0.10	**Hg**	Non-detected
Vitamin E (α-tocopherol)	3.07	**Al, Ni**	Non-detected

2.3. Experimental Design and Analytical Parameters

In relation to the treatments used, group I received 4.4 mL of 0.9% physiological saline intraperitoneally, group II received the same intraperitoneal volume with the antibiotic ceftriaxone (100 mg/kg) and group III the same volume with ceftriaxone (100 mg/kg) + TASE (0.5 mg/kg; referred to allicin content). In Figure 1, we showed the experimental scheme of the study.

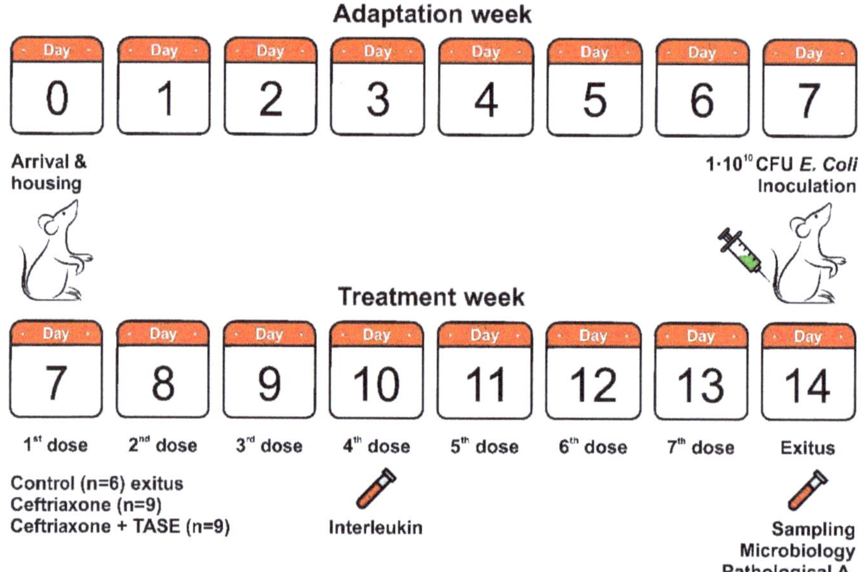

Figure 1. Chronological scheme of the study. Day zero started with the arrival of the animals, which included their respective housing, marking and the beginning of an adaptation process to the animal facility that lasted seven days. On the seventh day, *E. coli* inoculation and first treatment dose were performed. Blood samples for interleukin determination were taken 72 h after. On day 14, animals were sacrificed by lethal doses of anesthesia, with subsequent sampling for microbiology (blood and peritoneal fluid), interleukins (blood) and pathological anatomy (peritoneum, liver).

The following clinical parameters were evaluated daily: weight, mobility, appearance (normal, ocular secretions, nasal secretions, whisker position, lack of grooming, piloerection and dehydration), clinical signs (abdominal distension, hardening distension, temperature) and behavior (normal, hypoactive, lethargy; response to stimuli).

Interleukin (IL) 1β/IL-1F2, IL-6 and tumor necrosis factor alpha (TNF-α) were determined in blood samples on treatment day 3 (T3) and 7 (T7) with the corresponding Quantikine® Rat ELISA method (R&D Systems), following the manufacturer's instructions. All samples were diluted 1:3 in RD5Y diluent. Each diluted sample and standards were processed in duplicate. Internal quality control was performed with recombinant buffered IL control material of known concentration. The final reading was made at 450 nm and corrected to 550 nm in a microplate reader.

Peritoneal fluid was also sampled on T7 (last day of the experiment) directly from the peritoneal cavity. Peritoneal fluid study was performed to determine cellularity by Giemsa staining. During the exploratory laparotomy, the degree of peritoneal inflammation was evaluated macroscopically. Liver and peritoneum samples were taken for histopathological evaluation. Thus, samples were paraformaldehyde fixed, paraffin embedded, and 4 µm sections were made for hematoxylin/eosin staining to analyze the presence or absence of congestion and immune cells. A blinded expert pathologist evaluated the samples.

2.4. Statistical Analysis

A descriptive statistic of the quantitative variables was carried out to verify that the minimum and maximum values were in an adequate range. All data were expressed as mean ± standard error of the mean (x ± SEM).

To analyze the qualitative variables, proportions were compared with the chi-square test and Fisher's exact correlation. The means of the continuous variables were compared with U Mann–Whitney's test and normal distribution was verified by Shapiro Wilk's test.

In comparisons at different times within the same groups, tests were applied for paired variables (Student t for dependent variables or Wilcoxon test as the case may be). For the comparison between groups, the Kruskal–Wallis test was used as a function of normality. A significance level of 95% was used for statistical analysis. SPSS 25.0 for Windows (SPSS Inc., Chicago, IL, USA).

3. Results

Our model of abdominal sepsis is able to generate 100% lethality in rats if no antibiotic treatment is provided [17], so the control group could not finish the experimentation due to the fact that all members of the group died after 4–6 h post-inoculum. Unfortunately, one rat from group II also died after the inoculum despite the ceftriaxone administration, so we completed group II with eight rats.

In relation to clinical parameters, we found differences when comparing the body weights of group II and III on days 9, 10 and 11 ($p < 0.05$), corresponding to the third (T3), fourth (T4), and fifth (T5) dose of treatment (Figure 2). In relation to stress and suffering (nasal secretions, ocular secretions, whiskers position, lack of grooming, piloerection, lethargy, and diarrhea), we could only find differences in the level of activity of rats during the septic process 72 h post-inoculum (T3), showing greater hypoactivity in group II compared to group III ($p \leq 0.05$). Statistically significant differences were also observed in the position of the whiskers ($p \leq 0.05$) and in the presence of ocular secretions ($p \leq 0.05$) at that time point. No statistically significant differences were found in the rest of the clinical signs studied (Table 2), and from T4 to T7 (end of the study; data not shown).

In relation to the biochemical parameters studied (IL-1, IL-6, TNF- α), a comparison was made for the values of each interleukin between treated groups in T3 and T7 (Figure 3). Considering the levels of IL-1, in T3 we found statistically significant differences when comparing group II with respect to III ($p < 0.05$). It was also observed that IL-6 in T3 was lower in group III, although the values were not statistically significant. As for TNF-α, no differences between groups were assessed.

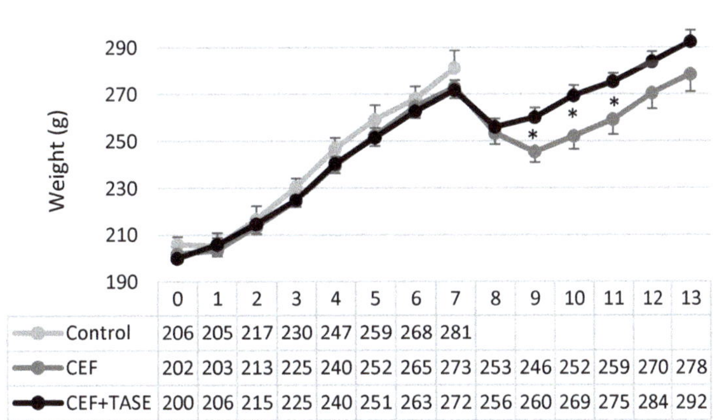

	0	1	2	3	4	5	6	7	8	9	10	11	12	13
Control	206	205	217	230	247	259	268	281						
CEF	202	203	213	225	240	252	265	273	253	246	252	259	270	278
CEF+TASE	200	206	215	225	240	251	263	272	256	260	269	275	284	292

Figure 2. Weight monitoring during experiment: control group (8 days), group treated with ceftriaxone (CEF; 14 days) and group treated with ceftriaxone + thiosulfinate-enriched *Allium sativum* extract (CEF + TASE; 14 days). On day 7, the bacterial inoculum was introduced. Mean ± SEM. * $p < 0.05$.

Table 2. Clinical parameters in relation to stress and suffering at T1, T2 and T3 (24, 48 and 72 h post-inoculum, respectively) for nasal secretions, eye secretions, position of whiskers, lack of grooming, hair erection, lethargy, and diarrhea. * $p \leq 0.05$.

		T1	
Variable, n/nt (%)	**CEF**	**CEF + TASE**	***p*-Value**
Nasal secretions 1	8/8 (100)	7/9 (78)	0.471
Ocular secretions 1	7/8 (88)	7/9 (78)	0.600
Whiskers separation 1	2/8 (25)	6/9 (67)	0.153
Lack of grooming 1	8/8 (100)	6/9 (67)	0.206
Piloerection 1	8/8 (100)	8/9 (89)	0.999
Hypoactivity 1	8/8 (100)	9/9 (100)	-
Diarrhea 1	2/8 (25)	0/9 (0)	0.206
		T2	
Variable, n/nt (%)	**CEF**	**CEF + TASE**	***p*-Value**
Nasal secretions 2	7/8 (88)	6/9 (67)	0.576
Ocular secretions 2	6/8 (75)	5/9 (56)	0.620
Whiskers separation 2	2/8 (25)	7/9 (78)	0.057
Lack of grooming 2	8/8 (100)	6/9 (67)	0.206
Piloerection 2	7/8 (88)	5/9 (56)	0.294
Hypoactivity 2	6/8 (75)	4/9 (44)	0.335
Diarrhea 2	2/8 (25)	0/9 (0)	0.206
		T3	
Variable, n/nt (%)	**CEF**	**CEF + TASE**	***p*-Value**
Nasal secretions 3	6/8 (75)	3/9 (33)	0.153
Ocular secretions 3	5/8 (63)	1/9 (11)	0.05 *
Whiskers separation 3	6/8 (75)	2/9 (22)	0.05 *
Lack of grooming 3	0/8 (0)	0/9 (0)	-
Piloerection 3	5/8 (63)	2/9 (22)	0.153
Hypoactivity 3	5/8 (63)	1/9 (11)	0.05 *
Diarrhea 3	1/8 (13)	0/9 (0)	0.471

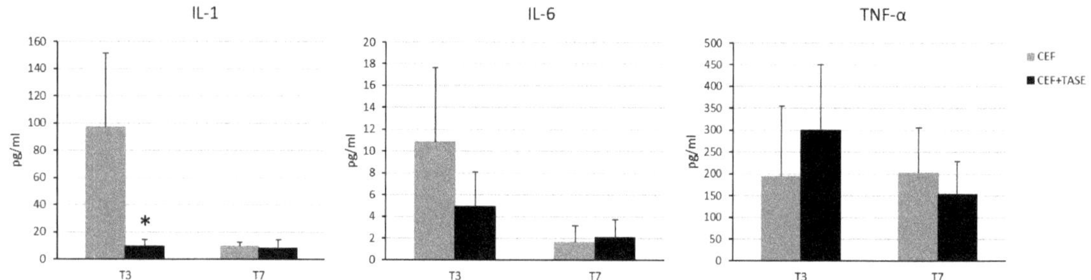

Figure 3. Interleukin levels in T3 and T7 (treatment day 3 and 7, respectively). CEF = ceftriaxone. CEF + TASE = ceftriaxone + thiosulfinate-enriched *Allium sativum* extract. Mean ± SEM. * $p < 0.05$.

The peritoneal liquid and blood cultures of the control group were positive for multi-sensitive *E. coli* ATCC 25922 and identical to the inoculum (Table 3). Additionally, as mentioned before, all the rats from the control group did not recover from the inoculum and died after 4–6 h, showing that our sepsis model is lethal if left untreated. In group II, only one rat died after inoculation, and of the remaining eight rats, six showed *Enterococcus faecalis* in blood cultures, and two in peritoneal liquid. In the blood cultures of the two remaining rats, the multi-sensitive bacteria *E. coli* ATCC 25922 appeared. In group III, *E. coli* ATCC 25922 was not detected neither in blood nor in peritoneal liquid. In fact, eight out of nine rats showed *Enterococcus faecalis* in blood samples, and only one out of nine rats showed *Enterococcus faecalis* in peritoneal liquid. There was also one rat that was negative for both bacteria.

Table 3. Results of blood and peritoneal fluid cultures stratified by treatment groups and their respective antibiogram.

<table>
<tr><th colspan="6">RESULTS OF MICROBIOLOGIC STUDIES</th></tr>
<tr><th></th><th colspan="2">Number of Rats</th><th>Blood Cultures</th><th>Peritoneal Liquid Culture</th><th>Antibiogram (Sensitive to Ampicilin, Vancomycin, Teicoplanin)</th></tr>
<tr><td>Control (Group I)</td><td colspan="2">3</td><td>Positive E. coli ATCC 25922</td><td>Positive E. coli ATCC 25922</td><td>Multisensitive</td></tr>
<tr><td rowspan="4">Ceftriaxone (Group II)</td><td rowspan="3">8</td><td rowspan="2">6</td><td rowspan="2">Enterococcus faecalis</td><td>2–E. faecalis</td><td rowspan="2">Sensitive</td></tr>
<tr><td>4–Negative</td></tr>
<tr><td>2</td><td>Positive E. coli ATCC 25922</td><td>Negative</td><td>Multisensitive</td></tr>
<tr><td colspan="2">1</td><td>Exitus before 24 h (no samples collected)</td><td>-</td><td>-</td></tr>
<tr><td rowspan="3">Ceftriaxone + TASE * (Group III)</td><td colspan="2" rowspan="2">8</td><td rowspan="2">E. faecalis</td><td>1–E. faecalis</td><td rowspan="2">Sensitive</td></tr>
<tr><td>7–Negative</td></tr>
<tr><td colspan="2">1</td><td>Negative</td><td>Negative</td><td>-</td></tr>
</table>

* TASE (0.5 mg/kg; referred to allicin content).

Regarding the histopathological analysis and organ evaluation, the inflammatory cell count, the presence of bacteria in the liver and on the peritoneal surface, as well as the congestion and hepatic vacuolization between treatment groups (group II and III), no statistically significant differences were found (Table 4).

Table 4. Histopathological analysis and organs evaluation in relation to inflammation, bacteria, congestion, and vacuolization.

Variable, n/nt (%)	CEF	CEF + TASE	*p*-Value
Liver—hepatic congestion	8/9 (89)	8/9 (89)	-
Liver—sinusoidal PMN leukocytes	4/9 (44)	2/9 (22)	0.62
Liver—serosa PMN leukocytes	3/9 (33)	1/9 (11)	0.576
Liver—bacteria	2/9 (22)	0/9 (0)	0.471
Liver—perinuclear vacuolization	6/9 (33)	7/9 (78)	0.599
Peritoneum—PMN leukocytes	3/9 (33)	2/9 (22)	0.999
Peritoneum—bacteria	3/9 (33)	1/9 (11)	0.576

4. Discussion

Until now, many models of sepsis have been described in animal experimentation, but most of them failed to replicate the human heterogeneous septic process, which is dependent on the genetic susceptibility of each individual and influenced by sex, age, comorbidities and drug consumption [18,19]. The murine model described here generates an efficient, controlled and easily reproducible intraperitoneal infection that could serve as a basis for future lines of research.

Scientific research based on the use of garlic derivatives has led to ambiguous conclusions on the beneficial effects of this plant, thus preventing the application of garlic products in the treatment of certain diseases [20]. This situation can be attributed to several factors, among which the following can be highlighted: the chemical instability of this type of compound [21], the great diversity of industrial processes for their production, the lack of coherence in terms of the medical properties investigated, and the chemical composition of the products used in these clinical investigations [22]. With the freeze-dried garlic used in this experimental model, these deficiencies could be overcome by using stable and known concentrations of allicin and other thiosulfinates over time.

At present, there are no studies in the scientific literature that have demonstrated a decrease in morbidity and mortality in relation to sepsis, except for antibiotics and the goal-guided resuscitation strategy [23]. Some of the drugs that have been tested are, among others, corticoids [12], immunoglobulins [13], antithrombin III [14], vasopressin [24] or anti-TNF [25]. However, the clinical results obtained in our animal experimentation study are encouraging in this sense and confirm our working hypothesis showing earlier recovery of weight, less ocular secretions, separation of whiskers and decrease in hypoactivity in the group where TASE was administered. We also found lower levels of IL-1 on the third day of treatment, and a tendency to decrease the pro-inflammatory cytokine IL-6 in the TASE group. All these data would support the immunomodulatory role of the lyophilized garlic, thanks to its action in the inflammatory cascade [26], thus achieving the attenuation of sepsis and septic shock. Previous work has described the ability to suppress inflammatory signals of lipopolysaccharide (LPS) through the expression of anti-inflammatory genes, and the reduction in pro-inflammatory cytokines (IL-6 and MCP-1) [27]. The work of Lee et al. [28] showed the immunomodulatory activity of garlic in an experimental sepsis model where clamping and blind puncture were performed to induce peritonitis. The authors described how the administration of methyl 3-formyl-4-methylpentanoate (a natural compound derived from garlic) led to the inhibition of apoptosis of lymphocytes in the spleen and significantly inhibited the production of pro-inflammatory cytokines such as TNF-α, IL-1β and IL-6. The production of TNF-α and IL-6 stimulated by LPS was also strongly inhibited by the compound methyl 3-formyl-4-methylpentanoate sucrose in macrophages derived from mouse bone marrow [28]. In our study, we could not assess any significant difference on TNF-α levels in the TASE group.

The microbiological analysis showed how the blood cultures of the control group were positive for *E. coli*. This bacterium corresponded to the same inoculum with which sepsis was generated. However, no blood culture or peritoneal fluid were positive for *E. coli* in the TASE-treated group. This result highlights the fact that the inoculum was sensitive to the antibiotic and coadjuvant treatments used. In fact, it is known that there is a 90% sensitivity of *E. coli* to ceftriaxone and only 65% sensitivity of *Klebsiella pneumoniae*. Despite these moderately high percentages, we are currently in a global state of alarm due to the increased resistance of several microorganisms to this antibiotic compared to previous studies [29]. Moreover, most of the blood and peritoneal fluid cultures in our antibiotic treatment group showed the presence of *Enterococcus faecalis*. This result could be explained by considering the broad-spectrum efficacy of ceftriaxone on Gram-negative and some Gram-positive bacteria. This change of gastrointestinal microbiota would favor *Enterococcus faecalis*. One likely explanation for this selection is based on related studies in mice where LPS and flagellin from Gram-negative and anaerobic bacteria stimulated the production of RegIIIγ in Paneth cells by interactions with Toll-like receptors. Paneth cells have an important role in the defense mechanisms of the gastrointestinal tract in several animal species, thanks to their secretions of lysozyme, phospholipase A2 and defensins [30]. RegIIIγ is a C-type lectin receptor, capable of recognizing carbohydrates present on the surface of pathogens and is responsible of the internalization of the pathogen for antigen presentation and the induction of an immunological response. Thus, the level of RegIIIγ maintains the balance between the bacteria that compose the intestinal microbiota and the host [31]. When ceftriaxone used in our study killed the Gram-negative bacteria, it decreased the production of RegIIIγ and facilitated the growth of Gram-positive coccus (i.e., *Enterococcus faecalis*). Then, those Gram-positive could cross the intestinal barrier and reach the systemic circulation, liver and more. Therefore, if the antibiotic treatment persists for a long time, there could be a potential risk of bacterial endocarditis.

Few studies with animal models associate histological, clinical, and microbiological findings in intraperitoneal organs such as liver, peritoneum, and intestine after induced peritonitis. In this sense, our study tried to correlate those findings with the clinical response to a treatment based on a thiosulfinate-enriched garlic extract. Only Lee et al. [28] examined the lung after peritonitis for these inflammatory changes, and also after therapy with a garlic derivative. Unfortunately, we could not obtain any histopathological difference between treatment groups because the tissue and organ evaluations were assessed at the end of the experiment and rats from both treatment groups were mostly recovered from the septic insult. Moreover, our study has several limitations that are inherent to the animal model of sepsis and septic shock that we use. First, the amount of blood collected was limited and did not allow the measurement of a greater number of inflammatory factors and biomarkers of endothelial damage. Secondly, no measurements were taken in relation to myocardial function and macrocirculation such as mean arterial pressure, contractility, peripheral vascular preload, and resistance to perform a target-guided therapy as it is usually performed in humans.

5. Conclusions

Thiosulfinate-enriched *Allium sativum* extract used as an adjuvant to antibiotic treatment and to sepsis management could improve the response profile and attenuate the outcome of the sepsis shock, mostly during the first days of the combined treatment. Further research would be necessary to clarify the immunomodulatory role of this plant extract.

6. Patents

Patent WO 2008/102036 A1. Method for obtaining a freeze-dried, stable extract from plants of the *Allium* genus.

National patent (Spanish Trademark number ES2675282A1). *Allium sativum* extract, its use for the manufacture of a medicinal product for the treatment of diseases, and its obtaining procedure.

Author Contributions: Conceptualization, F.J.R.-C., D.P.-V. and J.M.P.-O.; data curation, L.M.-P. and J.R.M.-R.; formal analysis, F.J.R.-C., O.M., V.B., N.V., S.I. and V.M.; investigation, O.M., N.B.-R., R.G. and L.M.-P.; methodology, D.P.-V., P.V. and S.I.; project administration, F.J.R.-C., D.P.-V. and J.M.P.-O.; resources, F.J.R.-C., D.P.-V. and L.A.G.; supervision, F.J.R.-C., D.P.-V. and J.M.P.-O.; visualization, J.R.M.-R.; writing—original draft, F.J.R.-C., O.M. and S.I.; writing—review and editing, F.J.R.-C., P.V., V.B., N.B.-R., R.G. and J.M.P.-O. All authors have read and agreed to the published version of the manuscript.

Funding: There are no sources of funding for the work.

Institutional Review Board Statement: The study (ref. PI-HGUCR 1/2014) was approved by the Animal Experimentation Committee of the University General Hospital, Ciudad Real. It was authorized by the Office of Agriculture of Castilla-La Mancha (Spain) and this experiment followed the ARRIVE guidelines developed by the National Center for the Replacement, Refinement and Reduction of Animals in Research (nc3rs).

Informed Consent Statement: Not applicable.

Data Availability Statement: Not applicable.

Acknowledgments: The authors are grateful to the Pathology, Microbiology and Clinical Analysis Departments of the Hospital General Universitario de Ciudad Real. We also thank Clara Villar-Rodríguez for her technical support (Translational Research Unit, Hospital General Universitario de Ciudad Real).

Conflicts of Interest: L.A.G. is part of the authors of the registered brand Aliben© (European Trade mark number 10543429) which entitles the lyophilized *Allium sativum* extract employed in this study (patent WO 2008/102036 A1. Method for obtaining a freeze-dried, stable extract from plants of the *Allium* genus). D.P., P.V., J.M.P-O., J.R.M-R., L.A.G. and F.J.R-C. are co-contributors of a national registered patent (Spanish Trade mark number ES2675282A1), which employs the lyophilized *Allium sativum* extract, its use for the manufacture of a medicinal product for the treatment of diseases, and its obtaining procedure.

References

1. Singer, M.; Deutschman, C.S.; Seymour, C.W.; Shankar-Hari, M.; Annane, D.; Bauer, M.; Bellomo, R.; Bernard, G.R.; Chiche, J.D.; Coopersmith, C.M.; et al. The Third International Consensus Definitions for Sepsis and Septic Shock (Sepsis-3). *JAMA* **2016**, *315*, 801–810. [CrossRef]
2. Derek, C.; Angus, D.C.; van der Poll, T. Severe sepsis and septic shock. *N. Engl. J. Med.* **2013**, *369*, 840–851.
3. Deutschman, C.S.; Tracey, K.J. Sepsis: Current dogma and new perspectives. *Immunity* **2014**, *40*, 463–475. [CrossRef]
4. Sprung, C.L.; Sakr, Y.; Vincent, J.L.; Le Gall, J.R.; Reinhart, K.; Ranieri, V.M.; Gerlach, H.; Fielden, J.; Groba, C.B.; Payen, D. An evaluation of systemic inflammatory response syndrome signs in the sepsis occurrence in acutely III patients (SOAP) study. *Intensive Care Med.* **2016**, *32*, 21–27.
5. Tattelman, E. Health effects of garlic. *Am. Fam. Physician* **2005**, *72*, 103–106. [PubMed]
6. Keiss, H.P.; Dirsch, V.M.; Hartung, T.; Haffner, T.; Trueman, L.; Auger, J.; Kahane, R.; Vollmar, A.M. Garlic (Allium sativum L.) modulates cytokine expression in lipopolysaccharide-activated human blood thereby inhibiting NF-κB activity. *J. Nutr.* **2003**, *133*, 2171–2175. [CrossRef] [PubMed]
7. Sela, U.; Ganor, S.; Hecht, I.; Brill, A.; Miron, T.; Rabinkov, A.; Wilchek, M.; Mirelman, D.; Lider, O.; Hershkoviz, R. Allicin inhibits SDF-1 a -induced T cell interactions with fibronectin and endothelial cells by down-regulating cytoskeleton rearrangement, Pyk-2 phosphorylation and VLA-4 expression. *Immunology* **2014**, *111*, 391–399. [CrossRef] [PubMed]
8. Bruk, R.; Aeed, H.; Brazovsky, E.; Noor, T.; Herskoviz, R. Allicin, the active component of garlic, prevents immune-mediated, concanavalin A-induced hepatic injury in mice. *Liver Int.* **2005**, *25*, 613–621. [CrossRef]
9. Apitz-Castro, R.; Escalante, J.; Vargas, R.; Jain, M.K. Ajoene, the antiplatelet principle of garlic, synergistically potentiates the antiaggregatory action of prostacyclin, forskolin, indomethacin and dypiridamole on human platelets. *Thromb. Res.* **1984**, *42*, 303–311. [CrossRef]
10. Rhodes, A.; Evans, L.E.; Alhazzani, W.; Levy, M.M.; Antonelli, M.; Ferrer, R.; Kumar, A.; Sevransky, J.E.; Sprung, C.L.; Nunnally, M.E.; et al. Surviving Sepsis Campaign: International Guidelines for Management of Sepsis and Septic Shock: 2016. *Crit. Care Med.* **2017**, *45*, 486–552. [CrossRef]
11. Ranieri, V.M.; Thompson, B.T.; Finfer, S.; Barie, P.S.; Dhainaut, J.F.; Douglas, I.S.; Gårdlund, B.; Marshall, J.C.; Rhodes, A. PROWESS-SHOCK Academic Steering Committee. Unblinding plan of PROWESS-SHOCK trial. *Intensive Care Med.* **2011**, *37*, 1384–1385. [CrossRef] [PubMed]
12. Annane, D.; Sébille, V.; Charpentier, C.; Bollaert, P.E.; François, B.; Korach, J.M.; Capellier, G.; Cohen, Y.; Azoulay, E.; Troché, G.; et al. Effect of treatment with low doses of hydrocortisone and fludrocortisone on mortality in patients with septic shock. *JAMA* **2002**, *88*, 862–871. [CrossRef]
13. Taccone, F.S.; Stordeur, P.; De backer, D.; Creteur, J.; Vincent, J.L. Gamma-globulin levels in patients whit community acquired septic shock. *Shock* **2009**, *32*, 379–385. [CrossRef] [PubMed]
14. Allingstrup, M.; Wetterslev, J.; Ravn, F.B.; Møller, A.M.; Afshari, A. Antithrombin III for critically ill patients: A systematic review with meta-analysis and trial sequential analysis. *Intensive Care Med.* **2016**, *42*, 505–520. [CrossRef]
15. Reynolds, P.; Wall, P.; van Griensven, M.; McConnell, K.; Lang, C.; Buchman, T. Shock supports the use of animal research reporting guidelines. *Shock* **2012**, *8*, 1–3. [CrossRef]
16. Fujisawa, H.; Suma, K.; Origuchi, K.; Kumagai, H.; Seki, T.; Ariga, T. Biological and chemical stability of garlic-derived allicin. *J. Agric. Food Chem.* **2008**, *56*, 4229–4235. [CrossRef] [PubMed]
17. Montenegro, O.; Illescas, S.; González, J.C.; Padilla, D.; Villarejo, P.; Baladrón, V.; Galán, R.; Bejarano, N.; Medina-Prado, L.; Villaseca, N.; et al. Development of animal experimental model for bacterial peritonitis. *Rev. Esp. Quimioter.* **2020**, *33*, 18–23. [CrossRef]
18. Guillon, A.; Preau, S.; Aboab, J.; Azabou, E.; Jung, B.; Silva, S.; Textoris, J.; Uhel, F.; Vodovar, D.; Zafrani, L.; et al. Preclinical septic shock research: Why we need an animal ICU. *Ann. Intensive Care.* **2019**, *9*, 66. [CrossRef]
19. Van der Poll, T. Preclinical Sepsis Models. *Surg. Infect.* **2012**, *13*, 287–292. [CrossRef]
20. Sivam, G.P. Protection against Helicobacter pylori and other bacterial infections by garlic. *J. Nutr.* **2001**, *131*, 1106S–1108S. [CrossRef]
21. Lawson, L.; Wood, S.C.; Hunghes, B.G. HPLC analysis of allicin and other thiosulfinates in garlic clove homogenates. *Plant Med.* **1991**, *57*, 363–370. [CrossRef] [PubMed]
22. Staba, E.J.; Lash, L.; Staba, J.E. A commentary on the effects of garlic extraction and formulation on product composition. *J. Nutr.* **2001**, *131*, 1118S–1119S. [CrossRef] [PubMed]
23. Gyawali, B.; Ramakrishna, K.; Dhamoon, A.S. Sepsis: The evolution in definition, pathophysiology, and management. *SAGE Open Med.* **2019**, *7*, 2050312119835043. [CrossRef] [PubMed]
24. Patel, B.M.; Chittock, D.R.; Russell, J.A.; Walley, K.R. Beneficial effects of short-term vasopressin infusion during severe septic shock. *Anesthesiology* **2002**, *96*, 576–582. [CrossRef]
25. Reinhart, K.; Karzai, W. Anti-tumor necrosis factor therapy in sepsis: Update on clinical trials and lesson learned. *Critical Care Med.* **2001**, *29*, S121–S125. [CrossRef] [PubMed]
26. Arreola, R.; Quintero-Fabián, S.; López-Roa, R.I.; Flores-Gutiérrez, E.O.; Reyes-Grajeda, J.P.; Carrera-Quintanar, L.; Ortuño-Sahagún, D. Immunomodulation and anti-inflammatory effects of garlic compounds. *J. Immunol. Res.* **2015**, *15*, 401630. [CrossRef]

27. You, S.; Nakanishi, E.; Kuwata, H.; Chen, J.; Nakasone, Y.; He, X.; He, J.; Liu, X.; Zhang, S.; Zhang, B.; et al. Inhibitory effects and molecular mechanisms of garlic organosulfur compounds on the production of inflammatory mediators. *Mol. Nutr. Food Res.* **2013**, *57*, 2049–2060. [CrossRef]
28. Lee, S.K.; Park, Y.J.; Ko, M.J.; Wang, Z.; Lee, H.Y.; Choi, Y.W.; Bae, Y.S. A novel natural compound from garlic (*Allium sativum* L.) with therapeutic effects against experimental polymicrobial sepsis. *Biochem. Biophys. Res. Commun.* **2015**, *464*, 774–779. [CrossRef]
29. Bushra, R.; Sial, A.A.; Rizvi, M.; Shafiq, Y.; Aslam, N.; Bano, N. Report: Sensitivity Pattern of Ceftriaxone against different Clinical Isolates. *Pak. J. Pharm. Sci.* **2016**, *29*, 249–253.
30. Wang, J.; Tian, F.; Wang, P.; Zheng, H.; Zhang, Y.; Tian, H.; Zhang, L.; Gao, X.; Wang, X. Gut Microbiota as a Modulator of Paneth Cells During Parenteral Nutrition in Mice. *J. Parenter. Enteral Nutr.* **2018**, *42*, 1280–1287. [CrossRef]
31. Vaishnava, S.; Yamamoto, M.; Severson, K.M.; Ruhn, K.A.; Yu, X.; Koren, O.; Ley, R.; Wakeland, E.K.; Hooper, L.V. The Antibacterial Lectin RegIIIγ Promotes the Spatial Segregation of Microbiota and Host in the Intestine. *Science* **2011**, *334*, 255–258. [CrossRef] [PubMed]

Article

Cellular Antioxidant Effects and Bioavailability of Food Supplements Rich in Hydroxytyrosol

Cecilia Bender *, **Sarah Straßmann and Pola Heidrich**

Institut Kurz GmbH, Stöckheimer Weg 1, 50829 Köln, Germany; s.strassmann@institut-kurz.de (S.S.); p.heidrich@institut-kurz.de (P.H.)
* Correspondence: c.bender@kurz-italia.com

Abstract: The present study evaluates the effect of olive (*Olea europaea* L.) vegetation water on human cells regarding its antioxidant properties and radical scavenger bioactivities. To this aim, two food supplements containing concentrated olive water in combination with 6% lemon juice or 70% grape juice, respectively, were assessed in different oxidation assays. From the investigated polyphenols, hydroxytyrosol, present in olives and in a lesser extent in grapes, was found to be the most abundant in both formulations, followed by tyrosol and oleuropein for the olive-derived concentrate with lemon juice, and by proanthocyanidins and tyrosol for the olive concentrate with grape juice. Cellular studies suggest that both formulations are effective antioxidants. In particular, the combination of olive and grape extracts showed a remarkable superoxides-, hydroxyl radicals-, and hydrogen peroxides-scavenging activity, while the formulation containing 94% olive concentrate wasmore potent in protecting the cells against lipoxidation. Both products showed a significant and similar effect in preventing advanced glycation end products' (AGEs) formation. In addition, preliminary data indicate that hydroxytyrosol is absorbed into the human body when administered via these hydrophilic matrices, as confirmed by the urinary excretion of free hydroxytyrosol. Since the availability of phytochemicals largely depends on the vehicle in which they are solved, these findings are of relevance and contribute to supporting the healthful effects here assessed in a cellular environment.

Keywords: hydroxytyrosol; olive extract; olive polyphenols; grape extract; oleuropein; antioxidant capacity; F2-isoprostanes; AGEs

Citation: Bender, C.; Straßmann, S.; Heidrich, P. Cellular Antioxidant Effects and Bioavailability of Food Supplements Rich in Hydroxytyrosol. *Appl. Sci.* **2021**, *11*, 4763. https://doi.org/10.3390/app11114763

Academic Editor: Luca Mazzoni

Received: 30 April 2021
Accepted: 20 May 2021
Published: 22 May 2021

1. Introduction

Many consumers associate the fruit of the olive tree (*Olea europaea* L.) mainly with the resulting oil, the precious olive oil that is considered particularly healthy compared to other oils. The raw olive fruit contains several types of phenols, thecontents of which vary with the olive cultivar, mainly tyrosol and its derivatives, phenolic acids, and flavonoids [1]. However, many of these substances that the olive has to offer from a health perspective are water-soluble, and thus remain largely in the residue of the olive pressing and the oil contains only a small part. This is the case of hydroxytyrosol, a polar phenol slightly soluble in fats, which can be found in olives as a simple phenol, or either esterified with elenolic acid to form oleuropein aglycone, and which is naturally present in significantly higher concentrations in the olive fruit's aqueous fraction.

The vegetation water, resulting from the pressing of the olive fruits duringthe production of the olive oil, is rich in bioactive compounds, particularly polar phenols, and typically contains 98% of the total phenols of the olive fruit [2].

The positive health effects of olive polyphenols are already known; in particular, hydroxytyrosol has potential antioxidant, anti-inflammatory, and health benefits mainly related with cardiovascular diseases [3–6].

Bioavailability and pharmacokinetic analyses, which were mainly reported with pure hydroxytyrosol and with olive oil, suggest that hydroxytyrosol can be rapidly absorbed from blood and distributed in the human body [7], metabolized, and quickly eliminated in urine mainly as glucuronide and sulfate [8]. Currently, hydroxytyrosol from different sources is available on the market. Its absorption and subsequent urine excretion may be dependent on the vehicle of administration [9]. Thus, the bioavailability of hydroxytyrosol and its precursors (oleuropein and tyrosol) from those specific sources would be a prerequisite for its health effects in humans.

The present study addresses the bioactivity of hydroxytyrosol-rich extracts, obtained from the vegetation water resulting from olive oil production. Herein, hydroxytyrosol is present as both a simple phenol and as oleuropein aglycone.

For the study, olive-derived concentrates combined with 6% lemon juice or 70% grape juice and marketed as liquid supplements were characterized. Bioactivities, mainly related to the antioxidant potential, were evaluated in cultured cells by means of the antioxidant capacity (cellular antioxidant activity assay, superoxide dismutase and catalase activities), the protection against lipoxidation (inhibition of F2-isoprostanes formation) and glycation (inhibition of AGEs formation). In addition, preliminary data on the bioavailability and urinary recovery of free hydroxytyrosol through acute administration of the food supplements are presented from an open-label cross-over study with four volunteers.

Despite the difference in the composition of both formulations, the main phytochemical in the ones that were investigated was hydroxytyrosol, present in both olive fruit and to a lesser extent in grapes. The treatment of the cells with the supplements gave positive results, although these were slightly different in magnitude, through antioxidant actions. The high bioactivity observed suggests a possible application in the maintenance of the cellular redox state and for related health benefits.

2. Materials and Methods

2.1. Standards and Reagents

2,2′-azobis [2-methylpropionamide] dihydrochloride (AAPH), quercetin dihydrate, and 2′-7′-dichlorodihydrofluorescin diacetate (DCFH2-DA) were purchased from Sigma-Aldrich (Milan, Italy). Hydroxytyrosol and oleuropein were procured from Cayman Chemical (Ann Arbor, MI, USA). Resveratrol was purchased from Sigma-Aldrich (Steinheim, Germany). Dulbecco's modified Eagle's medium (DMEM) high-glucose culture media, L-glutamine, trypan blue solution, and trypsin-EDTA solution 10X were culture grade and purchased from Merck (Milan, Italy). Fetal bovine serum (FBS), Dulbecco's phosphate buffered saline (PBS) without Mg^{2+} and Ca^{2+}, and Hank's balanced salts solution (HBSS) were culture grade and purchased from Euroclone SpA (Milan, Italy). Water, acetonitrile, formic acid, and methanol (all LC-MS-grade) were purchased from VWR Chemicals (Darmstadt, Germany). All other chemicals were analytical grade and purchased from common sources.

2.2. Sample Material

The food supplements analyzed are derived from olive fruit (*Olea europaea* L.) vegetation water subjected to filtration and concentration, and were supplied by Fattoria La Vialla (Castiglion Fibocchi, Arezzo, Italy). The commercial brands are Oliphenolia bitter™ and Oliphenolia™, hereinafter referred to as P-1 and P-2, respectively. P-1 consists of 94% concentrated olive aqueous fraction and 6% lemon juice (*Citrus limon* L. fructus); while P-2 is characterized by 30% concentrated olive extract and 70% grape juice (*Vitis vinifera* L. fructus).

2.3. Analysis of Polyphenols

The samples were diluted with methanol (50:50 v/v), ultrasonicated, centrifuged, and filtrated through 0.45 μm regenerated cellulose filters prior to measurement by UHPLC-MS, with an Acquity UPLC I-Class system coupled to a XEVO-TQS micro mass spectrometer (bothWaters, Milford, MA, USA). The instrument consisted of a sample manager cooled at 10 °C, a binary pump, a column oven, and a diode array detector measuring at 280 nm.

The column oven temperature was set at 40 °C. The gradient started with 2% A and raised linearly to 15% within 5.5 min, then to 100% A within 1 min before holding for 1.5 min as a washing step; it then decreased back to 2% B within 1 min and was equilibrated for 2 min. Eluent B was water with 0.1% formic acid, eluent A was acetonitrile with 0.1% formic acid, the flow was 0.4 mL/min on an HSS T3 RP column (150 mm x 2.1 mm, 1.7 µm particle size) combined with a precolumn (Acquity UPLC HSS T3 VanGuard, 100 Å, 2.1 mm × 5 mm, 1.8 µm), both from Waters (Milford, MA, USA). The injection volume was 2 µL.

The peaks were identified by MS/MS (MRM 153 > 123 for hydroxytyrosol, 539 > 377 for oleuropein and SIR 137 for tyrosol in negative mode and MRM 229 > 135 for resveratrol operating in positive ion mode). The source voltage was kept at 1.5 kV, and the cone voltage was 20 V. The source temperature was set at 150 °C and the desolvation temperature at 350 °C with a desolvation gas flow of 650 L/h and a cone gas flow of 50 L/h. Standard substances were used as reference.

Proanthocyanidin monomers were determined according to Kelm et al. [10].

Data were acquired and processed using MassLynx (Waters, Milford, MA, USA).

2.4. Cell Cultures

Human hepatocellular carcinoma (HepG2) and human keratinocytes (HaCat) cell lines were obtained from CLS (Cell Lines Service GmbH, Germany) and cultured at 37 °C under a humidified atmosphere of 5% CO_2 in DMEM containing 2 mM L-Glutamine, 4.5 g/L glucose, and 10% of heat-inactivated FBS. Experiments were performed with DMEM low-glucose (Lonza Ltd., Morristown, NJ, USA) supplemented with 2 mM L-Glutamine without FBS in either, 6-well culture plates for AGEs, catalase, and superoxide dismutase (SOD), 12-well plates for F2-isoprostanes, or 96-well black plates for the cellular antioxidant activity (CAA) and vitality assays. For each cell-based test, P-1 and P-2 samples were centrifuged, sterile-filtered and directly diluted into culture media before testing.

2.5. Cellular Viability Assay

To determine the optimal growth conditions of cells following 4 h treatment with the sample material, five serial dilutions were evaluated for P-1 and P-2 (range 1:150 to 1:750 and 1:250 to 1:1250 for HaCat and HepG2 cells, respectively), and the metabolic activity was monitored using the resazurin toxicity assay according to the manufacturer's instructions (Tox-8 kit, Sigma-Aldrich, Italy). Fluorescence was read at 37 °C (Em. 590 nm/Ex. 540 nm) in a multiwell fluorescence reader (Fluostar Optima, BMG LabTech, Offenburg, Germany). Data were processed using Mars 2.0 Optima Data Analysis software (BMG LabTech GmbH, Germany).

2.6. Cellular Antixidant Activity (CAA)

The intracellular reactive oxygen species (ROS) formation was detected with the CAA method by spectrofluorimetry using the cell-permeable probe DCFH2-DA, as previously described [11,12]. Briefly, HepG2 cells were cultured until confluence and pre-incubated for an hour with DCFH2-DA and increasing concentrations of the sample dilutions (1:750 to 1:250 v/v) or the quercetin standard. After the addition of AAPH, the absorbed probe was oxidized to a high fluorescent molecule within the cytoplasm, which was measured at 37 °C for an hour at excitation 485 nm and emission 540 nm (Fluostar Optima, BMG LabTech, Germany). Raw data were analyzed using MARS 2.0 Optima Data Analysis software (BMG LabTech, Germany). Results are expressed as µmol of quercetin equivalency (QE) per mL of product and as the mean of five independent measurements ± standard deviation.

2.7. Cellular Extract Preparation

At the indicated time points, the cells were harvested in ice-cold PBS and collected in 2 mL centrifuge tubes before homogenization for the AGEs, catalase and SOD experiments. For whole lysates preparation, samples were homogenized using the Cell Disruptor Genie® (Scientific Industries Inc., Bohemia, NY, USA) with 0.5 mm glass beads, according to the manufacturer's instructions. Whole protein lysates were obtained by centrifugation for 10 min at 10,000× g, clear supernatants were transferred to clean tubes and their total protein content was determined according to the Bradford method [13], and they were then preservedat −80 °C for further analysis.

2.8. Catalase Activity Assay

HepG2 cells were cultured without (untreated control) or with the specific sample material (dilution 1:750 *v/v*) for 72 h. After treatment cells were harvested and lysed, the catalase activity was immediately measured by fluorescence using a commercial assay (Arbor Assays Ltd., Ann Arbor, MI, USA, Cat. No: K033-F1) according to manufacturer's instructions. Raw data were analyzed using Mars 2.0 Optima Data Analysis software (BMG LabTech GmbH, Germany), and the results were normalized with the total protein content, expressed as the mean of two experiments and as units of catalase activity per mg of protein ± standard deviation, and then compared with the untreated control.

2.9. Superoxide Dismutase (SOD)

HepG2 cells were incubated for 72 h with 1:750 *v/v* dilution of the samples.After treatment cells were harvested and lysed for further absolute quantification of SOD activity using a commercial kit (Sigma-Aldrich, Italy; SOD assay, Cat. No: 19160) following the manufacturer's recommendations. In the presence of oxygen, xanthine oxidase generates $O_2^{\bullet-}$, which in turn converts a colorless substrate into a yellow product. Samples with increasing levels of SOD cause a decrease in the $O_2^{\bullet-}$ concentration, reducing the yellow color, which is read at 450 nm. Raw data were analyzed using Mars 2.0 Optima Data Analysis software (BMG LabTech GmbH, Germany), and the results were normalized, expressed as mean of three experiments in terms of units of SOD activity per mg of protein ± standard deviation, and compared with the untreated cells.

2.10. Endogenous F2-Isoprostanes Measurement

HaCat cells were seeded in 12-well plates (500,000 cells/mL) and pre-incubated overnight without (untreated control) or with diluted samples (1:750 *v*/*v*). After replacing the culture media, lipoxidation was provoked by incubating with AAPH 1 mM for 2.5 h. Supernatants were then removed, centrifuged and immediately investigated for 8-epi PGF2α concentrations using a commercial ELISA kit (item n. 516360, Cayman Chemical, USA) following manufacturer's protocol. Results, expressed as the mean of three experiments ± standard deviation, were determined using Mars 2.0 Optima Data Analysis software (BMG LabTech GmbH, Germany).

2.11. Endogenous AGEs Measurement

HaCat cells were plated and incubated overnight in complete culture medium. For the experiments the culture medium was replaced with serum-free medium without (untreated control) or with diluted samples (1:750 *v*/*v*). After an hour of incubation, the medium was replaced with an AGEs-inducer solution containing glyoxal and/or S-p-bromobenzylglutathione cyclopentyl diester at increasing concentrations and incubated for 4 h. Whole protein lysates were processed for quantitative determination of AGEs with a commercial ELISA kit (Cusabio Ltd., Wilmington, DE, USA, Cat. No: CSB-E09412h) according to the manufacturer's recommendations. Spectrophotometric measurements were recorded with a multiwell reader (Fluostar Optima, BMG Labtech, Germany), and the raw data were analyzed using Mars 2.0 Optima Data Analysis software (BMG LabTech GmbH, Germany) and expressed as the mean of two experiments ± standard deviation and as AGEs concentrations relative to protein content.

2.12. Bioavailability

A pilot, open-label, single-dose, two-period, cross-over design study was conducted in our laboratory to test the urinary excretion of free hydroxytyrosol and trans-resveratrol in self-reported healthy volunteers. In the investigation, two males and two females received, after an overnight fast, a single dose (50 mL) of one food supplement with 200 mL of water separated by one week wash-out period before administration of a single dose of the second food supplement. Urine samples were collected immediately before intake (baseline) and after 30 min of intake. The samples were centrifuged and filtered before being measured by LC-MS/MS, as described above. Freshly prepared urine-blank samples spiked with standards were used for the hydroxytyrosol calibration.

3. Results

3.1. Characterization of the Food Supplements

Table 1 shows the chemical characterization of a representative batch of P-1 and P-2. Of the selected phytochemicals identified, hydroxytyrosol is the main phenolic compound in both samples, as shown by the LC-MS analysis.

Table 1. Chemical analysis of selected phytochemicals of olive-derived food supplements. Values are expressed as mean ± standard deviation of 2 determinations. [a] single determination. * manufacture's data; – not determined.

	P-1	P-2
Parameters	**Content (mg/L)**	
Hydroxytyrosol	1196 ± 35	1399 ± 45
Oleuropein [a]	18.1	11.9
Tyrosol	27.0 ± 0.4	43.9 ± 3.5
Trans-resveratrol	–	3.34 ± 0.04
Total phenolic content [a]*	11,220	11,371
Proanthocyanidin monomers	–	165 ± 4

A comparison of the average amount of total phenolics gives similar results for both samples. Despite the ratio of olive water being lower in P-2, this compound contains higher amounts of hydroxytyrosol and derivatives. This result is consistent with P-2's combination of grape juice and further concentrated olive vegetation water, as both would contribute to the hydroxytyrosol content. In addition, P-2 comprises trans-resveratrol and proanthocyanidin monomers from grapes, phytonutrients that have considerable antioxidant properties, and that are also said to have positive effects on health [14–18]. It is very likely that P-2 contains additional polyphenols from grapes, and that the two products may also differ in the content of vitamins, or the phytocomplex they contain, though they will not be described in detail here, as hydroxytyrosol is the main polyphenol from the olive on which we are focused.

3.2. Bioassays

Since the aqueous olive extracts show a high content of natural phenols, we examined their possible connection with the prevention of oxidative stress. To this end, sample dilutions found to be non-toxic to the HepG2 and HaCat cell lines were further investigated for their antioxidant effects.

3.2.1. CAA

The CAA measures the ability of an antioxidant sample to inhibit the formation of induced reactive oxygen species (ROS) within cells [11]. While antioxidant samples inhibit the formation of free radicals in a dose-dependent manner, an increment in intracellular fluorescence denotes an increment in ROS formation [19]. Both purified extracts exert a strong antioxidant activity in the CAA assay (Table 2 and Figure 1) by inhibiting the production of peroxides at the intracellular level.

Table 2. CAA results. ⁊ HepG2 cells' viability > 90% in the dilution range tested; ** QE = quercetin equivalency; * $p < 0.05$ according to *t* test.

Treatment	Dilution Range ⁊ (*v*/*v*)	µmol QE/mL **
Sample P-1	1:250- 1:750	6.77 ± 0.77 *
Sample P-2	1:250- 1:750	9.20 ± 1.57 *

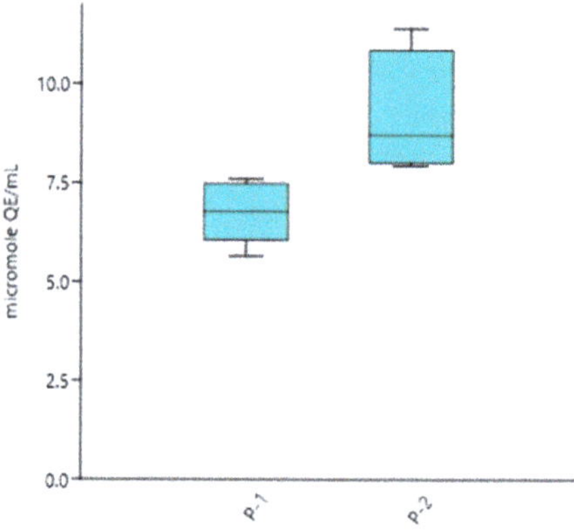

Figure 1. CAA of P-1 and P-2. Data are represented as box plots, showing median, 25–75% quartiles, standard error, and total range of values from 5 experiments. QE = quercetin equivalency.

We found that P-2, containing grape and olive concentrates, displayed a more potent cellular antioxidant potential compared to P-1 containing 94% of the olive extract ($p < 0.05$). In a preventive treatment, the CAA values were 9.20 and 6.77 µmoles quercetin equivalents per mL for P-2 and P-1, respectively.

Interestingly, the results for both purified extracts of olive water are even superior to those of extracts known to be highly antioxidant, such as pure chokeberry juice, which in the CAA test yields an average of 5.27 µmol QE/mL (unpublished results).

Similarly, data reported elsewhere have showed that polyphenols extracted from olive vegetation water are able to inhibit ROS production in human neutrophils and in endothelial cells exposed in vitro [20,21].

3.2.2. SOD and Catalase Activities

To minimize the harmful effects of excess ROS, aerobic organisms have developed several lines of antioxidant defense, which are employed in addition to the direct action of antioxidant molecules [22]. Among these, the enzymatic defense plays a key role in oxidative damage prevention [23]. Catalase and superoxide dismutase (SOD) directly scavenge hydrogen peroxide and superoxide radicals, respectively, converting them into less reactive species.

Given the performance of P-1 and P-2 samples in the CAA test, we evaluated the endogenous SOD and catalase defense in HepG2 human cells after treatment with both formulations, and then compared these to the untreated cells (baseline activity). Under the conditions tested, both supplements showed an increased SOD activity ($p < 0.01$) in liver cells compared to the untreated control (Table 3). A similar effect on catalase activity was found with the P-2 sample, but not with the P-1 sample.

Table 3. Antioxidant enzyme activity measured in cell-lysates after 72 h of treatment with the sample material. Average results are expressed in terms of activity units per mg of protein ± standard deviation. UTC: untreated control cells. * $p < 0.01$ according to *t*-test.

Treatment	U SOD/mg Protein	U Catalase/mg Protein
UTC	6.87 ± 0.44 *	5.52 ± 0.04
P-1	7.93 ± 0.13 *	5.51 ± 0.03
P-2	9.23 ± 0.24 *	7.39 ± 0.18

Similar results supporting the increase in catalase and SOD after treatment with phenolic-rich olive water were obtained by others through a series of cellular tests [20], and in vivo in rats' liver [24].

In accordance with the results obtained in the CAA dosage, the P-2 formulation containing grape and olive extracts exerts a stronger antioxidant effect than the formulation without grape extract. This suggests a potent synergism of olive water polyphenols in the presence of grape extract, and could be explained by the different phenolic compositions, which include, among other things, grape-derived phenolic compounds, such as trans-resveratrol, anthocyanins and proanthocyanidins, as well as a further concentration of the olive vegetation water, which results in a higher content of hydroxytyrosol and tyrosol.

3.2.3. Cellular Peroxidation

Isoprostanes are a family of eicosanoids produced by the random oxidation of tissue phospholipids by oxygen radicals. Several studies have shown that the content of F2-isoprostanes in the human body (i.e., measured in vivo) is directly related to oxidative damage to lipids. In particular, the 8-epi-prostaglandin F2-alpha isomer (8-isoprostane), an end product of the lipid peroxidation chain reaction, has been proposed as a reliable signaling molecule for antioxidant deficiency and oxidative stress [25–28]. The degradation of lipids by lipoxidation occurs as a result of oxidative damage, and consequently the levels of isoprostanes increase, contributing in turn to the development of many diseases related to oxidative stress.

To evaluate the potential of olive-derived supplements in protecting cellular lipids from oxidative damage, we measured the levels of free 8-isoprostanes in cultured cells. Modest amounts of 8-isoprostanes were present in the culture media under normal culture conditions (untreated control), which were increased by oxidative stress after induction by AAPH treatment, but these decreased after incubation with olive-derived extracts or quercetin treatments (Figure 2 and Table 4); compared with untreated cells (baseline level), treatment with P-1 caused a significantly ($p < 0.05$) greater reduction in the content of free 8-isoprostanes (~29% reduction), while P-2 showed a moderate yet not significant effect on lipid oxidation protection (~6% reduction).

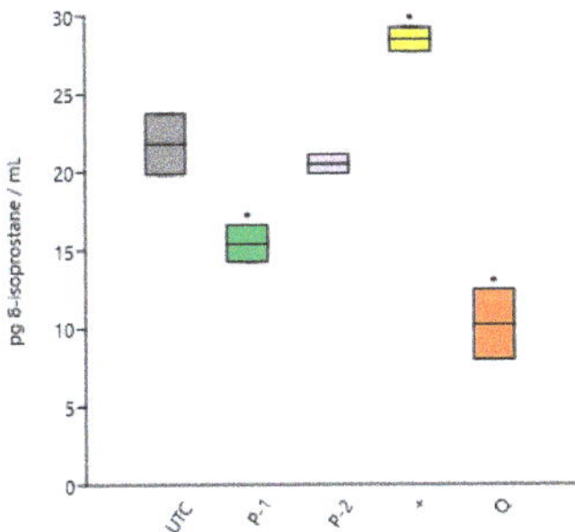

Figure 2. Box plots represent the free isoprostanes (8-epi PGF2α) released in culture, showing median, 25–75% quartiles, standard error, and total range of values from 3 experiments. UTC: untreated control cells; +: lipid peroxidation initiator; Q: quercetin control. * $p < 0.05$ versus UTC.

Table 4. Free isoprostanes released in culture after treatments. The relative values (third column) were normalized to the UTC, which represents the basal amount of isoprostanes under normal culture conditions, and multiplied by 100. UTC: untreated control cells; AAPH: lipid peroxidation initiator; Q: quercetin 100 μM. * $p < 0.05$ vs. UTC.

Treatment	8-epi PGF2α (pg/mL)	Relative Isoprostanes (%)
UTC	21.80 ± 1.95	100.0
AAPH	28.45 ± 0.76 *	130.5 *
P-1	15.42 ± 1.18 *	70.7 *
P-2	20.49 ± 0.60	93.9
Q	10.24 ± 2.23 *	47.0 *

The products derived from olives appear to cut down the number of oxidative attacks on the cell lipid membrane; in fact, the basal oxidative stress is effectively quenched compared to the untreated control cells, thus promoting cellular health. Similarly, in humans, oils rich in olive polyphenols have been shown to reduce the urinary excretion of F2-isoprostanes [29].

3.2.4. Antiglycation Activity

Advanced glycation end products (AGEs) are a heterogeneous group of substances that are formed in the human body during non-enzymatic glycosylation between the carbonyl group of a reducing sugar and a free amino group of a protein [30,31]. AGEs play an important, albeit complicated, role in cellular aging processes, in which they are produced in large quantities, causing oxidative stress, inflammatory reactions, and chronic diseases such as diabetes and cardiovascular disease [32,33]. Hyperglycemia, the accumulation of triosephosphates and ketone bodies, lipid peroxidation, and oxidative stress can increase AGEs formation [34]. The irreversible glycation of proteins in turn results in structural alterations and the accumulation of defective proteins in cells, which impactnormal physiological functions [35].

Since increased intracellular and extracellular stress is a source of AGEs accumulation in vivo, and the antioxidant activities of natural phenolics may inhibit the AGEs production [36,37], we investigated cellular AGEs to further support the potential of P-1 and P-2 in preventing the cellular stress.

We have found that both products reduce the formation of new AGEs (Table 5).

Table 5. AGEs content in cell lysates after treatment with the extract products and an AGEs-inducer mix. UTC: untreated control cells; +: AGEs inducer solution. Relative AGEs values are normalized to the UTC, which represents the basal amount of AGEs in normal culture conditions, multiplied by 100.

Treatment	μg AGEs/mg Protein	Relative AGEs (%)
UTC	1.02 ± 0.006	100
+	2.64 ± 0.042	258
P-1	0.52 ± 0.008	50.9
P-2	0.47 ± 0.007	46.0

Relative values greater than 100 indicate cellular accumulation of AGEs (e.g., treatment with the AGE-inducer alone), while lower values indicate a potential for the sample to reduce the AGEs formation. Pre-incubation of the cells with each product, followed by the induction of AGEs formation, effectively cleanses the cellular AGEs, with respect to both the AGEs-inducer treatment alone and with the cells under normal culture conditions. The strong positive influence is similar for both pre-treatments with each olive-derived concentrates, and both show a nearly 50% reduction in AGEs formation in vitro.

Our results are consistent with data from a recent study conducted with an olive leaf extract concentrated in hydroxytyrosol (54.5 mg/g), in which a reduction in AGEs production was demonstrated in HepG2 cells subjected to carbonyls-induced stress [38].These authors reported that, likely due to a synergistic effect of hydroxytyrosol and other minor compounds with similar polarities, the olive leaf extract exerts a wide antiglycative activity.

3.3. Bioavailability

Several cellular effects have been evaluated in vitro to test the antioxidant potential of olive water concentrates. Although, the bioavailability of such components is a prerequisite to any health claim made based on cellular tests. To investigate whether the hydroxytyrosol and trans-resveratrol consumed with the food supplements here studied are bioavailable, we conducted a pilot internal study with four volunteers (two male, two female). Volunteers in a fasted state received a single dose of 50 mL of P-2 in period 1 and 50 mL of P-1 in period 2, separated by a one-week wash-out period. Trans-resveratrol and hydroxytyrosol in its free form were undetectable in urinary samples collected immediately before intake. Free hydroxytyrosol, but not free trans-resveratrol, was detectable in urinary samples collected 30 min after the intake of both supplements and in all the volunteers (Figure 3).

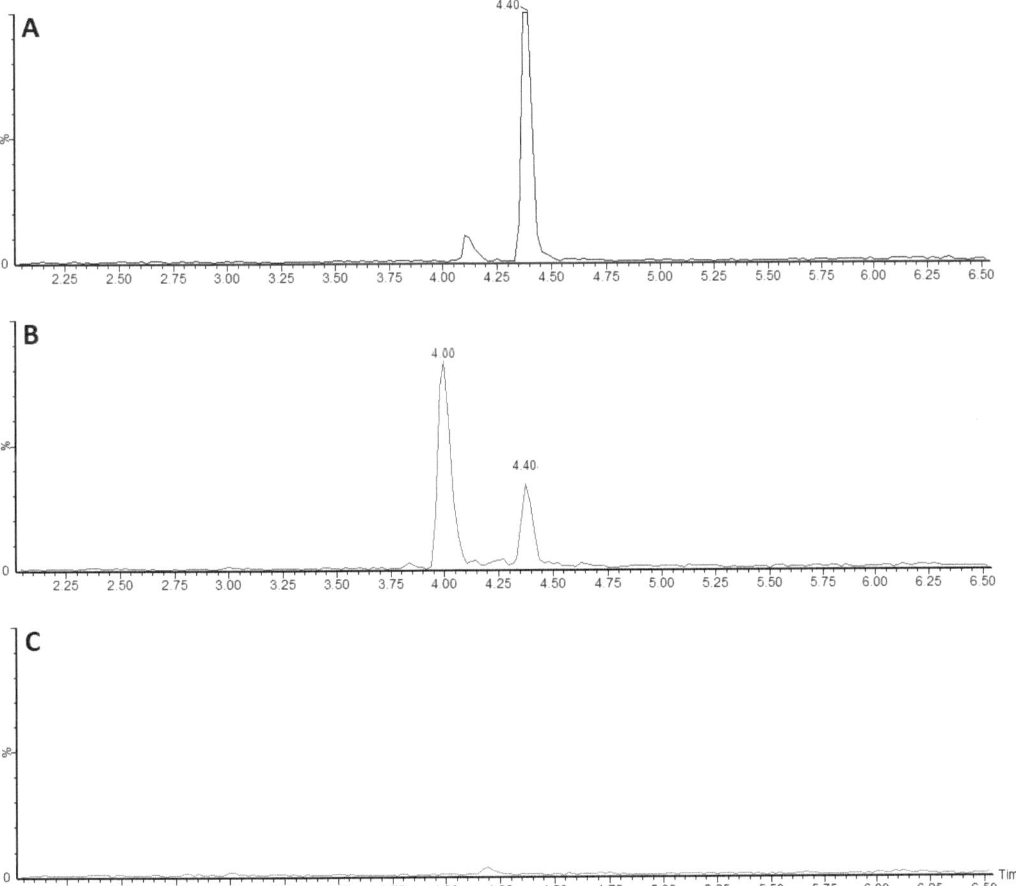

Figure 3. UHPLC-MS/MS chromatograms of the MRM 153 > 123. (**A**) hydroxytyrosol standard in urine; (**B**) representative urine sample taken 30 min after ingestion; (**C**) representative urine sample before intake of the supplement/product.

Similarly, previous human studies have showed that plasma and/or urinary hydroxytyrosol increase following the oral administration of hydroxytyrosol, consumed with olive oil [39–41], liquid or encapsulated olive leaf extract [42], and encapsulated extract from oil mill wastewater [8]. In these experiments, the absorption and excretion of orally administered hydroxytyrosol, collected mainly in conjugated forms and in a dose-dependent manner, have been shown. Furthermore, it has been shown that human absorption may differ depending on the composition of the food matrix through which the hydroxytyrosol is dispensed. Besides this, the bioavailability is likely influenced by wide interindividual variability in the absorption and metabolism [43].

Our preliminary results show that hydroxytyrosol consumed together with a hydrophilic vehicle is bioavailable in the human body, independently of the interactions of the combined fruit extracts used. Moreover, hydroxytyrosol can be detected in urine in its free form. This is relevant as the absorption of hydroxytyrosol is dependent on the vehicle of administration [9]. This is likely because the interaction between olive water and grape or lemon juices influences the hydroxytyrosol absorption and recovery yield. Further studies should be done to verify this hypothesis.

4. Discussion

Our study explores the antioxidant potential of two food supplements derived from olive vegetation water, mainly characterized by a high content of hydroxytyrosol. The antioxidant activity was determined by measuring the cellular antioxidant activity, the catalase and SOD activities in the HepG2 cell line, as well as the inhibition of lipid peroxidation and glycoxidation in the HaCat cell line. The tests carried out have shown that both olive-derived products have a strong positive influence on cells; this influence is complex and not one-dimensional. This reflects the complex nature of the sample materials and suggests a powerful synergy of hydroxytyrosol with other olive phenols, which is further potentiated in the presence of the grape phytocomplex. In particular, both supplements are able to reduce oxidative parameters in vitro.

On one hand, P-1, containing 94% olive water concentrate, showed a good capacity in the CAA and SOD assays, but a better performance in preventing isoprostanes formation in vitro when compared to P-2 (Table 6). On the other hand, P-2, containing olive and grape concentrates, showed a greater antioxidant potential for scavenging reactive species, as indicated by its greater potential in removing the hydroxyl radicals and by its higher SOD and catalase activities. Regarding the prevention of AGEs accumulation, both products showed an excellent capacity in vitro. The overall better antioxidant performance of P-2 in vitro could be explained by the higher concentration of olive-derived polyphenols, as well as the presence of grape-derived antioxidants, which include, among others, the trans-resveratrol, anthocyanins and proanthocyanidin monomers.

Table 6 describes the cellular effects measured, and their links to the attributed in vivo effects.

The mechanism of this positive influence needs to be better understood and has led us to further investigations on the mechanisms and dynamics of the effects of food supplements derived from olive vegetation water on human cells. However, understanding the absorption and bioavailability of these key molecules after oral administration remains a prerequisite before any potential health effect can be derived. In this sense, the pilot trial shows that the hydroxytyrosol supplied with ahydrophilic matrix combining olive fruit concentrate and lemon or grape juices is effectively absorbed, and then urinarily excreted as hydroxytyrosol in its free form.

In subsequent studies, the exact excreted fraction will be determined, and further focus will be placed on the metabolites to obtain a broader picture of the entire ADME properties.

Overall, preliminary data obtained in vitro indicate that the aqueous extracts of olives can actually improve the cellular redox status and related markers, and that their main active ingredient is bioavailable to the human body. Aqueous olive concentrates, with or without grape concentrate, are valid candidates for the prevention of cellular oxidative damage, and thus merit further attention.

Table 6. Results of bioassays on human cell lines and the effects' connections in vivo. Direction of in vitro effect: increased (↑), decreased (↓), or no effect (~). Effect magnitude compares P-1 to P-2.

In Vitro Test	Cellular Effect	Magnitude and Direction of the Effect In Vitro		Associated In Vivo Effect
		P-1	P-2	
CAA	Removal of hydroxyl radicals	↑	↑↑	Hydroxyl radical damage protection. Consequences of hydroxyl radicals are related to artherosclerosis, cancer and neurological disorders [44].
AGEs	Glycation ability	↓	↓	An increase in AGEs is associated with cellular aging processes, oxidative stress, inflammatory reactions, and chronic diseases such as diabetes, cardiovascular disease [32,33], and chronic kidney disease [45].
SOD activity	Superoxide removal	↑	↑↑	Oxidative damage protection. SOD deficiency is associated with diabetes [46].
Catalase activity	Hydrogen peroxide removal	~	↑↑	Oxidative damage protection. Catalase deficiency is associated with diabetes, Alzheimer's disease, Parkinson's disease, and acatalasemia [47].
8-isoprostane	Lipid peroxidation	↓	~	Reliable biomarker of lipoxidation (an increase on F2-isoprostanes is correlated with oxidative stress) [28,48].

Author Contributions: Conceptualization, investigation, writing—original draft preparation, methodology, formal analysis: C.B.; writing—review and editing, formal analysis: S.S.; visualization, methodology: P.H. All authors have read and agreed to the published version of the manuscript.

Funding: This research was funded by the German Federal Ministry for Economy and Energy under the program ZIM (Zentrales Innovationsprogramm Mittelstand), grant numbers 16KN038023 and ZF4226901SK6.

Institutional Review Board Statement: Ethical review and approval were waived for this study, as only urine was taken, there was no medical intervention and there was no danger to the health of the test persons.

Informed Consent Statement: Informed consent was obtained from all subjects involved in the study.

Conflicts of Interest: The authors declare no conflict of interest.

References

1. Rothwell, J.A.; Perez-Jimenez, J.; Neveu, V.; Medina-Remon, A.; M'Hiri, N.; Garcia-Lobato, P.; Manach, C.; Knox, C.; Eisner, R.; Wishart, D.S.; et al. Phenol-Explorer 3.0: A major update of the phenol-explorer database to incorporate data on the effects of food processing on polyphenol content. *Database* **2013**, *2013*, bat070. [CrossRef] [PubMed]
2. Obied, H.K.; Allen, M.S.; Bedgood, D.R.; Prenzler, P.D.; Robards, K.; Stockmann, R. Bioactivity and analysis of biophenols recovered from olive mill waste. *J. Agric. Food Chem.* **2005**, *53*, 823–837. [CrossRef] [PubMed]
3. Visioli, F.; Galli, C. biological properties of olive oil phytochemicals. *Crit. Rev. Food Sci. Nutr.* **2002**, *42*, 209–221. [CrossRef]
4. Covas, M.I.; Nyyssönen, K.; Poulsen, H.E.; Kaikkonen, J.; Zunft, H.J.; Kiesewetter, H.; Gaddi, A.; de la Torre, R.; Mursu, J.; Bäumler, H.; et al. The effect of polyphenols in olive oil on heart disease risk factors: A randomized trial. *Ann. Intern. Med.* **2006**, *145*, 333–341. [CrossRef] [PubMed]
5. Covas, M.I.; de la Torre, R.; Fitó, M. Virgin olive oil: A key food for cardiovascular risk protection. *Br. J. Nutr.* **2015**, *113*, S19–S28. [CrossRef]

6. EFSA Panel on Dietetic Products, Nutrition and Allergies (NDA). Scientific opinion on the substantiation of health claims related to polyphenols in olive and protection of LDL particles from oxidative damage (ID 1333, 1638, 1639, 1696, 2865), maintenance of normal blood HDL-cholesterol concentrations (ID 1639), maintenance of normal blood pressure (ID 3781), "anti-inflammatory properties" (ID 1882), "contributes to the upper respiratory tract health" (ID 3468), "can help to maintain a normal function of gastrointestinal tract" (3779), and "contributes to body defences against external agents" (ID 3467) pursuant to Article 13(1) of Regulation (EC) No 1924/2006. *EFSA J.* **2011**, *9*. [CrossRef]
7. D'Angelo, S.; Manna, C.; Migliardi, V.; Mazzoni, O.; Morrica, P.; Capasso, G.; Pontoni, G.; Galletti, P.; Zappia, V. Pharmacokinetics and metabolism of hydroxytyrosol, a natural antioxidant from olive oil. *Drug Metab. Dispos.* **2001**, *29*, 1492–1498.
8. Khymenets, O.; Crespo, M.C.; Dangles, O.; Rakotomanomana, N.; Andres-Lacueva, C.; Visioli, F. Human hydroxytyrosol's absorption and excretion from a nutraceutical. *J. Funct. Foods* **2016**, *23*, 278–282. [CrossRef]
9. Visioli, F.; Galli, C.; Grande, S.; Colonnelli, K.; Patelli, C.; Galli, G.; Caruso, D. Hydroxytyrosol excretion differs between rats and humans and depends on the vehicle of administration. *J. Nutr.* **2003**, *133*, 2612–2615. [CrossRef] [PubMed]
10. Kelm, M.A.; Johnson, J.C.; Robbins, R.J.; Hammerstone, J.F.; Schmitz, H.H. High-performance liquid chromatography separation and purification of cacao (*Theobroma cacao* L.) procyanidins according to degree of polymerization using a diol stationary phase. *J. Agric. Food Chem.* **2006**, *54*, 1571–1576. [CrossRef]
11. Wolfe, K.L.; Liu, R.H. Cellular antioxidant activity (caa) assay for assessing antioxidants, foods, and dietary supplements. *J. Agric. Food Chem.* **2007**, *55*, 8896–8907. [CrossRef] [PubMed]
12. Bender, C.; Graziano, S.; Zimmermann, B.F. Study of stevia rebaudiana bertoni antioxidant activities and cellular properties. *Int. J. Food Sci. Nutr.* **2015**, *66*, 553–558. [CrossRef] [PubMed]
13. Bradford, M.M. A rapid and sensitive method for the quantitation of microgram quantities of protein utilizing the principle of protein-dye binding. *Anal. Biochem.* **1976**, *72*, 248–254. [CrossRef]
14. King, R.E.; Bomser, J.A.; Min, D.B. Bioactivity of resveratrol. *Compr. Rev. Food Sci. Food Saf.* **2006**, *5*, 65–70. [CrossRef]
15. Yu, W.; Fu, Y.C.; Wang, W. Cellular and molecular effects of resveratrol in health and disease. *J. Cell. Biochem.* **2012**, *113*, 752–759. [CrossRef] [PubMed]
16. Smoliga, J.M.; Baur, J.A.; Hausenblas, H.A. Resveratrol and health—A comprehensive review of human clinical trials. *Mol. Nutr. Food Res.* **2011**, *55*, 1129–1141. [CrossRef]
17. Ramírez-Garza, S.; Laveriano-Santos, E.; Marhuenda-Muñoz, M.; Storniolo, C.; Tresserra-Rimbau, A.; Vallverdú-Queralt, A.; Lamuela-Raventós, R. Health effects of resveratrol:Results from human intervention trials. *Nutrients* **2018**, *10*, 1892. [CrossRef] [PubMed]
18. Khan, N.; Mukhtar, H. Tea polyphenols in promotion of human health. *Nutrients* **2019**, *11*, 39. [CrossRef]
19. Wang, H.; Joseph, J.A. Quantifying cellular oxidative stress by dichlorofluorescein assay using microplate reader. *Free Radic. Biol. Med.* **1999**, *27*, 612–616. [CrossRef]
20. Bedouhene, S.; Hurtado-Nedelec, M.; Sennani, N.; Marie, J.; El-Benna, J.; Moulti-Mati, F. Polyphenols extracted from olive mill wastewater exert a strong antioxidant effectin human neutrophils. *Int. J. Waste Resour.* **2014**, *4*. [CrossRef]
21. Rossi, T.; Bassani, B.; Gallo, C.; Maramotti, S.; Noonan, D.M.; Albini, A.; Bruno, A. Effect of a purified extract of olive mill waste water on endothelial cell proliferation, apoptosis, migration and capillary-like structure in vitro and in vivo. *J. Bioanal. Biomed.* **2015**. [CrossRef]
22. Sies, H. Strategies of antioxidant defense. *Eur. J. Biochem.* **1994**, *215*, 213–219. [CrossRef] [PubMed]
23. Ighodaro, O.M.; Akinloye, O.A. First line defence antioxidants-superoxide dismutase (SOD), catalase (CAT) and glutathione peroxidase (GPX): Their fundamental role in the entire antioxidant defence grid. *Alex. J. Med.* **2018**, *54*, 287–293. [CrossRef]
24. Fki, I.; Sashnoun, Z.; Sayadi, S. Hypocholesterolemic effects of phenolic extracts and purified hydroxytyrosol recovered fromolive mill wastewater in rats fed a cholesterol-rich diet. *J. Agric. Food Chem.* **2007**, *55*, 624–631. [CrossRef] [PubMed]
25. Morrow, J.D.; Hill, K.E.; Burk, R.F.; Nammour, T.M.; Badr, K.F.; Roberts, L.J. A series of prostaglandin F2-like compounds are produced in vivo in humans by a non-cyclooxygenase, free radical-catalyzed mechanism. *Proc. Nati. Acad. Sci. USA* **1990**, *87*, 9383–9387. [CrossRef] [PubMed]
26. Milne, G.L.; Musiek, E.S.; Morrow, J.D. F2-Isoprostanes as markers of oxidative stress in vivo: An overview. *Biomarkers* **2005**, *10*, 10–23. [CrossRef]
27. Milne, G.L.; Yin, H.; Hardy, K.D.; Davies, S.S.; Roberts, L.J. Isoprostane generation and function. *Chem. Rev.* **2011**, *111*, 5973–5996. [CrossRef]
28. EFSA Panel on Dietetic Products, Nutrition and Allergies (NDA). Guidance for the scientific requirements for health claims related to antioxidants, oxidative damage and cardiovascular health. *EFSA J.* **2018**, *21*. [CrossRef]
29. Visioli, F.; Caruso, D.; Galli, C.; Viappiani, S.; Galli, G.; Sala, A. Olive oils rich in natural catecholic phenols decrease isoprostane excretion in humans. *Biochem. Biophys. Res. Commun.* **2000**, *278*, 797–799. [CrossRef]
30. Singh, R.; Barden, A.; Mori, T.; Beilin, L. Advanced glycation end-products: A review. *Diabetologia* **2001**, *44*, 129–146. [CrossRef]
31. Rowan, S.; Bejarano, E.; Taylor, A. Mechanistic targeting of advanced glycation end-products in age-related diseases. *Biochim. Biophys. Acta (BBA) Mol. Basis Dis.* **2018**. [CrossRef] [PubMed]
32. Nagai, R.; Mori, T.; Yamamoto, Y.; Kaji, Y.; Yonei, Y. Significance of advanced glycation end products in aging-related disease. *Anti-Aging Med.* **2010**, *7*, 112–119. [CrossRef]

33. Rungratanawanich, W.; Qu, Y.; Wang, X.; Essa, M.M.; Song, B.J. Advanced glycation end products (AGEs) and other adducts in aging-related diseases and alcohol-mediated tissue injury. *Exp. Mol. Med.* **2021**, *53*, 168–188. [CrossRef] [PubMed]
34. Rahbar, S.; Figarola, J.L. Novel inhibitors of advanced glycation endproducts. *Arch. Biochem. Biophys.* **2003**, *419*, 63–79. [CrossRef]
35. Moldogazieva, N.T.; Mokhosoev, I.M.; Mel'nikova, T.I.; Porozov, Y.B.; Terentiev, A.A. Oxidative stress and advanced lipoxidation and glycation end products (ALEs and AGEs) in aging and age-related diseases. *Oxid. Med. Cell. Longev.* **2019**, *19*, 1–14. [CrossRef]
36. Crascì, L.; Lauro, M.L.; Puglisi, G.; Panico, A. Natural antioxidant polyphenols on inflammation management: Anti-glycation activity vs metalloproteinases inhibition. *Crit. Rev. Food Sci. Nutr.* **2017**, *58*, 893–904. [CrossRef]
37. Yeh, W.J.; Hsia, S.M.; Lee, W.H.; Wu, C.H. Polyphenols with antiglycation activity and mechanisms of action: A review of recent findings. *J. Food Drug Anal.* **2017**, *25*, 84–92. [CrossRef]
38. Navarro, M.; Morales, F.J.; Ramos, S. Olive leaf extract concentrated in hydroxytyrosol attenuates protein carbonylation and the formation of advanced glycation end products in a hepatic cell line (HepG2). *Food Funct.* **2017**, *8*, 944–953. [CrossRef]
39. Visioli, F.; Galli, C.; Bornet, F.; Mattei, A.; Patelli, R.; Galli, G.; Caruso, D. Olive oil phenolics are dose-dependently absorbed in humans. *FEBS Lett.* **2000**, *468*, 159–160. [CrossRef]
40. Miró-Casas, E.; Covas, M.I.; Fitò, M.; Farré-Albadalejo, M.; de la Torre, R. Tyrosol and hydroxytyrosol are absorbed from moderate and sustained doses of virgin olive oil in humans. *Eur. J. Clin. Nutr.* **2003**, *57*, 186–190. [CrossRef]
41. Suárez, M.; Valls, M.R.; Romero, M.P.; Macià, A.; Fernàndez, S.; Giralt, M.; Solà, R.; Motilva, M.J. Bioavailability of phenols from a phenol-enriched olive oil. *Br. J. Nutr.* **2011**, *106*, 1691–1701. [CrossRef] [PubMed]
42. De Bock, M.; Thorstensen, E.B.; Derraik, J.G.B.; Henderson, H.V.; Hofman, P.L.; Cutfield, W.S. Human absorption and metabolism of oleuropein and hydroxytyrosol ingested as olive (*Olea europaea* L.) leaf extract. *Mol. Nutr. Food Res.* **2013**, *57*, 2079–2085. [CrossRef] [PubMed]
43. Cassidy, A.; Minihane, A.M. The role of metabolism (and the microbiome) in defining the clinical efficacy of dietary flavonoids. *Am. J. Clin. Nutr.* **2016**, *105*, 10–22. [CrossRef]
44. Lipinski, B. Hydroxyl radical and its scavengers in health and disease. *Oxid. Med. Cell. Longev.* **2011**, *2011*, 809696. [CrossRef] [PubMed]
45. Liguori, I.; Russo, G.; Curcio, F.; Bulli, G.; Aran, L.; Della-Morte, D.; Gargiulo, G.; Testa, G.; Cacciatore, F.; Bonaduce, D.; et al. Oxidative stress, aging, and diseases. *Clin. Interv. Aging.* **2018**, *13*, 757–772. [CrossRef] [PubMed]
46. Kim, C.H. Expression of the extracellular superoxide dismutase protein in diabetes. *Arch. Plast. Surg.* **2013**, *40*, 517–521. [CrossRef]
47. Nandi, A.; Yan, L.J.; Jana, C.K.; Das, N. Role of Catalase in oxidative stress-and age-associated degenerative diseases. *Oxid. Med. Cell. Longev.* **2019**, *2019*, 9613090. [CrossRef]
48. Montuschi, P.; Corradi, M.; Ciabattoni, G.; Nightingale, J.; Kharitonov, S.A.; Barnes, P.J. Increased 8-isoprostane, a marker of oxidative stress, in exhaled condensate of asthma patients. *Am. J. Respir. Crit. Care Med.* **1999**, *160*, 216–220. [CrossRef]

 *applied sciences*

MDPI

Article

Phenolic Profiling of Five Different Australian Grown Apples

Heng Li [1], Vigasini Subbiah [1], Colin J. Barrow [2], Frank R. Dunshea [1,3] and Hafiz A. R. Suleria [1,2,*]

[1] Faculty of Veterinary and Agricultural Sciences, School of Agriculture and Food, The University of Melbourne, Parkville, VIC 3010, Australia; hengl2@student.unimelb.edu.au (H.L.); vsubbiah@student.unimelb.edu.au (V.S.); fdunshea@unimelb.edu.au (F.R.D.)
[2] Centre for Chemistry and Biotechnology, School of Life and Environmental Sciences, Deakin University, Waurn Ponds, VIC 3217, Australia; colin.barrow@deakin.edu.au
[3] Faculty of Biological Sciences, The University of Leeds, Leeds LS2 9JT, UK
* Correspondence: hafiz.suleria@unimelb.edu.au; Tel.: +61-470-439-670

Abstract: Apples (*Malus domestica*) are one of the most widely grown and consumed fruits in the world that contain abundant phenolic compounds that possess remarkable antioxidant potential. The current study characterised phenolic compounds from five different varieties of Australian grown apples (Royal Gala, Pink Lady, Red Delicious, Fuji and Smitten) using LC-ESI-QTOF-MS/MS and quantified through HPLC-PDA. The phenolic content and antioxidant potential were determined using various assays. Red Delicious had the highest total phenolic (121.78 ± 3.45 mg/g fw) and total flavonoid content (101.23 ± 3.75 mg/g fw) among the five apple samples. In LC-ESI-QTOF-MS/MS analysis, a total of 97 different phenolic compounds were characterised in five apple samples, including Royal Gala (37), Pink Lady (54), Red Delicious (17), Fuji (67) and Smitten (46). In the HPLC quantification, phenolic acid (chlorogenic acid, 15.69 ± 0.09 mg/g fw) and flavonoid (quercetin, 18.96 ± 0.08 mg/g fw) were most abundant in Royal Gala. The obtained results highlight the importance of Australian apple varieties as a rich source of functional compounds with potential bioactivity.

Keywords: apple; royal gala; pink lady; red delicious; smitten; fuji; phenolic compounds; antioxidant activity; LC-ESI-QTOF-MS/MS; HPLC

Citation: Li, H.; Subbiah, V.; Barrow, C.J.; Dunshea, F.R.; Suleria, H.A.R. Phenolic Profiling of Five Different Australian Grown Apples. *Appl. Sci.* **2021**, *11*, 2421. https://doi.org/10.3390/app11052421

Academic Editor: Luca Mazzoni

Received: 19 February 2021
Accepted: 4 March 2021
Published: 9 March 2021

1. Introduction

Apples (*Malus domestica*) are widely grown and consumed fruits. In 2018, apple production across the globe was 86 million tonnes, mainly from China, America and New Zealand, whereas the apple production in Australia was over 2.6 million tonnes [1]. Apples are usually supplied to the market in the form of fresh fruit or processed products, including dried apples, apple cider, apple juice and sauce [2]. Apples are enriched with bioactives compounds [3], vitamins (water and fat soluble) and minerals like calcium, potassium and phosphorus [4]. These compounds are required by the human body to perform various functions like strengthening of the bones, building muscles, filtering out waste [3], and have positive health benefits against several chronic diseases, including type 2 diabetes, asthma and rheumatoid arthritis [5].

The varieties of apples are due to the difference of agroclimatic regions and zones, cultivation practices, nutritional composition and sensory characteristics [6]. Royal Gala, one of the variety of apples having bright shiny red colour, with stripes ranging from straw yellow to amber orange, has a sensory profile that is sweet, soft, crunchy and slightly acidic [7,8]. Pink Lady is a variety that has been originated from a cross between 'Golden Delicious' and 'Lady Williams', known for its sweet taste, firmness and possesses a scald-free surface [6]. A consumer panel in New Zealand appreciated the Pink Lady variety for its dense flesh, excellent crispness, juiciness, good sugar-acid balance and sweet flavour [9]. The Red Delicious variety when compared to the previous two varieties has a darker crimson red surface with traces of yellow and orange [10]. The physical characteristics of Red Delicious is an elongated form with a thick peel, grainy and tender with a melting texture, usually exhibiting small but evident humps on the skin surface [11]. While different varieties exhibit different appearances, taste and shapes, apples have one common characteristics, which are the high concentrations of phenolic compounds that exhibit high antioxidant potential [12].

Phenolic compounds are important plant secondary metabolites which exhibit excellent abilities to reduce and eliminate free radicals thereby providing antioxidant and anti-lipid peroxidation properties [13,14]. The phenolic compounds exhibiting antioxidation potential have made the food and nutrition market interested in phenolic compounds, thus replacing the existing chemical anti-oxidation ingredients in food to increase the nutritional value and health benefits [14]. One of the polyphenol mechanisms is the removal of free radicals by supplying hydrogen atoms or separate electrons from the phenol group and eliminating related enzymes, thereby preventing the production of free radicals and their intermediate products [15]. Additionally, phenolic compounds can react with metal ions to inactivate the Fenton reaction [16]. The antioxidant potential are often determined by using a series of different in vitro spectrophotometric-based assays including the total antioxidant capacity (TAC), 2,2′-azino-bis-3-ethylbenzothiazoline-6-sulfonic acid (ABTS) assay, the ferric reducing ability of plasma (FRAP) and 2,2′-diphenyl-1-picrylhydrazyl (DPPH) [17].

Liquid chromatography coupled with mass spectrometry (LC-ESI-QTOF-MS/MS) is an effective tool used for the identification and characterisation of phenolic compounds. High pressure liquid chromatography (HPLC) combined with photodiode array detector (HPLC-PDA) is used for the quantification of the phenolics [18,19]. According to a previous study, few phenolic compounds have been identified in apples through HPLC and LC-ESI-QTOF-MS analysis including flavanols (catechin), dihydrochalcones (chlorogenic acid), phenolic acids and anthocyanins [20].

Although there are many studies that have isolated and identified phenolic compounds in different apples, only a few have focused on Australian grown apples. The novelty of this study will encourage the Australian producers to utilise the low-grade produce of the apples to a better use as it is rich in phenolics, since premature or overripe fruits compromise the quality and do not meet the standards of the supermarkets. Therefore, in the current research we extracted phenolics from five popular varieties of Australian grown apples (Royal Gala, Pink Lady, Red Delicious, Fuji and Smitten) and estimated their antioxidant potential. The outcome of the current research will add adequate information on the phenolics and antioxidant potential of Australian grown apples for their further application in the food, nutraceutical and pharmaceutical industries.

2. Materials and Methods

2.1. Chemicals and Reagents

The chemicals used for the extraction and characterisation were of analytical grade and purchased from Sigma-Aldrich (St. Louis, MO, USA). The chemicals used for phenolic estimation and antioxidant assays were procured from Sigma-Aldrich (St. Louis, MO, USA) including ferric (III) chloride anhydrous, 50% acetic acid, 2,4,6-tripyridyl-s-triazine (TPTZ), acetonitrile, catechin, ascorbic acid, vanillin, aluminium chloride hexahydrate, 2,2′-diphenyl-1-picrylhydrazyl, 2,2′-azino-bis(3-ethylbenz-thiazoline-6-sulphonate), potassium persulfate and Folin-Ciocalteu´s phenol. The standards for HPLC including protocatechuic acid, epicatechin, gallic acid, epicatechin gallate, caffeic acid, quercetin, chlorogenic acid, *p*-hydroxybenzoic acid and kaempferol were procured from Sigma-Aldrich (Castle Hill, NSW, Australia). Ammonium molybdate and sodium acetate hydrated were procured from Sigma-Aldrich (Castle Hill, NSW, Australia). Moreover, 99% ethanol was procured from Thermo Fisher (Waltham, MA, USA), and 98% sulfuric acid was purchased from RCI Labscan Ltd. (Rongmuang, Thailand).

2.2. Sample Preparation and Extraction

Australian grown apple varieties (Royal Gala, Pink Lady, Red Delicious, Fuji and Smitten) were bought from a local market in Melbourne, VIC, Australia. All the samples were fully matured and ripen before harvested, transported and distributed to the local retailers within 2–3 days using refrigerated trucks. The apple peels were removed by a peeler and the core was separated to obtain the pulp. Subsequently, the pulps were blended into a slurry using a blender. 5 g of slurry samples were macerated in 20 mL of 70% ethanol (w/v) by slightly modifying the protocol of our earlier published study of Gu et al. [21]. The slurry samples were homogenised to prepare the sample extracts of the apples in a homogeniser at 10,000 rpm for 30 s. The homogenised extract samples were incubated in a shaking incubator at 120 rpm, 4 °C for 12 h. The samples were centrifuged for 15 min at 5000 rpm (4 °C). A syringe filter was used to filter the extracts used for LC-ESI-QTOF-MS/MS and HPLC-PDA studies and the samples were stored at −20 °C for further analysis.

2.3. Estimation of Phenolic Compounds and Antioxidant Assays

The estimation of phenolic compounds present in the samples and their potential antioxidant activities were analysed following our previously published protocols of Tang et al. [22] and Wang et al. [23].

2.3.1. Determination of Total Phenolic Content (TPC)

The spectrophotometric method of Yunfeng et al. [24] was used for the determination of TPC with some modifications. For this, 25 μL of the apple extract with 200 μL water and 25 μL Folin–Ciocalteu reagent solution were added to 96-well plates. The reaction mixture was incubated for 5 min (25 °C). Then, 5 μL of 10% sodium carbonate was added to the reaction mixture and incubated for 60 min in the dark at room temperature. The absorbance of the reaction mixture was measured at 765 nm using spectrophotometer. The standard used was gallic acid (0–200 μg/mL) to construct the standard curve and the values of TPC was expressed in mg of gallic acid equivalent per gram of sample (mg GAE/g of sample) (fw).

2.3.2. Determination of Total Flavonoids Content (TFC)

The Total Flavonoids Content (TFC) was determined by improvising the aluminium protocol described in Rajurkar and Hande [25]. For this, 80 μL of the apple extract with 120 μL of 50 g/L sodium acetate solution and 80 μL of 2% aluminium chloride were added into the 96-well plate subsequently incubate the reaction mixture at 25 °C for 2.5 h. The absorbance was measured at 440 nm. Quercetin calibration curve (0–50 μg/mL) was constructed and TFC was expressed in quercetin equivalent (mg QE/g fw).

2.3.3. Determination of Total Tannin Content (TTC)

The vanillin-sulfuric acid method with some modifications of Mesfin and Won Hee [26] was used to determine TTC. 25 μL of the apple extract was added to 25 μL of 32% sulfuric acid and 150 μL of 4% vanillin solution in the 96-well plate. The reaction mixture was incubated for 15 min at 25 °C. The absorbance was measured at 500 nm and expressed in mg of catechin equivalent per g of sample weight (mg CE/g fw) based on a calibration curve with concentration from 0–1000 μg/mL.

2.3.4. 2,2′-Diphenyl-1-picrylhydrazyl (DPPH) Assay

The DPPH method was used to determine the free radical scavenging activity [27]. For this, 40 μL of DPPH methanolic solution (0.1 mM) and 40 μL of extract were added into the 96-well plate. The reaction mixture was shaken vigorously and incubated for 30 min at 25 °C. The absorbance was measured at 517 nm. The standard used was ascorbic acid to construct the standard curve (0 to 50 μg/mL). The obtained values were expressed in mg of ascorbic acid equivalent per gram (mg AAE/g) (fw).

2.3.5. Ferric Reducing Antioxidant Power (FRAP) Assay

The ferric reducing ability was assessed by modifying the FRAP method of Faiza et al. [28]. The FRAP solution was prepared at the ratio of 10:1:1, 300 mM sodium acetate solution, 20 mM Fe [III] solution and 10 mM TRTZ. 20 μL of the apple extract and 280 μL of FRAP dye solution added to the 96-well plate. The reaction mixture was incubated for 10 min at 37 °C. The absorbance was measured at 593 nm. The ascorbic acid standard curve (0–150 μg/mL) was constructed and the values obtained were expressed in mg of ascorbic acid equivalent per gram of sample (mg AAE/g fw).

2.3.6. 2,2′-Azino-bis-3-ethylbenzothiazoline-6-sulfonic Acid (ABTS) Assay

In ABTS assay, the free radical scavenging activity of the apple samples were determined by following the protocol as in Rajurkar and Hande [25]. First, 88 μL of 140 mM potassium persulfate and 5 mL of 7 mM ABTS solution were mixed to form the $ABTS^+$ stock solution and incubated for 16 h in a dark area. 290 μL of prepared diluted ABTS solution was mixed with 10 μL of extract. Subsequently, incubation of the reaction mixture in the dark area for 6 min (25 °C). The absorbance was measured at 734 nm. The standard curve used to calculate the antioxidant potential was of ascorbic acid (0 to 150 μg/mL). The values were expressed in ascorbic acid equivalents (mg AAE/g) of sample.

2.3.7. Total Antioxidant Capacity (TAC)

The phosphomolybdate [29] method was used to determine the TAC. The formulation for phosphomolybdate reagent was 0.6 M sulphuric acid, 0.004 M ammonium molybdate and 0.028 M sodium phosphate. Then, 260 μL phosphomolybdate reagent was mixed with 40 μL extracts in the 96-well plate. The incubation of the reaction mixture was at 95 °C for 10 min. The absorbance was read at 695 nm after the reaction mixture cools down to room temperature. Ascorbic acid standard curve (0–200 μg/mL) constructed to determine the values of TAC and expressed in mg ascorbic acid equivalents (AAE) per gram (fw).

2.4. *LC-ESI-QTOF-MS/MS Analysis of Phenolic Compounds*

The identification and characterisation of phenolics in five varieties of apples were conducted using LC-ESI-QTOF-MS/MS and following the protocol described in Suleria et al. [18]. The separation of compounds was carried out through LC column 250 × 4.6 mm, 4 μm with column temperature at 25 °C. The HPLC buffers were sonicated at room temperature for 10 min. The binary solvent delivery system was used as follows: Mobile phase A: 2% acetic acid and 98% water; Mobile phase B: acetonitrile, water and acetic acid (50:49.5:0.5, *v*/*v*/*v*). The injected sample volume was 6 μL and the flow rate was at 0.8 mL/min. The program set was carried out as following: 0 min (10% B), 20 min (25% B), 30 min (35% B), 40 min (40% B), 70 min (55% B), 75 min (80% B), 77 min (100% B), 79 min (100% B), 82–85 min (isocratic 10% B). Negative and positive modes were performed for peak identification. Nitrogen gas was used as a nebulizer and drying gas at 45 psi, temperature at 300 °C with the flow rate of 5 L/min. The range of mass spectra were 50–1300 amu. Agilent LC-ESI-QTOF-MS/MS Mass Hunter workstation software (Qualitative Analysis, version B.03.01, Agilent, Santa Clara, CA, USA) was used for data acquisition and analysis.

2.5. *HPLC-PDA Analysis*

The HPLC-PDA analysis of polyphenols in apples was carried out using Agilent 1200 series HPLC [30,31]. The volume of the injected sample was 20 μL. 280 nm, 320 nm and 370 nm were the wavelengths used for detection. The column and the conditions used were as followed in LC-ESI-QTOF-MS/MS analysis. The wavelengths were used for the identification of hydroxybenzoic acids, hydroxycinnamic acids and flavanol group, respectively. The acquisition of the data and analysis were carried out using Agilent LC-ESI-QTOF-MS/MS Mass Hunter workstation software (Qualitative Analysis, version B.03.01, Agilent, Santa Clara, CA, USA).

2.6. Statistical Analysis

The experiments were performed in triplicates (n = 3) and the data was expressed in mean ± standard deviation. One-way analysis of variance (ANOVA) followed by Tukey's honestly significant differences (HSD) multiple rank test were performed to see the significant difference between the phenolic compounds and antioxidant activities at $p < 0.05$.

3. Results and Discussion

3.1. Phenolic Compound Estimation (TPC, TFC and TTC)

The Folin–Ciocalteu's reagent method determined the total phenolic content in the apple extracts and were expressed as gallic acid equivalents (GAE/g fw) as shown in Table 1. Red Delicious apple showed the highest TPC with 121.78 ± 3.45 mg GAE/g and significantly higher than other samples ($p < 0.05$). The total polyphenol content of five different varieties of apples were in the order of Red Delicious > Royal Gala > Fuji > Pink Lady > Smitten. According to the study of Ting et al. [32], Praveen et al. [33] and Almeida et al. [34], Red Delicious had more phenolic content than Gala, Fuji and Pink Lady, which is consistent to the result of our study. Almeida et al. [34] reported that Fuji apple contains 14.7 ± 0.4 mg (GAE)/g and Ting et al. [32] study showed that Fuji has 489.59 ± 4.21 mg (GAE)/g, the difference in the phenolic content might be due to the geographical location, soil nutrients, growth period and harvest season [35]. Additionally, due to the lack of research on Smitten apple variety, there is no valid data for Smitten for comparison.

Flavonoids have attracted a lot of attention due to their strong antioxidant activity [36]. In TFC, Red Delicious apple had the highest flavonoid content of 101.23 ± 3.75 mg QE/g and the lowest flavonoid content was present in Smitten. In a previous study, TFC of Red Delicious (98 mg QE/g) and Royal Gala (89 mg QE/g) were similar to that of our apple samples [37]. In another study, the values of total flavonoid content of Fuji apple (108 mg QE/g) was reported more than our value which may be due to the difference of varieties or solvent extraction ratio [38]. The TTC in our selected apples ranged between 4.65 ± 0.03 to 2.17 ± 0.05 mg CE/g. Fuji apple showed higher level of tannin content followed by Pink Lady, Smitten, Royal Gala and Red Delicious. Previously, the total tannin content of different varieties ranged from 0.75 mg CE/g to 14.79 mg CE/g, which is consistent with our results [39]. Overall, the variety of Red Delicious had the highest content of TPC and TFC and Fuji variety had a high content of TTC.

Table 1. Phenolic content and antioxidant potential in five varieties of apples.

Antioxidant Assays	Royal Gala	Pink Lady	Red Delicious	Fuji	Smitten
TPC (mg GAE/g)	104.21 ± 3.10 ^{b}	94.23 ± 2.24 ^{c}	121.78 ± 3.45 ^{a}	102.26 ± 2.14 ^{b}	83.98 ± 1.05 ^{d}
TFC (mg QE/g)	93.73 ± 1.10 ^{b}	81.23 ± 2.25 ^{d}	101.23 ± 3.75 ^{a}	87.26 ± 1.54 ^{c}	72.19 ± 1.75 ^{e}
TTC (mg CE/g)	3.45 ± 0.09 ^{d}	4.25 ± 0.01 ^{b}	2.17 ± 0.05 ^{e}	4.65 ± 0.03 ^{a}	3.95 ± 0.08 ^{c}
DPPH (mg AAE/g)	3.39 ± 0.05 ^{b}	2.56 ± 0.03 ^{c}	3.53 ± 0.07 ^{a}	1.98 ± 0.01 ^{d}	1.17 ± 0.02 ^{e}
FRAP (mg AAE/g)	4.12 ± 0.07 ^{b}	3.15 ± 0.12 ^{c}	4.42 ± 0.01 ^{a}	2.12 ± 0.04 ^{d}	2.15 ± 0.02 ^{d}
ABTS (mg AAE/g)	3.22 ± 0.12 ^{a}	2.94 ± 0.01 ^{b}	3.24 ± 0.09 ^{a}	1.87 ± 0.10 ^{c}	1.49 ± 0.09 ^{d}
TAC (mg AAE/g)	2.68 ± 0.09 ^{b}	2.19 ± 0.11 ^{c}	3.12 ± 0.01 ^{a}	1.96 ± 0.08 ^{d}	1.32 ± 0.01 ^{e}

All values are expressed as the mean ± SD and performed in triplicates. Different letters (a, b, c, d, e) within the same column are significantly different ($p < 0.05$) from each other. The five varieties of apples are reported based on fresh weight. CE (catechin equivalents), QE (quercetin equivalents), GAE (gallic acid equivalents), AAE (ascorbic acid equivalents). TFC (total flavonoids content), TPC (total phenolic content), TTC (total tannins content), FRAP (ferric reducing ability of plasma), DPPH (2,2′-diphenyl-1-picrylhydrazyl), TAC (total antioxidant capacity), ABTS (2,2′-azino-bis-3-ethylbenzothiazoline-6-sulfonic acid).

3.2. Antioxidant Activities (DPPH, FRAP, ABTS and TAC)

The antioxidant potential of five varieties of apple samples were estimated by four assays including DPPH, FRAP, ABTS and TAC assays, and the antioxidant activities were expressed in ascorbic acid (AAE) per gram (fw) as mentioned in Table 1.

In the DPPH assay, the free radical scavenging activity is determined which is attributed to the phenolic compounds [40]. The apple varieties in the current study varied from 1.17 to 3.53 mg AAE/g. Red Delicious had the highest antioxidant potential followed by Royal Gala, Pink Lady, Fuji and Smitten. Previous studies reported that antioxidant potential for over ten varieties of apples ranged from 0.26 to 9.30 mg AAE/g [41,42]. The values of Fuji and Red Delicious apples are slightly higher than ours which might be because of the cultivar, location, maturity and storage of apples which may change the concentration of antioxidant potential [43].

FRAP assay can provide comprehensive information about the antioxidant activities of five varieties of apples since various antioxidant assays can help us to understand the antioxidant properties of apples better [44]. In FRAP assay, the electron transfer method was used to measure the capacity to reduce Fe^{3+} to Fe^{2+} [20]. The FRAP values were significantly different ($p < 0.05$) from 2.12 $\pm$ 0.04 mg AAE/g to 4.42 $\pm$ 0.01 mg AAE/g among the apple varieties. The highest FRAP capacity was recorded in Red Delicious, followed by Royal Gala, Pink Lady, Fuji, and Smitten.

In the ABTS assay, the antiradical scavenging activities were determined based of the hydrogen atom donating tendency of polyphenols [40]. The highest antioxidant ability was demonstrated in the order of Red Delicious > Royal Gala > Pink Lady > Fuji > Smitten. Upon comparison with the previous studies' Royal Gala and Fuji showed higher antioxidant ability than the previous reported values [41,42]. The reason might be because of the cultivar, location, maturity and storage of apples which may change the concentration of antioxidant potential [43]. In the TAC assay, the mechanism very similar to FRAP where reduction of molybdenum (VI) to molybdenum (V) in the presence of phenolics. In the current study, Red Delicious had the highest total antioxidant followed by Royal Gala, Pink Lady, Fuji and Smitten. Previously Khanizadeh et al.'s [35] study showed the values ranging from 0.323 to 1.246 mg AAE/g and the values were lower than our study. A difference in the concentration might be because of the difference between cultivars, location, harvesting time and maturity of samples [6].

3.3. Correlation between Phenolic Compounds and Antioxidant Activities

The correlation between the polyphenols and antioxidant activities was performed with a Pearson's correlation test (Table 2). TPC shows a strong positive correlation with TFC with r^2 = 0.975, $p \leq 0.01$, this indicates that TFC contributes largely to the total phenolic content. Additionally, TPC was strongly correlated with TAC with r^2 value of 0.920 ($p \leq 0.05$). A previous study by Vasantha Rupasinghe and Clegg [45] reported a similar correlation between TPC and TAC.

Table 2. Correlation coefficients (r^2) between phenolic contents and antioxidant assays.

Variables	TPC	TFC	TTC	DPPH	ABTS	FRAP
TFC	0.975 **					
TTC	−0.736	−0.702				
DPPH	0.832	0.903 *	−0.685			
ABTS	0.754	0.815	−0.830	0.952 **		
FRAP	0.681	0.756	−0.614	0.961 **	0.938 **	
TAC	0.920 *	0.952 **	−0.751	0.980 **	0.931 *	0.912 *

** Significant correlation with $p \leq 0.01$; * Significant correlation with $p \leq 0.05$.

TFC had a significantly strong correlation with DPPH and TAC with r^2 value of 0.903 ($p \leq 0.01$) and 0.952 ($p \leq 0.05$) respectively indicating that flavonoids were one of the significant contributors for the antioxidant activities. The results confirm with the previous studies of Maleeha et al. [46] and Ruiz-Torralba et al. [47], on phenolic compounds contributing towards antioxidant potential. A non-significant correlation were observed between TTC and antioxidant assays indicating the contribution of tannins to antioxidant activity is limited, which confirms with Kam et al. [48] study.

The correlation among the antioxidant assays had strong correlation with each other. Significant positive correlation was observed between DPPH with ABTS, FRAP and TAC (r^2 = 0.952, r^2 = 0.961, and r^2 = 0.980, $p \leq 0.01$). The correlation displayed in our study was similar to Kriengsak et al. [49], where a high correlation was observed between the four assays. Similarly, ABTS was observed to have high significant correlation with FRAP and TAC with r^2 = 0.938, $p \leq 0.01$ and r^2 = 0.931 ($p \leq 0.05$), respectively. On the other hand, FRAP was correlated with TAC with r^2 = 0. 912 ($p \leq 0.05$).

Overall, phenolic compounds were highly correlated with antioxidant assays, which indicated that both classes of phenolic compounds including phenolic acids and flavonoids have strong antioxidant potential. The four antioxidants' assays were strongly correlated with each other.

3.4. Phenolic Compounds Profile by LC-MS/MS Analysis

LC- MS/MS has been a useful and reliable tool for identification and characterisation of phenolics in several plant samples. Qualitative analyses of phenolics from five varieties of apples (Royal Gala, Pink Lady, Red Delicious, Fuji and Smitten) were achieved using mass spectrometry in both negative and positive modes of ionisation (ESI^-/ESI^+). The compounds in the apples were identified based on their precursor ions and MS spectra. The basis for the compounds to be further analysed were the PCDL library score more than 80 and mass error < 5 ppm (Table 3). In our current study, 97 different phenolic compounds were characterised in five apple samples, including 27 phenolic acids, 52 flavonoids, 5 lignans and 13 other polyphenols.

Table 3. Identification and characterisation of polyphenols in apples by using LC-ESI-QTOF-MS/MS.

No.	Proposed Compounds	Molecular Formula	RT (min)	Ionization (ESI^+/ESI^-)	Molecular Weight	Theoretical (*m*/*z*)	Observed (*m*/*z*)	Error (ppm)	MS^2 Product Ions	Samples
					Phenolic acid					
					Hydroxybenzoic acids					
1	Gallic acid 4-*O*-glucoside	$C_{13}H_{16}O_{10}$	6.866	$[M-H]^-$	332.0743	331.0670	331.0674	1.2	169, 125	RG
2	Protocatechuic acid 4-*O*-glucoside	$C_{13}H_{16}O_9$	7.379	** $[M-H]^-$	316.0794	315.0721	315.0718	−1.0	153	RD, F, * RG, S, PL
3	2-Hydroxybenzoic acid	$C_7H_6O_3$	7.608	** $[M-H]^-$	138.0317	137.0244	137.0242	−1.5	93	PL, * RD, RG, S, F
4	3-*O*-Methylgallic acid	$C_8H_8O_5$	12.930	$[M+H]^+$	184.0372	185.0445	185.0452	3.8	170, 142	F, * PL
5	2,3-Dihydroxybenzoic acid	$C_7H_6O_4$	15.580	$[M-H]^-$	154.0266	153.0193	153.0196	2.0	109	RG, * PL, F
					Hydroxycinnamic acids					
6	m-Coumaric acid	$C_9H_8O_3$	5.256	** $[M-H]^-$	164.0473	163.04	163.0393	−4.3	119	S,* RD, RG, PL, F
7	Caffeic acid	$C_9H_8O_4$	5.898	$[M+H]^+$	180.0423	181.0496	181.0494	−1.1	143, 133	S
8	*p*-Coumaroyl tartaric acid	$C_{13}H_{12}O_8$	8.632	$[M-H]^-$	296.0532	295.0459	295.0468	3.1	115	F
9	Cinnamic acid	$C_9H_8O_2$	9.314	** $[M-H]^-$	148.0524	147.0451	147.0449	−1.4	103	RG, * RD, F
10	3-Caffeoylquinic acid	$C_{16}H_{18}O_9$	12.979	** $[M-H]^-$	354.0951	353.0878	353.088	0.6	253, 190, 144	PL, S, * RG, F
11	3- *p*-Coumaroylquinic acid	$C_{16}H_{18}O_8$	18.131	** $[M-H]^-$	338.1002	337.0929	337.0924	−1.5	265, 173, 162	PL,* RG, F, S
12	*p*-Coumaric acid 4-*O*-glucoside	$C_{15}H_{18}O_8$	20.881	$[M-H]^-$	326.1002	325.0929	325.0925	−1.2	163	PL,* RG, F
13	Rosmarinic acid	$C_{18}H_{16}O_8$	22.273	$[M-H]^-$	360.0845	359.0772	359.0755	−4.7	179	* PL, F
14	Caffeic acid 3-*O*-glucuronide	$C_{15}H_{16}O_{10}$	22.737	$[M-H]^-$	356.0743	355.067	355.0677	2.0	179	PL
15	Ferulic acid	$C_{10}H_{10}O_4$	23.366	** $[M-H]^-$	194.0579	193.0506	193.0505	−0.5	178, 149, 134	S,* PL, F
16	Caffeoyl glucose	$C_{15}H_{18}O_9$	24.244	$[M-H]^-$	342.0951	341.0878	341.0886	2.3	179, 161	RD,* PL
17	Ferulic acid 4-*O*-glucuronide	$C_{16}H_{18}O_{10}$	25.785	$[M -H]^-$	370.09	369.0827	369.0814	−3.5	193	* PL, F
18	1-Sinapoyl-2,2′-diferuloylgentiobiose	$C_{43}H_{48}O_{21}$	26.763	$[M-H]^-$	900.2688	899.2615	899.2579	−4.0	613, 201	PL
19	Sinapic acid	$C_{11}H_{12}O_5$	30.185	** $[M-H]^-$	224.0685	223.0612	223.0603	−4.0	205, 163	* F, PL, S
20	3-Feruloylquinic acid	$C_{17}H_{20}O_9$	33.605	** $[M-H]^-$	368.1107	367.1034	367.1019	−4.1	298, 288, 192, 191	* RG, F
21	1,2,2′-Triferuloylgentiobiose	$C_{42}H_{46}O_{20}$	34.101	$[M-H]^-$	870.2582	869.2509	869.2498	−1.3	693, 517	S
22	Ferulic acid 4-*O*-glucoside	$C_{16}H_{20}O_9$	35.526	** $[M-H]^-$	356.1107	355.1034	355.1039	1.4	193, 178, 149, 134	* PL, RG, S, F
23	*p*-Coumaroyl malic acid	$C_{13}H_{12}O_7$	41.506	$[M-H]^-$	280.0583	279.051	279.0524	5.0	163, 119	S

Table 3. *Cont.*

No.	Proposed Compounds	Molecular Formula	RT (min)	Ionization (ESI^+/ESI^-)	Molecular Weight	Theoretical (*m/z*)	Observed (*m/z*)	Error (ppm)	MS^2 Product Ions	Samples
				Hydroxyphenylacetic acids						
24	2-Hydroxy-2-phenylacetic acid	$C_8H_8O_3$	31.517	** $[M-H]^-$	152.0473	151.04	151.0402	1.3	136, 92	PL
25	3,4-Dihydroxyphenylacetic acid	$C_8H_8O_4$	20.749	** $[M-H]^-$	168.0423	167.035	167.0343	−4.2	149, 123	* RG, PL, F
				Hydroxyphenylpropanoic acids						
26	Dihydroferulic acid 4-sulfate	$C_{10}H_{12}O_7S$	4.076	$[M-H]^-$	276.0304	275.0231	275.0229	−0.7	195, 151, 177	F
27	Dihydroferulic acid 4-*O*-glucuronide	$C_{16}H_{20}O_{10}$	6.866	$[M-H]^-$	372.1056	371.0983	371.0986	0.8	195	* RG, PL
				Flavonoids **Anthocyanins**						
28	Cyanidin 3-*O*-diglucoside-5-*O*-glucoside	$C_{33}H_{41}O_{21}$	21.567	$[M+H]^+$	773.214	774.2213	774.2216	0.4	610, 464	S
29	Cyanidin 3-*O*-(6″-p-coumaroyl-glucoside)	$C_{30}H_{27}O_{13}$	22.205	** $[M+H]^+$	595.1452	596.1525	596.1553	4.7	287	RG,* PL
30	Peonidin 3-*O*-sambubioside-5-*O*-glucoside	$C_{33}H_{41}O_{20}$	22.561	** $[M+H]^+$	757.2191	758.2264	758.2228	−4.7	595, 449, 287	* S, F
31	Cyanidin 3,5-*O*-diglucoside	$C_{27}H_{31}O_{16}$	37.067	** $[M+H]^+$	611.1612	612.1685	612.1693	1.3	449, 287	* F, S, PL
32	Delphinidin 3-*O*-xyloside	$C_{20}H_{19}O_{11}$	37.212	$[M+H]^+$	435.0927	436.1	436.0996	−0.9	303	PL
33	Delphinidin 3-*O*-glucosyl-glucoside	$C_{27}H_{31}O_{17}$	37.232	** $[M+H]^+$	627.1561	628.1634	628.1648	2.2	465, 303	F
34	Cyanidin 3-*O*-(2-*O*-(6-*O*-(E)-caffeoyl-D glucoside)-D-glucoside)-5-*O*-D-glucoside	$C_{43}H_{49}O_{24}$	38.918	$[M+H]^+$	949.2614	950.2687	950.2679	−0.8	787, 463, 301	RG
35	Delphinidin 3-*O*-galactoside	$C_{21}H_{21}O_{12}$	45.301	** $[M-H]^-$	465.1033	464.096	464.0964	0.9	303	S, F,* PL
				Dihydrochalcones						
36	3-Hydroxyphloretin 2′-*O*-glucoside	$C_{21}H_{24}O_{11}$	24.659	$[M-H]^-$	452.1319	451.1246	451.1249	0.7	289, 273	* PL, RG, F, S
37	3-Hydroxyphloretin 2′-*O*-xylosyl-glucoside	$C_{26}H_{32}O_{15}$	37.564	$[M-H]^-$	584.1741	583.1668	583.1665	−0.5	289	RG
38	Phloridzin	$C_{21}H_{24}O_{10}$	51.613	** $[M-H]^-$	436.1369	435.1296	435.1284	−2.8	273	* RG, PL, S, F

Table 3. *Cont.*

No.	Proposed Compounds	Molecular Formula	RT (min)	Ionization (ESI+/ESI−)	Molecular Weight	Theoretical (m/z)	Observed (m/z)	Error (ppm)	MS² Product Ions	Samples
					Dihydroflavonols					
39	Dihydromyricetin 3-O-rhamnoside	$C_{21}H_{22}O_{12}$	23.549	** [M-H]−	466.1111	465.1038	465.1031	−1.5	301	RG, F,* PL, F, PL
40	Dihydroquercetin 3-O-rhamnoside	$C_{21}H_{22}O_{11}$	32.081	** [M-H]−	450.1162	449.1089	449.1081	−1.8	303	S,* PL
41	Dihydroquercetin	$C_{15}H_{12}O_7$	38.674	** [M-H]−	304.0583	303.051	303.0518	2.6	285, 275, 151	S,* PL
					Flavanols					
42	(+)-Gallocatechin	$C_{15}H_{14}O_7$	4.494	** [M-H]−	306.074	305.0667	305.068	4.3	261, 219	S, PL, F,* RD
43	(+)-Gallocatechin 3-O-gallate	$C_{22}H_{18}O_{11}$	11.106	[M-H]−	458.0849	457.0776	457.0781	1.1	305, 169	F,* S
44	Procyanidin dimer B1	$C_{30}H_{26}O_{12}$	21.362	** [M-H]−	578.1424	577.1351	577.1333	−3.1	451	* PL, RG, S, F
45	(+)-Catechin 3-O-gallate	$C_{22}H_{18}O_{10}$	22.306	** [M-H]−	442.09	441.0827	441.0805	−5.0	289, 169, 125	* PL, F
46	(+)-Catechin	$C_{15}H_{14}O_6$	26.597	** [M-H]−	290.079	289.0717	289.0706	−3.8	245, 205, 179	* RG, S, PL, F
47	4′-O-Methyl-(-)-epigallocatechin 7-O-glucuronide	$C_{22}H_{24}O_{13}$	27.607	[M-H]−	496.1217	495.1144	495.116	3.2	451, 313	RG,* PL, F
48	Procyanidin trimer C1	$C_{45}H_{38}O_{18}$	28.966	** [M-H]−	866.2058	865.1985	865.1961	−2.8	739, 713, 695	* RG, S, PL, F
49	Cinnamtannin A2	$C_{60}H_{50}O_{24}$	35.444	** [M-H]−	1154.269	1153.2617	1153.263	1.1	739	RG,* PL, F
50	Prodelphinidin dimer B3	$C_{30}H_{26}O_{14}$	67.792	** [M+H]+	610.1323	611.1396	611.1407	1.8	469, 311, 291	PL,* F
					Flavanones					
51	Hesperetin 3′,7-O-diglucuronide	$C_{28}H_{30}O_{18}$	21.163	** [M-H]−	654.1432	653.1359	653.1361	0.3	477, 301, 286, 242	S,* PL
52	6-Prenylnaringenin	$C_{20}H_{20}O_5$	35.742	[M+H]+	340.1311	341.1384	341.1375	−2.6	323, 137	F
53	Narirutin	$C_{27}H_{32}O_{14}$	38.326	[M-H]−	580.1792	579.1719	579.171	−1.6	271	RG
54	Hesperetin 3′-O-glucuronide	$C_{22}H_{22}O_{12}$	52.421	** [M+H]+	478.1111	479.1184	479.1199	3.1	301, 175, 113, 85	RD, RG, PL,* F
					Flavones					
55	Apigenin 7-O-apiosyl-glucoside	$C_{26}H_{28}O_{14}$	14.031	** [M+H]+	564.1479	565.1552	565.1552	0.0	296	PL,* S
56	Apigenin 7-O-glucuronide	$C_{21}H_{18}O_{11}$	15.812	** [M+H]+	446.0849	447.0922	447.093	1.8	271, 253	* PL, S
57	7,4′-Dihydroxyflavone	$C_{15}H_{10}O_4$	18.251	[M+H]+	254.0579	255.0652	255.0643	−3.5	227, 199, 171	F
58	Cirsilineol	$C_{18}H_{16}O_7$	26.744	** [M+H]+	344.0896	345.0969	345.0962	−2.0	330, 312, 297, 284	* PL, RD
59	Apigenin 6,8-di-C-glucoside	$C_{27}H_{30}O_{15}$	43.578	** [M-H]−	594.1585	593.1512	593.1527	2.5	503, 473	PL, S,* RG, F
60	6-Hydroxyluteolin 7-O-rhamnoside	$C_{21}H_{20}O_{11}$	46.758	** [M-H]−	448.1006	447.0933	447.0928	−1.1	301	* RG, PL, RD, S, F
61	Chrysoeriol 7-O-glucoside	$C_{22}H_{22}O_{11}$	54.226	** [M+H]+	462.1162	463.1235	463.1255	4.3	445, 427, 409, 381	RG, PL,* F

Table 3. *Cont.*

No.	Proposed Compounds	Molecular Formula	RT (min)	Ionization (ESI⁺/ESI⁻)	Molecular Weight	Theoretical (m/z)	Observed (m/z)	Error (ppm)	MS² Product Ions	Samples
					Flavonols					
62	Myricetin 3-*O*-galactoside	$C_{21}H_{20}O_{13}$	19.288	[M-H]⁻	480.0904	479.0831	479.081	−4.4	317	RD
63	Quercetin 3-*O*-glucosyl-xyloside	$C_{26}H_{28}O_{16}$	21.146	[M-H]⁻	596.1377	595.1304	595.1291	−2.2	265, 138, 116	PL
64	Quercetin 3-*O*-xylosyl-rutinoside	$C_{32}H_{38}O_{20}$	23.124	** [M+H]⁺	742.1956	743.2029	743.2022	−0.9	479, 317	F,* S
65	Kaempferol 3-*O*-glucosyl-rhamnosyl-galactoside	$C_{33}H_{40}O_{20}$	24.867	** [M-H]⁻	756.2113	755.204	755.2068	3.7	285	RG,* F
66	Kaempferol 3-*O*-(2″-rhamnosyl-galactoside) 7-*O*-rhamnoside	$C_{33}H_{40}O_{19}$	25.198	** [M-H]⁻	740.2164	739.2091	739.2115	3.2	593, 447, 285	S,* F
67	Kaempferol 3-*O*-xylosyl-glucoside	$C_{26}H_{28}O_{15}$	28.135	** [M+H]⁺	580.1428	581.1501	581.1479	−3.8		* PL, RG, F
68	Kaempferol 3,7-*O*-diglucoside	$C_{27}H_{30}O_{16}$	37.879	** [M-H]⁻	610.1534	609.1461	609.1451	−1.6	447, 285	* RG, S
69	Myricetin 3-*O*-rhamnoside	$C_{21}H_{20}O_{12}$	39.996	** [M-H]⁻	464.0955	463.0882	463.0862	−4.3	317	* RD, RG, S
70	Quercetin 3-*O*-xylosyl-glucuronide	$C_{26}H_{26}O_{17}$	43.207	[M+H]⁺	610.117	611.1243	611.1255	2.0	479, 303, 285, 239	F,* PL
71	Quercetin 3-*O*-arabinoside	$C_{20}H_{18}O_{11}$	45.665	** [M-H]⁻	434.0849	433.0776	433.0781	1.2	301	* RG, S
					Isoflavonoids					
72	6″-*O*-Malonylglycitin	$C_{25}H_{24}O_{13}$	7.256	[M+H]⁺	532.1217	533.129	533.1286	−0.8	285, 270, 253	S
73	6″-*O*-Malonyldaidzin	$C_{24}H_{22}O_{12}$	16.246	[M+H]⁺	502.1111	503.1184	503.12	3.2	255	F
74	Dihydrobiochanin A	$C_{16}H_{14}O_{5}$	22.255	[M+H]⁺	286.0841	287.0914	287.0925	3.8	269, 203, 201, 175	F,* PL
75	Violanone	$C_{17}H_{16}O_{6}$	24.926	[M+H]⁺	316.0947	317.102	317.1016	−1.3	300, 285, 135	F
76	3′-Hydroxygenistein	$C_{15}H_{10}O_{6}$	27.116	[M+H]⁺	286.0477	287.055	287.0547	−1.0	269, 259	* S, F
77	Formononetin 7-*O*-glucuronide	$C_{22}H_{20}O_{10}$	42.45	** [M-H]⁻	444.1056	443.0983	443.0973	−2.3	267, 252	* S, F
78	5,6,7,3′,4′-Pentahydroxyisoflavone	$C_{15}H_{10}O_{7}$	42.893	** [M+H]⁺	302.0427	303.05	303.0487	−4.3	285, 257	* PL, S, RD, RG, F
79	6″-*O*-Malonylgenistin	$C_{24}H_{22}O_{13}$	64.297	** [M+H]⁺	518.106	519.1133	519.1157	4.6	271	* F, S

Table 3. *Cont.*

No.	Proposed Compounds	Molecular Formula	RT (min)	Ionization (ESI$^+$/ESI$^-$)	Molecular Weight	Theoretical (*m*/*z*)	Observed (*m*/*z*)	Error (ppm)	MS^2 Product Ions	Samples
					Lignans					
80	Enterolactone	$C_{18}H_{18}O_4$	4.234	$[M+H]^+$	298.1205	299.1278	299.1279	0.3	281, 187, 165	PL
81	7-Hydroxymatairesinol	$C_{20}H_{22}O_7$	47.587	$[M-H]^-$	374.1366	373.1293	373.1283	−2.7	343, 313,	S, F,* RG
82	Schisandrin C	$C_{22}H_{24}O_6$	59.344	$[M+H]^+$	384.1573	385.1646	385.1663	4.4	370, 315, 300	S,* F
83	Secoisolariciresinol-sesquilignan	$C_{30}H_{38}O_{10}$	59.607	$[M-H]^-$	558.2465	557.2392	557.2387	−0.9	539, 521, 509, 361	F
84	Schisandrol B	$C_{23}H_{28}O_7$	63.253	$[M+H]^+$	416.1835	417.1908	417.1929	5.0	224, 193, 165	F
					Other polyphenols					
					Curcuminoids					
85	Demethoxycurcumin	$C_{20}H_{18}O_5$	81.976	$[M-H]^-$	338.1154	337.1081	337.108	−0.3	217	RD
					Furanocoumarins					
86	Isopimpinellin	$C_{13}H_{10}O_5$	4.478	$[M+H]^+$	246.0528	247.0601	247.0605	1.6	232, 217, 205, 203	* RD, F
					Hydroxybenzaldehydes					
87	*p*-Anisaldehyde	$C_8H_8O_2$	26.251	** $[M+H]^+$	136.0524	137.0597	137.0596	−0.7	122, 109	PL,* F, S
88	4-Hydroxybenzaldehyde	$C_7H_6O_2$	44.568	** $[M-H]^-$	122.0368	121.0295	121.0301	5.0	77	S, F,* RD
					Hydroxycoumarins					
89	Coumarin	$C_9H_6O_2$	25.364	* $[M-H]^-$	146.0368	145.0295	145.0302	4.8	103, 91	F
					Hydroxyphenylpropenes					
90	2-Methoxy-5-prop-1-enylphenol	$C_{10}H_{12}O_2$	25.903	$[M+H]^+$	164.0837	165.091	165.0906	−2.4	149, 137, 133, 124	F
					Other polyphenols					
91	Salvianolic acid C	$C_{26}H_{20}O_{10}$	9.665	$[M-H]^-$	492.1056	491.0983	491.0963	−4.1	311, 267, 249	S
92	Salvianolic acid B	$C_{36}H_{30}O_{16}$	28.598	$[M-H]^-$	718.1534	717.1461	717.1436	−3.5	519, 339, 321, 295	RD
					Phenolic terpenes					
93	Rosmanol	$C_{20}H_{26}O_5$	22.23	$[M+H]^+$	346.178	347.1853	347.1844	−2.6	301, 241, 231	S
94	Carnosic acid	$C_{20}H_{28}O_4$	80.419	** $[M-H]^-$	332.1988	331.1915	331.1905	−3.0	287, 269	* RD, F
					Tyrosols					
95	Hydroxytyrosol 4-*O*-glucoside	$C_{14}H_{20}O_8$	14.338	** $[M-H]^-$	316.1158	315.1085	315.109	1.6	153, 123	F,* PL
96	3,4-DHPEA-AC	$C_{10}H_{12}O_4$	25.537	** $[M-H]^-$	196.0736	195.0663	195.0658	−2.6	135	* PL, F, S
97	Demethyloleuropein	$C_{24}H_{30}O_{13}$	51.646	* $[M-H]^-$	526.1686	525.1613	525.1599	−2.7	495	* RG, F

* Data presented in the table are from the sample indicated with an asterisk; ** Compounds were detected in both negative $[M-H]^-$ and positive $[M+H]^+$ mode of ionization while only single mode data was presented. Apple samples mentioned in abbreviations are Royal Gala "RG"; Red Delicious "RD"; Fuji "F"; Smitten "S"; Pink Lady "PL".

3.4.1. Phenolic Acids

In our research, 27 phenolic acids including hydroxyphenylacetic acids (2), hydroxycinnamic acids (18), hydroxybenzoic acids (5), and hydroxyphenylpropanoic acids (2) were identified and characterised in five varieties of apples.

Compound **1** was tentatively characterised as protocatechuic acid 4-*O*-glucoside present in negative mode of ionisation and identified in Royal Gala, Red Delicious and Fuji apples. The compound had precursor ion at *m/z* 315.0718 and on further MS/MS analysis showed product ions at *m/z* 125 (loss of CO_2, 44 Da) and *m/z* 169 (loss of hexosyl moiety, 162 Da) [50]. In previous study of Gu et al. [21] reported tentatively characterised protocatechuic acid 4-*O*-glucoside from fresh apples. Compound **12** (($[M-H]^-$ *m/z* at 325.0925) was tentatively characterised as *p*-Coumaric acid 4-*O*-glucoside based on the product ions at *m/z* 163, due to the loss of hexosyl moiety (162 Da) from the precursor ions [50]. Identified in Pink Lady, Royal Gala and Fuji apples.

Compound **7** was tentatively characterised as caffeic acid in Smitten variety based on the precursor ion at $[M+H]^+$ at *m/z* 181.0494 and confirmed based on the MS^2 fragmentation with product ions at *m/z* 143 (loss of two water molecules, 36 Da) and *m/z* 133 (loss of HCOOH, 46 Da) [51]. Compound 15 was observed in Smitten, Pink Lady and Fuji and tentatively characterised as ferulic acid based on the precursor ion at ($[M-H]^-$ at *m/z* 193.0505. Upon further MS/MS analysis, the product ions at *m/z* 178 (loss of CH_3, 15 Da), *m/z* 149 (loss of CO_2, 44 Da) and *m/z* 134 (loss of CH_3-CO_2, 59 Da) confirmed the compound [52]. Compounds **19** (($[M-H]^-$ *m/z* at 223.0603) identified in Fuji, Pink Lady and Smitten apples. MS/MS analysis confirmed the compound as sinapic acid by fragments at *m/z* 205 and *m/z* 163 due to the consecutive loss of H_2O and 2CHO from the precursor ion respectively [53]. Previously, Lee et al. [54] reported the presence of caffeic acid, ferulic acid and sinapic acid in apples. Caffeic acid abundantly present in both pulp and peel [54]. Other phenolic compounds to our best knowledge were first time detected in Australian grown apples.

3.4.2. Flavonoids

A total of 52 Flavonoids were identified in the five apple samples including an thocyanins (8), dihydrochalcones (3), dihydroflavonols (3), flavanols (9), flavones (4), flavanones (7), flavonols (10), and Isoflavonoids (8).

Compound **31** (Cyanidin 3,5-*O*-diglucoside) and compound **33** (Delphinidin 3-*O*-glucosyl-glucoside) were both detected in the positive mode of ionization with the precursor ions at *m/z* 612.1693 and *m/z* 628.1648, respectively. The MS/MS experiment allowed the further identification of these compounds based on the peaks after removal of the sugar moieties for both compounds [55].

Compound **36** and compound **37** were tentatively characterised as 3-hydroxyphloretin 2′-*O*-glucoside and 3-hydroxyphloretin 2′-*O*-xylosyl-glucoside present in negative mode of ionisation with precursor ions at *m/z* 451.1249 and *m/z* 583.1665, respectively. 3-hydroxyphloretin 2′-*O*-glucoside was confirmed by fragment ions at *m/z* 289 [M-H-glucoside] and *m/z* 273 [M-H-phloretin aglycon] [56] identified in Pink Lady, Royal Gala, Fuji and Smitten apples. Whereas, 3-Hydroxyphloretin 2′-*O*-xylosyl-glucoside was identified by fragment ions at *m/z* 289, due to the loss of xylosyl-glucoside disaccharide (132 + 162 Da) [57] observed in Royal Gala apples. Phloridzin (compound **38)** with precursor ion at [($[M-H]^-$, *m/z* 435.1284], and confirmed by product ions at *m/z* 273 due to the loss of glucoside (162 Da) [58] identified in Pink Lady, Royal Gala, Fuji and Smitten apples. Kelebek et al. [58] reported the presence of phloridzin in apples.

Three flavanols derivatives (Compound **44**, **46**, **48**) were all detected in four samples including Pink Lady, Royal Gala, Fuji and Smitten apples. Compound **44**, **46**, **48** with negative mode of ionisation with precursor ions at *m/z* 577.1333, *m/z* 289.0706 and *m/z* 865.1961 were tentatively characterised as procyanidin dimer B1, (+)-catechin and procyanidin trimer C1 respectively. The compound procyanidin trimer C1 was confirmed by product ions at *m/z* 739, *m/z* 713 and *m/z* 695, due to the loss of heterocyclic ring fission

(HRF) reaction (126 Da), loss of retro-Diels-Alder (RDA) (152 Da) and loss of H_2O [59]. While the loss of phloroglucinol (126 Da) from the precursor ion confirmed the presence of procyanidin dimer B1 [60]. Whereas, (+)-catechin compound confirmed based on the fragment ions at *m/z* 245, *m/z* 205 and *m/z* 179, due to corresponding loss of CO_2 (44 Da), flavonoid A ring (84 Da) and flavonoid B ring (110 Da) from the precursor ion, respectively [50]. Previously Nicoli et al. [61] reported the presence of (+)-catechin in apple varieties. (+)-catechin has a positive health benefit including scavenging free radicals, delaying aging and benefitting the intestinal microbes [62].

Compound **51** (hesperetin 3′,7-*O*-diglucuronide) and compound **53** (narirutin) were found both in negative ionization modes based on the precusor ions at *m/z* 653.1361 and *m/z* 579.1710, respectively. Compound **51** was confirmed by the product ion at *m/z* 477 [M-H-glucuronide, loss of 176 Da], *m/z* 301 [M-H-2 glucuronide, loss of 352 Da], *m/z* 286 [M-H-2glucuronide-CH_3, loss of 367 Da] and *m/z* 242 [M-H-2glucuronide-OCH_2-CHO] [63], while compound **53** was confirmed by loss of neohesperidose moiety (308 Da) [64] from the precursor ion. In our study compound **51** was identified in Smitten and Pink Lady whereas compound **53** was identified in Royal Gala and Red Delicious. To our best knowledge it was first time detected in Australian grown apples.

Apigenin 7-*O*-glucuronide (Compound **56**) and cirsilineol (compound **58**) were tentatively characterised in negative mode of ionisation at *m/z* 447.0930 and *m/z* 345.0962, respectively. The MS/MS analysis confirmed the compound **56** at product ions *m/z* 271 due to the corresponding loss of glucuronide (176 Da) and loss of glucuronide and *m/z* 253 due to the loss of H_2O-CH_2O (194 Da) from the precursor ion [65]. The presence of cirsilineol was confirmed by the product ions at *m/z* 330 [M+H-CH_3], *m/z* 312 [M+H-CH_3-H_2O], *m/z* 297 [M+H-2CH_3-H_2O] and *m/z* 284 [M+H-CH_3-H_2O-CO] [66]. According to previous reports, compounds have been characterised in several plants including Ocimum species [66].

Compound **62** (Myricetin 3-*O*-galactoside with ($[M\text{-}H]^-$ *m/z* at 479.081) identified in Red Delicious and compound **63** (Quercetin 3-*O*-glucosyl-xyloside with ($[M\text{-}H]^-$ *m/z* at 595.1291) identified in Pink Lady were only detected in the negative ionization mode, and identified according to the fragment peaks at *m/z* 317 [M-H-glucoside, loss of 162 Da] [67] and *m/z* 265 [M-H-glucose-xylose, loss of 330 Da] [51], respectively. Compound **65, 66** and **68** present in the negative mode of ionisation were identified as kaempferol 3-*O*-glucosyl-rhamnosyl-galactoside, kaempferol 3-*O*-(2″-rhamnosyl-galactoside) 7-*O*-rhamnoside and kaempferol 3,7-*O*-diglucoside according to the ($[M\text{-}H]^-$ at *m/z* 755.2068, *m/z* 739.2115 and *m/z* 609.1451, respectively Kaempferol 3-*O*-glucosyl-rhamnosyl-galactoside exhibited the product ions at *m/z* 285, corresponding to the loss of the sugar units from the precursor ion [68]. The presence of kaempferol 3-*O*-(2″-rhamnosyl-galactoside) 7-*O*-rhamnoside was confirmed by the product ions at *m/z* 593 [M-H-$C_6H_{10}O_4$], *m/z* 447 [M-H-2$C_6H_{10}O_4$], and *m/z* 285 [M-H-2$C_6H_{10}O_4$-$C_6H_{10}O_5$] [69]. Whereas, kaempferol 3,7-*O*-diglucoside exhibited the product ions at *m/z* 447 and *m/z* 285, corresponding to the loss of glucoside and consecutive loss of glucoside from the parent ion [70]. It worth noted that these compounds were first time detected in Australian grown apple samples to the best of our knowledge.

Compound **73** and **75** detected in positive mode were identified as 6″-*O*-Malonyldaidzin and violanone with precursor ion at *m/z* 503.1200 and *m/z* 317.1016, respectively. 6″-*O*-Malonyldaidzin was confirmed by the product ion at *m/z* 255 [71], corresponding to the loss of malonyl-glucoside from precursor, while the compound violanone was confirmed by the intensive peaks at *m/z* 300 [M+H-CH_3, loss of 15 Da], *m/z* 285 [M+H-2CH_3, loss of 30 Da] and *m/z* 135 [M+H-$C_{10}H_{12}O_3$] [72]. Previously, several studies had discovered the existence of the above isoflavonoids in fruits [71,73–76].

3.4.3. Lignans

Compound **82** (Schisandrin C) was detected only in the positive ionization mode with precursor ions at *m/z* 385.1663. The fragmentation peaks confirmed the compound schisantherin C based on product ions at m/z 370 [M+H-CH_3OH], m/z 315 [M+H-C_5H_{10}] and m/z 300 [M+H-CH_3-C_5H_{10}] [77].

3.4.4. Other Polyphenols

In other polyphenols, curcuminoids (1), furanocoumarins (1), hydroxybenzaldehydes (2), hydroxycoumarins (1), hydroxyphenylpropenes (1), phenolic terpenes (2), tyrosols (3) and other polyphenols (2), while tyrosols was the dominant subclass were identified in apple samples.

Compound **88** was tentatively characterised as 4-hydroxybenzaldehyde based on the precursor ion at ([M-H]$^-$ at m/z 121.0301 and confirmed based on the MS^2 fragmentation, which exhibited the loss of CO_2 (44 Da) from the precursor, resulting in the product ion at m/z 77 [78]. Rosmanol (compound **93**) was found in positive modes, and tentatively characterised according to the precursors [M+H]$^+$ at m/z 347.1844. In the MS^2 experiment, peaks at m/z 301 (loss of H_2O) and m/z 231(loss of CO_2) achieved the identification of coumarin [79]. Meanwhile, compound **94** (carnosic acid with ([M-H]$^-$ at m/z 331.1905) was confirmed by the fragments at m/z 287 and m/z 296, resulting from the loss of CO_2 and further loss of H_2O from the precursor [80]. To best of our knowledge, this is the first time it has been detected in apple samples.

Compounds **95** and **96** detected in negative mode were detected as hydroxytyrosol 4-*O*-glucoside and 3,4-DHPEA-AC, precursor ion at m/z 315.1090 and m/z 195.0658, respectively. On further analysis, hydroxytyrosol 4-*O*-glucoside was confirmed by the product ions at m/z 153 and m/z 123, corresponding to the loss of glucoside (162 Da) and glucoside-CH_2O (192 Da) from the precursor ion, respectively [78] and 3,4-DHPEA-AC was confirmed by the product ions at m/z 135 [M-H-$C_2H_4O_2$] [81].

Compounds **91** and **92** were found in negative ionization mode and identified as salvianolic acid C and salvianolic acid B with precursor ions at m/z 491.0963 and m/z 717.1436, respectively. Salvianolic acid C was confirmed by the product ion at m/z 311 [M-H-caffeic acid], m/z 267 [M-H-caffeic-CO_2] and m/z 249 [M-H-CO_2-H_2O][82], while salvianolic acid B was confirmed by the intensive peaks at m/z 519 [M-H-Danshensu, loss of 198 Da], m/z 339 [M-H-Danshesu-caffeic acid, loss of 378], m/z 321 [M-H-2 $\times$ Danshensu, loss of 396 Da] and m/z 295 [M-H-Danshensu-caffeic acid-CO_2, loss of 422 Da][82]. Previously, both compounds were detected in *Salvia miltiorrhiza* [83]. Salvianolic acid, known for its antioxidant potential, can effectively remove oxygen free radicals in the human body. This compound is one of the natural products with the strongest antioxidant effect [84]. However, these compounds have been discovered for the first time in apple varieties to the best of our knowledge.

3.5. Quantitative Analysis of Phenolic Compounds by HPLC-PDA

The most effective way of quantification of phenolic compounds is by HPLC-PDA analysis [85]. In our study, 10 phenolic compounds (mainly phenolic acids and flavonoids) were chosen to be quantified since it is difficult to complete the qualification of all the identified compounds. Since a few compounds have too low UV absorption to be detected, the content of phenolic compounds in five apple samples are shown in Table 4.

In phenolic acids, chlorogenic acid, *p*-hydroxybenzoic acid and caffeic acid were the major phenolic acids in Royal Gala, while Pink Lady contained high content in chlorogenic acid, *p*-hydroxybenzoic acid and protocatechuic acid. It was observed that Red Delicious had highest content in caffeic acid when compared to other samples. Caffeic acid, chlorogenic acid and protocatechuic acid were detected in Fuji. Whereas Smitten apples had gallic acid and *p*-hydroxybenzoic acid, these compounds were not observed in Fuji.

According to previous studies, chlorogenic acid and caffeic acid have been identified and quantified in several apple cultivars [86,87]. While Soares et al.'s [88] study indicated that apples, including gala, showed a low concentration of gallic acid and *p*-hydroxybenzoic acid, only few studies focused on identification of Fuji. Hence, further studies are required to analyse the quantitation of Fuji and Smitten.

Table 4. Quantitative analysis in phenolic compounds of five kinds of apple samples.

No.	Compound Name	Molecular Formula	RT (min)	Royal Gala (mg/g)	Pink Lady (mg/g)	Red Delicious (mg/g)	Fuji (mg/g)	Smitten (mg/g)	Phenolic Class
1	Gallic acid	$C_7H_6O_5$	6.836	2.34 ± 0.06 [c]	1.23 ± 0.05 [d]	4.56 ± 0.09 [a]	-	3.25 ± 0.07 [b]	Phenolic acids
2	Protocatechuic acid	$C_7H_6O_4$	12.569	3.69 ± 0.07 [b]	4.59 ± 0.08 [a]	1.25 ± 0.05 [d]	2.59 ± 0.07 [c]	-	Phenolic acids
3	*p*-Hydroxybenzoic acid	$C_7H_6O_3$	20.24	4.6 ± 0.08 [b]	6.37 ± 0.09 [a]	2.13 ± 0.06 [c]	-	1.29 ± 0.05 [d]	Phenolic acids
4	Chlorogenic acid	$C_{16}H_{18}O_9$	20.579	11.25 ± 0.07 [b]	15.69 ± 0.09 [a]	4.59 ± 0.06 [c]	3.18 ± 0.05 [d]	1.24 ± 0.05 [e]	Phenolic acids
5	Caffeic acid	$C_9H_8O_4$	25.001	4.56 ± 0.06 [c]	2.14 ± 0.05 [e]	10.25 ± 0.09 [a]	5.69 ± 0.07 [b]	3.69 ± 0.05 [d]	Phenolic acids
6	Catechin	$C_{15}H_{14}O_6$	19.704	15.64 ± 0.08 [b]	10.25 ± 0.08 [c]	3.68 ± 0.05 [e]	18.61 ± 0.09 [a]	4.59 ± 0.07 [d]	Flavonoids
7	Epicatechin	$C_{15}H_{14}O_6$	24.961	7.13 ± 0.08 [a]	2.14 ± 0.06 [b]	2.14 ± 0.05 [b]	2.39 ± 0.06 [b]	7.59 ± 0.09 [a]	Flavonoids
8	Epicatechin gallate	$C_{22}H_{18}O_{10}$	38.015	3.21 ± 0.07 [a]	0.26 ± 0.02 [c]	-	1.21 ± 0.05 [b]	3.67 ± 0.07 [a]	Flavonoids
9	Quercetin	$C_{15}H_{10}O_7$	70.098	18.96 ± 0.08 [b]	7.45 ± 0.06 [d]	19.67 ± 0.09 [a]	4.98 ± 0.05 [e]	14.79 ± 0.07 [c]	Flavonoids
10	Kaempferol	$C_{15}H_{10}O_6$	80.347	14.25 ± 0.09 [a]	3.69 ± 0.05 [e]	9.67 ± 0.07 [c]	11.59 ± 0.08 [b]	6.97 ± 0.07 [d]	Flavonoids

Experiments performed in triplicates are expressed as the mean ± SD. Means followed by different letters (a, b, c, d, e) within the same column are significantly different ($p < 0.05$) from each other. Data of five kinds of apples are reported (fw).

In flavonoids, a total of four flavonoids (catechin, epicatechin, quercetin, kaempferol) were detected among five apple samples. In general, Fuji was detected the highest catechin content while Red Delicious was the lowest. In contrast, the highest quercetin was detected in Red Delicious while Fuji contained the lowest quercetin. Epicatechin was detected in Royal Gala and Smitten the compounds were 7.13 ± 0.08 mg/g and 7.59 ± 0.09 mg/g respectively. Smitten contained the highest Kaempferol (14.25 ± 0.09 mg/g) among five samples. Compound epicatechin gallate was negligible in all the samples.

Previous studies showed that catechin and quercetin are main flavonoids that contribute to the antioxidant potential of apples [61,89]. Previously reported that epicatechin and kaempferol have been successfully synthesised and characterised [90,91]. However, to the best of our knowledge epicatechin gallate was not detected in apples hence more further studies are needed to verify the detection of this flavonoids.

In conclusion, Royal Gala, Red Delicious and Smitten had abundant quercetin content. Pink Lady had a high concentration of compounds including chlorogenic acid and catechin. Fuji had most abundant amount kaempferol and catechin content among five samples. Finally, phenolic acids were more abundant in Pink Lady and Royal Gala while flavonoids were more abundant in Royal Gala, which is consistent with the previous study.

4. Conclusions

In conclusion, various methods have been successfully utilized for the determination, characterisation, and quantitation of phenolic compounds among five different varieties of Australian grown apples. In phenolic compound estimation, Red Delicious showed higher TPC, TFC, DPPH, FRAP, ABTS and TAC values than other apple samples while Fuji exhibited the highest TTC value. The correlation between flavonoids and phenolic acids exhibited a major contribution towards the antioxidant activities of apples. The LC-ESI-QTOF-MS/MS qualification identified a total of 97 different phenolic compounds in five apple samples, including phenolic acids, flavonoids, lignans, other polyphenols and stilbenes. 10 phenolic compounds were quantification through HPLC-PDA based on the difference of UV spectra and retention times. The analysis showed that phenolic acids were more abundant in Pink Lady and Royal Gala whereas flavonoids were more abundant in Royal Gala.

Author Contributions: Conceptualization, methodology, formal analysis, validation and investigation, H.L. and H.A.R.S.; resources, H.A.R.S., C.J.B. and F.R.D.; writing—original draft preparation, H.L. and H.A.R.S.; writing—review and editing, H.L., V.S., C.J.B., H.A.R.S. and F.R.D.; supervision, H.A.R.S. and F.R.D.; ideas sharing, H.A.R.S.; C.J.B. and F.R.D.; funding acquisition, H.A.R.S., F.R.D. and C.J.B. All authors have read and agreed to the published version of the manuscript.

Funding: This research was funded by the University of Melbourne under the "McKenzie Fellowship Scheme" (Grant No. UoM-18/21), the "Richard WS Nicholas Agricultural Science Scholarship" and the "Faculty Research Initiative Funds" funded by the Faculty of Veterinary and Agricultural Sciences, The University of Melbourne, Australia and "The Alfred Deakin Research Fellowship" funded by Deakin University, Australia.

Institutional Review Board Statement: Not applicable.

Informed Consent Statement: Not applicable.

Data Availability Statement: Data is contained within the article.

Acknowledgments: We would like to thank Nicholas Williamson, Shuai Nie and Michael Leeming from the Mass Spectrometry and Proteomics Facility, Bio21 Molecular Science and Biotechnology Institute, the University of Melbourne, VIC, Australia for providing access and support for the use of HPLC-PDA and LC-ESI-QTOF-MS/MS and data analysis.

Conflicts of Interest: The authors declare no conflict of interest.

References

1. FAO; FAOSTAT. Faostat Database. Rome, Italy. 2018. Available online: www.Fao.Org/faostat/zh/#data/qc (accessed on 16 April 2020).
2. Bouayed, J.; Hoffmann, L.; Bohn, T. Total phenolics, flavonoids, anthocyanins and antioxidant activity following simulated gastro-intestinal digestion and dialysis of apple varieties: Bioaccessibility and potential uptake. *Food Chem.* **2011**, *128*, 14–21. [CrossRef]
3. Guerra-Valle, M.E.; Moreno, J.; Lillo-Pérez, S.; Petzold, G.; Simpson, R.; Nuñez, H. Enrichment of apple slices with bioactive compounds from pomegranate cryoconcentrated juice as an osmodehydration agent. *J. Food Qual.* **2018**, *2018*, 7241981. [CrossRef]
4. Azam, H.M.; Alam, S.T.; Hasan, M.; Yameogo, D.D.S.; Kannan, A.D.; Rahman, A.; Kwon, M.J. Phosphorous in the environment: Characteristics with distribution and effects, removal mechanisms, treatment technologies, and factors affecting recovery as minerals in natural and engineered systems. *Environ. Sci. Pollut. Res.* **2019**, *26*, 20183–20207. [CrossRef] [PubMed]
5. O'Kane, G.M.; Richardson, A.; D'Almeida, M.; Wei, H. The cost, availability, cultivars, and quality of fruit and vegetables at farmers' markets and three other retail streams in canberra, act, australia. *J. Hunger Environ. Nutr.* **2019**, *14*, 643–661. [CrossRef]
6. Ribeiro, J.A.; Seifert, M.; Vinholes, J.; Rombaldi, C.V.; Nora, L.; Antillano, R.F.F. Erythorbic acid and sodium erythorbate effectively prevent pulp browning of minimally processed "royal gala" apples. *Ital. J. Food Sci.* **2019**, *31*, 573–590.
7. Belie, N.d.; Harker, F.R.; Baerdemaeker, J.d. Crispness judgement of royal gala apples based on chewing sounds. *Biosyst. Eng.* **2002**, *81*, 297–303.
8. Daillant-Spinnler, B.; MacFie, H.; Beyts, P.; Hedderley, D. Relationships between perceived sensory properties and major preference directions of 12 varieties of apples from the southern hemisphere. *Food Qual. Prefer.* **1996**, *7*, 113–126. [CrossRef]
9. Corrigan, V.K.; Hurst, P.L.; Boulton, G. Sensory characteristics and consumer acceptability of pink lady and other late-season apple cultivars. *N. Z. J. Crop Hortic. Sci.* **1997**, *25*, 375–383. [CrossRef]
10. Ioannides, Y.; Howarth, M.; Raithatha, C.; Defernez, M.; Kemsley, E.; Smith, A. Texture analysis of red delicious fruit: Towards multiple measurements on individual fruit. *Food Qual. Prefer.* **2007**, *18*, 825–833. [CrossRef]
11. Hampson, C.R.; Quamme, H.A.; Hall, J.W.; MacDonald, R.A.; King, M.C.; Cliff, M.A. Sensory evaluation as a selection tool in apple breeding. *Euphytica* **2000**, *111*, 79–90. [CrossRef]
12. Sethi, S.; Joshi, A.; Arora, B.; Bhowmik, A.; Sharma, R.; Kumar, P. Significance of frap, dpph, and cuprac assays for antioxidant activity determination in apple fruit extracts. *Eur. Food Res. Technol.* **2020**, *246*, 591–598. [CrossRef]
13. Hagiwara, K.; Goto, T.; Araki, M.; Miyazaki, H.; Hagiwara, H. Olive polyphenol hydroxytyrosol prevents bone loss. *Eur. J. Pharmacol.* **2011**, *662*, 78–84. [CrossRef] [PubMed]
14. Hongmin, G.S.W.Y.D.; Jianbo, G.J.L.L.Y. Study on the antioxidant and free radical scavenging ability of apple polypbenol. *J. Chinese Cereals Oils Assoc.* **2013**, *4*.
15. Suárez, B.; Álvarez, Á.L.; García, Y.D.; del Barrio, G.; Lobo, A.P.; Parra, F. Phenolic profiles, antioxidant activity and in vitro antiviral properties of apple pomace. *Food Chem.* **2010**, *120*, 339–342. [CrossRef]
16. Rodrigo, R.; Gil, D.; Miranda-Merchak, A.; Kalantzidis, G. Antihypertensive role of polyphenols. *Adv. Clin. Chem.* **2012**, *58*, 225.
17. Hong, Y.; Wang, Z.; Barrow, C.J.; Dunshea, F.R.; Suleria, H.A. High-throughput screening and characterisation of phenolic compounds in stone fruits waste by lc-esi-qtof-ms/ms and their potential antioxidant activities. *Antioxidants* **2021**, *10*, 234. [CrossRef]
18. Suleria, H.A.; Barrow, C.J.; Dunshea, F.R.J.F. Screening and characterisation of phenolic compounds and their antioxidant capacity in different fruit peels. *Foods* **2020**, *9*, 1206. [CrossRef]
19. Imran, M.; Butt, M.S.; Akhtar, S.; Riaz, M.; Iqbal, M.J.; Suleria, H.A.R. Quantification of mangiferin by high pressure liquid chromatography; physicochemical and sensory evaluation of functional mangiferin drink. *J. Food Process. Preserv.* **2016**, *40*, 760–769. [CrossRef]
20. Wojdylo, A.; Oszmianski, J.; Laskowski, P. Polyphenolic compounds and antioxidant activity of new and old apple varieties. *J. Agric. Food Chem.* **2008**, *56*, 6520–6530. [CrossRef]
21. Gu, C.; Howell, K.; Dunshea, F.R.; Suleria, H.A. Lc-esi-qtof/ms characterisation of phenolic acids and flavonoids in polyphenol-rich fruits and vegetables and their potential antioxidant activities. *Antioxidants* **2019**, *8*, 405. [CrossRef]
22. Tang, J.; Dunshea, F.R.; Suleria, H.A. Lc-esi-qtof/ms characterisation of phenolic compounds from medicinal plants (hops and juniper berries) and their antioxidant activity. *Foods* **2020**, *9*, 7.
23. Wang, Z.; Barrow, C.J.; Dunshea, F.R.; Suleria, H.A.R. A comparative investigation on phenolic composition, characterisation and antioxidant potentials of five different australian grown pear varieties. *Antioxidants* **2021**, *10*, 151. [CrossRef] [PubMed]
24. Yunfeng, P.; Tian, D.; Wenjun, W.; Yanju, X.; Xingqian, Y.; Mei, L.; Donghong, L. Effect of harvest, drying and storage on the bitterness, moisture, sugars, free amino acids and phenolic compounds of jujube fruit (*Zizyphus jujuba* cv. *Junzao). J. Sci. Food Agric.* **2018**, *98*, 628–634.
25. Rajurkar, N.S.; Hande, S.M. Estimation of phytochemical content and antioxidant activity of some selected traditional indian medicinal plants. *Indian J. Pharm. Sci.* **2011**, *73*, 146. [CrossRef] [PubMed]
26. Mesfin, H.; Won Hee, K. Antioxidant activity, total polyphenol, flavonoid and tannin contents of fermented green coffee beans with selected yeasts. *Fermentation* **2019**, *5*, 29.
27. Ouyang, H.; Hou, K.; Peng, W.; Liu, Z.; Deng, H. Antioxidant and xanthine oxidase inhibitory activities of total polyphenols from onion. *Saudi J. Biol. Sci.* **2018**, *25*, 1509–1513. [CrossRef]

28. Faiza, A.; Masood, S.B.; Ahmad, B.; Saima, T.; Hafiz, A.R.S. Comparative assessment of free radical scavenging ability of green and red cabbage based on their antioxidant vitamins and phytochemical constituents. *Curr. Bioact. Compd.* **2020**, *16*, 1231–1241.
29. Subbiah, V.; Zhong, B.; Nawaz, M.A.; Barrow, C.J.; Dunshea, F.R.; Suleria, H.A.R. Screening of phenolic compounds in australian grown berries by lc-esi-qtof-ms/ms and determination of their antioxidant potential. *Antioxidants* **2021**, *10*, 26. [CrossRef]
30. Zhong, B.; Robinson, N.A.; Warner, R.D.; Barrow, C.J.; Dunshea, F.R.; Suleria, H.A. Lc-esi-qtof-ms/ms characterisation of seaweed phenolics and their antioxidant potential. *Mar. Drugs* **2020**, *18*, 331. [CrossRef]
31. Ma, C.; Dunshea, F.R.; Suleria, H.A.R. Lc-esi-qtof/ms characterisation of phenolic compounds in palm fruits (jelly and fishtail palm) and their potential antioxidant activities. *Antioxidants* **2019**, *8*, 483. [CrossRef]
32. Ting, Z.; Lijun, S.; Zichao, W.; Tanzeela, N.; Tian, G.; Dan, L.; Pengfei, N.; Yurong, G. The antioxidant property and α-amylase inhibition activity of young apple polyphenols are related with apple varieties. *LWT Food Sci. Technol.* **2019**, *111*, 252–259.
33. Praveen, D.; Amit, B.; Sandeep, R.; Indra, D.B.; Ranbeer, S.R. Diversity of bioactive compounds and antioxidant activity in delicious group of apple in western himalaya. *J. Food Sci. Technol.* **2018**, *55*, 2587–2599.
34. Almeida, D.P.F.; Gião, M.S.; Pintado, M.; Gomes, M.H. Bioactive phytochemicals in apple cultivars from the portuguese protected geographical indication "maçã de alcobaça": Basis for market segmentation. *Int. J. Food Prop.* **2017**, *20*, 2206–2214. [CrossRef]
35. Khanizadeh, S.; Tsao, R.; Rekika, D.; Yang, R.; Charles, M.T.; Rupasinghe, H.P.V. Polyphenol composition and total antioxidant capacity of selected apple genotypes for processing. *J. Food Compos. Anal.* **2008**, *21*, 396–401. [CrossRef]
36. Ravishankar, D.; Rajora, A.K.; Greco, F.; Osborn, H.M. Flavonoids as prospective compounds for anti-cancer therapy. *Int. J. Biochem. Cell Biol.* **2013**, *45*, 2821–2831. [CrossRef]
37. Boyer, J.; Liu, R.H. Apple phytochemicals and their health benefits. *Nutr. J.* **2004**, *3*, 5. [CrossRef] [PubMed]
38. Oszmianski, J.; Lachowicz, S.; Gamsjager, H. Phytochemical analysis by liquid chromatography of ten old apple varieties grown in austria and their antioxidative activity. *Eur. Food Res. Technol.* **2020**, *246*, 437–448. [CrossRef]
39. Bahukhandi, A.; Dhyani, P.; Jugran, A.K.; Bhatt, I.D.; Rawal, R.S. Total phenolics, tannins and antioxidant activity in twenty different apple cultivars growing in West Himalaya, India. *Proc. Natl. Acad. Sci. USA India Sect. B Biol. Sci.* **2019**, *89*, 71. [CrossRef]
40. Yang, D.; Dunshea, F.R.; Suleria, H.A.R. Lc-esi-qtof/ms characterisation of australian herb and spices (garlic, ginger, and onion) and potential antioxidant activity. *J. Food Process. Preserv.* **2020**, *44*, e14497. [CrossRef]
41. Valavanidis, A.; Vlachogianni, T.; Psomas, A.; Zovoili, A.; Siatis, V. Polyphenolic profile and antioxidant activity of five apple cultivars grown under organic and conventional agricultural practices. *Int. J. Food Sci. Technol.* **2009**, *44*, 1167–1175. [CrossRef]
42. Jincan, L.; Pei, Z.; Siqian, L.; Nagendra, P.S. Antioxidant, antibacterial, and antiproliferative activities of free and bound phenolics from peel and flesh of fuji apple. *J. Food Sci.* **2016**, *81*, M1735–M1742.
43. Juan, K.; Tong-bin, H.; Wang-jing, X.; Chang-hai, J. Effect of 1-mcp and uv-c on antioxidant capacity in apple (malus pumila mil.) fruit during storage. *Sci. Technol. Food Ind.* **2019**, 281–285.
44. Pello-Palma, J.; Mangas-Alonso, J.J.; De la Fuente, E.D.; Gonzalez-Alvarez, J.; Diez, J.; Alvarez, M.D.G.; Abrodo, P.A. Characterisation of volatile compounds in new cider apple genotypes using multivariate analysis. *Food Anal. Methods* **2016**, *9*, 3492–3500. [CrossRef]
45. Vasantha Rupasinghe, H.P.; Clegg, S. Total antioxidant capacity, total phenolic content, mineral elements, and histamine concentrations in wines of different fruit sources. *J. Food Compos. Anal.* **2007**, *20*, 133–137. [CrossRef]
46. Maleeha, M.; Farooq, A.; Nazamid, S.; Muhammad, A. Variations of antioxidant characteristics and mineral contents in pulp and peel of different apple (malus domestica borkh.) cultivars from pakistan. *Molecules* **2012**, *17*, 390–407.
47. Ruiz-Torralba, A.; Guerra-Hernandez, E.J.; Garcia-Villanova, B. Antioxidant capacity, polyphenol content and contribution to dietary intake of 52 fruits sold in spain. *CyTA J. Food* **2018**, *16*, 1131–1138. [CrossRef]
48. Kam, A.; Li, K.M.; Razmovski-Naumovski, V.; Nammi, S.; Chan, K.; Li, G.Q. Variability of the polyphenolic content and antioxidant capacity of methanolic extracts of pomegranate peel. *Nat. Prod. Commun.* **2013**, *8*, 707. [CrossRef]
49. Thaipong, K.; Boonprakob, U.; Crosby, K.; Cisneros-Zevallos, L.; Byrne, D.H. Comparison of abts, dpph, frap, and orac assays for estimating antioxidant activity from guava fruit extracts. *J. Food Compos. Anal.* **2006**, *19*, 669–675. [CrossRef]
50. Escobar-Avello, D.; Lozano-Castellón, J.; Mardones, C.; Pérez, A.J.; Saéz, V.; Riquelme, S.; von Baer, D.; Vallverdú-Queralt, A. Phenolic profile of grape canes: Novel compounds identified by lc-esi-ltq-orbitrap-ms. *Molecules* **2019**, *24*, 3763.
51. Lin, H.; Zhu, H.; Tan, J.; Wang, H.; Wang, Z.; Li, P.; Zhao, C.; Liu, J. Comparative analysis of chemical constituents of moringa oleifera leaves from china and india by ultra-performance liquid chromatography coupled with quadrupole-time-of-flight mass spectrometry. *Molecules* **2019**, *24*, 942. [CrossRef]
52. Wang, J.; Jia, Z.; Zhang, Z.; Wang, Y.; Liu, X.; Wang, L.; Lin, R. Analysis of chemical constituents of melastoma dodecandrum lour. By uplc-esi-q-exactive focus-ms/ms. *Molecules* **2017**, *22*, 476. [CrossRef]
53. Geng, C.-A.; Chen, H.; Chen, X.-L.; Zhang, X.-M.; Lei, L.-G.; Chen, J.-J. Rapid characterisation of chemical constituents in saniculiphyllum guangxiense by ultra fast liquid chromatography with diode array detection and electrospray ionization tandem mass spectrometry. *Int. J. Mass Spectrom.* **2014**, *361*, 9–22. [CrossRef]
54. Lee, J.; Chan, B.L.S.; Mitchell, A.E. Identification/quantification of free and bound phenolic acids in peel and pulp of apples (*Malus domestica*) using high resolution mass spectrometry (hrms). *Food Chem.* **2017**, *215*, 301–310. [CrossRef]
55. Kim, I.; Lee, J. Variations in anthocyanin profiles and antioxidant activity of 12 genotypes of mulberry (morus spp.) fruits and their changes during processing. *Antioxidants* **2020**, *9*, 242. [CrossRef] [PubMed]

applied sciences

Review

Environmental Conditions and Agronomical Factors Influencing the Levels of Phytochemicals in *Brassica* Vegetables Responsible for Nutritional and Sensorial Properties

Francesca Biondi [1,2], Francesca Balducci [1], Franco Capocasa [1], Marino Visciglio [2], Elena Mei [2], Massimo Vagnoni [2], Bruno Mezzetti [1] and Luca Mazzoni [1,*]

1 Department of Agricultural, Food and Environmental Sciences, Università Politecnica delle Marche, via Brecce Bianche 10, 60131 Ancona, Italy; fra90ap@hotmail.it (F.B.); francesca.balducci@staff.univpm.it (F.B.); f.capocasa@staff.univpm.it (F.C.); b.mezzetti@staff.univpm.it (B.M.)

2 Valli di Marca ss Agricultural Company, c/da Valle 2, 63068 Montalto delle Marche, Italy; marino.visciglio@gmail.com (M.V.); maddymei@hotmail.it (E.M.); massimo.agv@gmail.com (M.V.)

* Correspondence: l.mazzoni@staff.univpm.it; Tel.: +39-071-220-4640

Abstract: Recently, the consumption of healthy foods has been related to the prevention of cardiovascular, degenerative diseases and different forms of cancers, underlying the importance of the diet for the consumer's health. Fruits and vegetables contain phytochemicals that act as protective factors for the human body, through different mechanisms of action. Among vegetables, *Brassica* received a lot of attention in the last years for the phytochemical compounds content and antioxidant capacity that confer nutraceutical value to the product. The amount of healthy bioactive compounds present in the *Brassica* defines the nutritional quality. These molecules could belong to the class of antioxidant compounds (e.g., phenols, vitamin C, etc.), or to non-antioxidant compounds (e.g., minerals, glucosinolates, etc.). The amount of these compounds in *Brassica* vegetables could be influenced by several factors, depending on the genotypes, the environmental conditions and the cultivation techniques adopted. The aim of this study is to highlight the main phytochemical compounds present in brassicas used as a food vegetable that confer nutritional and sensorial quality to the final product, and to investigate the main factors that affect the phytochemical concentration and the overall quality of *Brassica* vegetables.

Keywords: phytochemical compounds; antioxidant capacity; *Brassica* spp.; vegetables; cultivation techniques; glucosinolates

Citation: Biondi, F.; Balducci, F.; Capocasa, F.; Visciglio, M.; Mei, E.; Vagnoni, M.; Mezzetti, B.; Mazzoni, L. Environmental Conditions and Agronomical Factors Influencing the Levels of Phytochemicals in *Brassica* Vegetables Responsible for Nutritional and Sensorial Properties. *Appl. Sci.* **2021**, *11*, 1927. https://doi.org/10.3390/app11041927

Academic Editors: Alessandro Genovese and Carmela Spagnuolo

Received: 14 December 2020
Accepted: 19 February 2021
Published: 22 February 2021

1. Introduction

In recent years, the increasing incidence of cardiovascular, degenerative diseases and different forms of cancers has stimulated the interest of consumers in distinguishing healthy from unhealthy foods, as a consequence of the abandonment of Mediterranean diet which, in contrast to other eating regimes, was considered a model of healthy eating for years. The interest for consuming healthy food led to coining the new definition "functional food" for several foodstuffs. This term defines a food product that, in addition to carrying out the traditional alimentary function, also performs preventive and/or therapeutic effects against various human diseases, in particular chronic-degenerative diseases [1,2]. Fruits and vegetables contain phytochemicals that are responsible for these positive effects on human body.

Among vegetables, brassicas received a lot of attention in the last few years. They comprise a large and diverse group of widely consumed vegetables. *Brassica* is the Latin name of a genus that is taxonomically placed within the *Brassicaceae* (*Cruciferae*). The main cultivated and most consumed as food vegetables *Brassica* species in the world are indicated in Table 1. Other closely related vegetables within the *Brassicaceae* family are also reported.

The healthy potential of *Brassica* is bound to their phytochemical compounds. The main compounds responsible for healthy function are phenolic compounds, vitamins (C, B9, K), provitamin A (β-carotene), lutein and different types of glucosinolates [3]. The increasing interest for *Brassica* vegetables has been underlined by their economic importance (among the top 10 economic crops in the world [4]) and by the fact that there was an increase of about 11.5% of both the cultivated area and the production quantity from 2009 to 2019, with a slight increase in yield of about 1.5%. The Organization for Food and Agriculture of the United Nations (FAO) also reported that, in 2019, the global production of cauliflower, broccoli, cabbage and other Brassica crops was about 97 million tonnes, occupying a cultivated area of almost 4 million hectares. Asia accounts for more than 75% of the global *Brassica* vegetable production, with China producing almost a half of all of these vegetables (45 million tonnes). India, with more than 18 million tonnes, then Korea, Russia, and USA with more than 2 million tonnes, are also the biggest producers of cauliflower, broccoli, cabbage and other Brassica crops [5].

Table 1. Main species, subspecies (ssp.) and varieties (var.) of *Brassicaceae* family crops consumed as food vegetable in the world [4,6–8].

Species	ssp./var.	Common Name/Italian Name
Brassica oleracea **L.**	*italica*	Broccoli
	capitata *capitata f. rubra* *capitata f. alba*	Cabbage Red cabbage White cabbage
	botrytis	Cauliflower
	acephala sabellica *acephala laciniata* *acephala*	Curly Kale, Red, Green and Russian curly kale Black Cabbage, Italian or Tuscan cabbage Collards
	gemmifera	Brussels sprouts
	gongylodes	Kohlrabi
	sabauda	Savoy cabbage
	alboglabra	Chinese kale
	costata	Tronchuda cabbage
Brassica rapa **L.**	*rapa*	Turnip broccoli
	sylvestris	Turnip top, broccoli raab/cima di rapa, friarielli
	peckinensis	Chinese cabbage
	chinensis	Pak-choi, Chinese mustard
	japonica	Mizuna, curled mustard, Japanese greens
	perviridis	Tendergreen, Spinach mustard
	rapifera	Turnip
Brassica napus **L.**	*napus*	Rapeseed
	napobrassica	Swede/rutabaga
Brassica kaber/Brassica arvensis/Sinapis arvensis		Charlock/kaber
Brassica alba/Sinapis alba, Brassica hirta		White or yellow mustard
Brassica nigra/Sinapis nigra		Black mustard
Brassica campestris		Field mustard

Table 1. *Cont.*

Species	ssp./var.	Common Name/Italian Name
Brassica carinata		Ethiopian mustard, Abyssinian mustard, Texsel greens
Brassica juncea		Brown mustards
Raphanus sativus L.		Radish
Raphanus raphanistrum L.		Wild radish
Nasturtium officinale R. BR.		Watercress/Crescione d'acqua
Eruca sativa Mill.		Rocket/Rucola
Eruca vesicaria L.		Rocket Ruca, ruchetta
Diplotaxis tenuifolia L.		Wild rocket/Rughetta selvatica
Diplotaxis muralis L.		Wall rocket

The present review summarizes the main chemical compounds responsible for the sensorial and nutritional quality of *Brassica* spp., with particular emphasis on the factors affecting their level.

2. Nutritional Quality

Nutritional quality could be defined as the value of the product for the consumer's physical, psychological, or emotional well-being. The first term of this extended definition concerns the effects of food determined by its phytochemicals, i.e., the sum of all beneficial and harmful compounds and their nutritional (or biological) aspects [9]. In the case of *Brassica* spp., these molecules could belong to the class of antioxidant compounds (e.g., phenols, vitamin C, etc.), exerting their health effects through the ability to scavenge free radicals, or to non-antioxidant compounds (e.g., minerals, glucosinolates, etc.) that exert their function through direct mechanisms in the human metabolism, different from the scavenger activity.

2.1. Antioxidant Compounds

Total antioxidant capacity (TAC) is the ability of food to preserve an oxidizable substrate, inactivate the radical species or reduce an oxidized antioxidant. TAC is considered a fundamental parameter for the description of fruits and vegetables nutritional quality; it is an indicator of the presence of bioactive substances belonging to the antioxidants group. Each antioxidant compound performs its protecting activity through different mechanisms and with different efficiency, depending on its chemical structure and the matrix it acts on. For this reason, TAC analysis is usually preferred to the measurement of the single concentration of each antioxidant, mainly if the objective of the study is a general screening of the health effects of different fruit and vegetables.

Brassica vegetables, i.e., broccoli and kale, showed higher antioxidant potential than other vegetable crops, such as spinach, carrots, potatoes, beans and onions. In general, among *Brassica* vegetables, Brussels sprouts, broccoli, and red cabbage belong to the group that has the highest antioxidant capacity. Common cabbage possesses the lowest antioxidant capacity [10,11]. Contrasting results were reported in cauliflower by Azuma et al. [12] and Wu et al. [13]. The analysis of TAC is influenced by the extraction method and the type of reactive species in the reaction mixture [12].

Many researchers studied and identified the main antioxidant molecules present in *Brassicaceae* [13–16]. These antioxidant compounds belong to two main groups: water-soluble antioxidants and lipo-soluble antioxidants [14,15]. Kurlich et al. [16] and Wu et al. [13] reported that hydrophilic antioxidants are responsible for 80–95% of TAC in *Brassicaceae*, while lipo-soluble antioxidants account for only 5–20%.

2.1.1. Water-Soluble Antioxidants

- Phenolic compounds

Phenolic compounds are the most widespread antioxidant family present in vegetables. This large group of compounds is particularly present in *Brassica* vegetables and constitutes the main source of antioxidants in these plants [14,17]. These plants produce them as secondary metabolites for protection from pest and insect attack.

Their importance in human health is related to antioxidant and anti-inflammatory properties that could have preventive and/or therapeutic effects against obesity, cancer, and neurodegenerative and cardiovascular diseases [18]. Among *Brassica* species, kale and broccoli have the highest quantity of total polyphenols with about 13 mg gallic acid/g of dry weight [19,20].

Flavonoids represent common phenolic compounds in *Brassica*; they possess a lot of biological properties, e.g., antioxidant activity, a capillary protective effect, and an inhibitory effect elicited in various stages of tumours [14,21]. They are characterized by numerous subclasses, but the most important in *Brassica* are the following:

1. Flavonols: together with anthocyanins, they are the main represented flavonoids in *Brassica* species; they can be found in internal and external parts of leaves, seeds, shoots and sprouts leaves [22,23]. The most represented flavonols in *Brassica* vegetables are quercetin (up to 23 mg/100 g fresh product in kale), kaempferol (up to 47 mg/100 g fresh product in kale) and isorhamnetin (up to 24 mg/100 g fresh product in kale) [24]. Quercetin, found mainly in kale, is characterized by a strong antioxidant power (higher than vitamin C); it exerts its activity against free oxygen radicals and acts on the prevention of cardiovascular diseases and cancer, atherosclerosis and chronic inflammation, and the induction of enzymes that detoxify carcinogens [25,26]. Kaempferol 3-O-sophoroside is the main represented flavonol in broccoli florets; its high intake is linked with a lower risk of coronary heart disease [27]. Kaempferol and quercetin, and in less amounts, myricetin, are the main represented flavonols in *B. rapa* subsp. *sylvestris*.
2. Flavones: Apigenin and luteolin are the only flavones detected in hydrolysed extracts of different *Brassica* vegetables (up to 45 and 12 µg/g of fresh weight in Chinese cabbage, respectively), excluding broccoli, where they were not detected [28].
3. Anthocyanins were detected in *Brassica* vegetables and described by several authors [13,29]. They are present only in bright coloured species and varieties with red, orange and purple pigmentation, such as some kales, purple broccoli, and red and black cabbage. These compounds show an interesting antioxidant activity. The 80% of anthocyanins present in *Brassica* species are in the acylated form, more stable and easily absorbable by the organism. The main represented anthocyanins in cruciferous are cyanidin derivatives. In particular, red cabbage possesses eight main types of anthocyanins (for a total of up to 190 mg Cyanidin−3-Glucoside equivalents/100 g of fresh weight) [30]; cyanidin−3-diglucoside is the most represented [31]. In broccoli, more than 17 anthocyanins were detected [29].

Among phenolic compounds, even if they are not hydro-soluble, it is worth mentioning the lignans, diphenolic compounds that possess several biological activities, through their antioxidant and oestrogenic properties. Lignans may reduce the risk of certain cancers and cardiovascular diseases [16]. Some studies reported that lignans are mainly present in the kale family, broccoli and Brussel sprouts with lariciresinol (972, 599 and 493 µg/100 g fresh edible weight of Broccoli, Curly kale and Brussel sprouts, respectively) and pinoresinol (315, 1691 and 220 µg/100 g fresh edible weight of Broccoli, Curly kale and Brussel sprouts, respectively) being the most abundant [15,32].

- Vitamin C and vitamin B9 (Folic Acid)

Vitamin C, or ascorbic acid, is a powerful antioxidant, widely present and studied in fruits; however, many recent works have been focused on the importance of vitamin C in vegetables, mostly in *Brassicaceae* family. In *Brassica* vegetables vitamin C concentration varies a lot among species and subspecies, and it is strictly genotype- and environment-dependent [33,34]. Vitamin C performs countless biological activities in the human body and represents a nutritional compound fundamental for health. Ascorbic acid is a radical scavenger, an enzyme cofactor and a donator/acceptor in electrons transport at the plasma membrane level; its role is fundamental in the regeneration of α-tocopherol, and in the prevention and treatment of malignant and degenerative diseases [33,35].

Among *Brassica* genotypes, Brussel sprouts (76–192 mg/100 g edible portion) and kale (92–186 mg/100 g edible portion) seem to possess the highest content of vitamin C, followed by broccoli (34–146 mg/100 g edible portion) and cauliflower (17–81 mg/100 g edible portion), while white cabbage (19–47 mg/100 g edible portion) possesses the lowest amount [14].

Vitamin B9 (Folic acid) is an important vitamin present in *Brassica*, mainly in raw broccoli (63 μg/100 g of edible portion), Brussel sprouts (61 μg/100 g of edible portion) and kale (141 μg/100 g of edible portion) [36], that act as a coenzyme in many single carbon transfer reactions, in the synthesis of DNA and RNA and of protein components. Furthermore, it reduces the level of homocysteine in the blood, a risk factor for cardiovascular diseases. Among the several health activities that folic acid performs, it is strongly important in the prevention of megaloblastic anaemia, neuropsychiatric disorders and various forms of cancer in the foetus during pregnancy, also reducing the risk of neural tube defects [33,37]. These beneficial effects of folic acid, in particular on the pathogenesis of cancer, and neurological, haematological, and cardiovascular diseases may, in part, be due to its antioxidant activity, via its electron-accepting capacity [38,39].

2.1.2. Lipo-Soluble Antioxidants

Despite the low incidence of lipo-soluble antioxidants on the TAC of *Brassica*, several studies confirm the high content of lipo-soluble antioxidant in kale and broccoli, moderate in Brussels sprouts, and low amount in cauliflower and cabbage [33]. Among lipo-soluble antioxidants, carotenoids and vitamin E are the most important found in *Brassica* vegetables.

- Carotenoids

Carotenoids are responsible for the orange, yellow and red pigmentation of several fruits and vegetables, mainly carotenes and xanthophylls. The most represented carotenoids in *Brassica* vegetables are β-carotene, which the organism transforms to vitamin A, and lutein and zeaxanthin [14]. β-carotene prevents the insurgence of cancer and cardiovascular diseases, and decreases the risk of myocardial infarction, of immune dysfunction and age-related macular degeneration among smokers [33,40]. Muller [41] analysed the total carotenoid content of several *Brassica* species and reported them in decreasing order: Brussel sprouts (6.1 mg/100 g), broccoli (1.6 mg/100 g), red cabbage (0.43 mg/100 g) and finally white cabbage (0.26 mg/100 g). In the *Brassica oleracea* genus, kale possesses the highest content of carotenoids with over 10 mg/100 g of the edible portion [41].

The *Brassica* vegetable with the highest content of lutein and zeaxanthin is kale (3.04–39.55 mg/100 g); interesting contents were also found in broccoli and Brussels sprouts [14]. In *B. rapa* species, 16 carotenoids were identified by Wills and Rangga [42]; in *B. chinensis, parachinensis* and *pekinensis*, lutein and β-carotene are the most abundant carotenoids [15].

- Vitamin E

Vitamin E is formed by groups of compounds known as tocopherols and tocotrienols; in detail, α-tocopherol is the main compound found in *Brassica* vegetables, with the exception of cauliflower, that contains mainly γ-tocopherol [14,43]. Vitamin E performs a protective activity against coronary heart disease through the inhibition of LDL oxidation [44]. A high intake of vitamin E helps in the prevention of cancers, cardiovascular diseases, neurological disorders, and inflammatory diseases [33]. The content of vitamin E in *Brassica* species has been studied in the literature, as reported here in decreasing order: broccoli (0.82 mg/100 g), Brussels sprouts (0.40 mg/100 g), cauliflower (0.35 mg/100 g),

Chinese cabbage (0.24 mg/100 g), Red cabbage (0.05 mg/100 g), and white cabbage (0.04 mg/100 g) [43].

2.2. Micro- and Macro-Elements

Macro-elements, also called macronutrients, are those nutrients that the plants need in greater quantities for essential structural and energetic role. They are indispensable elements for the growth and development of the metabolic functions of plants. The fundamental nutrients are represented by nitrogen (N), phosphorus (P) and potassium (K).

Minerals, such as Boron (B), Copper (Cu), Cobalt (Co), Iron (Fe), Manganese (Mn), Zinc (Zn), and Selenium (Se), are required by plants in very small quantities and are known as microelements. Although trace elements are present in small quantities in plants, they play key roles in plant life; this is also demonstrated by the symptoms associated with deficiency phenomena. Their availability depends on the conditions of the soil. The high capacity of *Brassicaceae* to accumulate the metals present in the soil led this family to be considered a good heavy metal hyperaccumulator, giving the significant number of genera (11) and species (90) of those kinds of plants belong to *Brassicaceae* family [45].

However, micro- and macro-elements also play important roles in the human body. The elements K, Ca, Mg, Fe, Zn, Se, and Mn are fundamental in the regulation of many metabolic activities, in bones and teeth health, in cancer prevention, in the production of red blood cells, and participating as enzyme co-factors. Among *Brassicaceae*, kale is the richest in almost all the main macro and micro elements, with a particularly high amount of calcium (95–539 mg/100 g of edible portion), magnesium (20–67 mg/100 g of edible portion), phosphorus (13–92 mg/100 g of edible portion), potassium (20–491 mg/100 g of edible portion), zinc (0.3–0.9 mg/100 g of edible portion), iron (0.4–3.1 mg/100 g of edible portion), manganese (0.4–1.9 mg/100 g of edible portion), copper (0.02–1.03 mg/100 g of edible portion), and selenium (0–0.94 mg/100 g of edible portion). Mustard green is the richest in iron (1.64 mg/100 g of edible portion), Turnip in sodium (67 mg/100 g of edible portion) and Broccoli in selenium (2.5 mg/100 g of edible portion) [36,46].

2.3. Glucosinolates (GLS) and Isothiocyanates (ITCS)

Glucosinolates (GLS) are one of the most important secondary metabolites in *Brassicaceae* derived from amino acid biosynthesis [14,47]. GLS are glucosidic compounds containing sulphur, present in *Brassica* leaves, compartmentalized in the vacuole, at concentrations that are able to prevent the development of pathogens, diseases and pests [48]. Their concentrations vary among *Brassica* species [49], according to the developmental stage, tissue type, exposure to salt stress, environmental factors, or plant signalling molecules, including treatment with salicylic acid (SA), jasmonic acid (JA) and methyl-jasmonic acid (MeJA) [50–52]. However, their amount generally ranges from the 4.7–32.2 mg/100 g of Mustard spinach, 8.7–12.8 mg/100 g of Rocket, 9.7–33.7 mg/100 g of Chinese cabbage, up to the 65.4–151.1 mg/100 g of Kale, 149.4 mg/100 g of Chinese broccoli and 87.6–332.8 mg/100 g of Radish [53,54]. GLS can be divided into three chemical classes: arylaliphatic, indole and aliphatic, based on their amino acid precursor (aromatic amino acid, tryptophan and methionine, respectively) [55], as reported in Table 2. In *Brassica* vegetables, the most important GLS belong to the methionine-derived ones [56]. Some authors declared that the most popular food processing methods, such as boiling, blanching, and steaming, can significantly affect the final content of GLS. A mild-processing technique, such as blanching, is recommended in order to minimize the loss of GLS or their derivatives [57].

Table 2. Principal glucosinolates identified in leaves of *Brassica* vegetable crops.

Crop	Aliphatic Glucosinolates										Indole Glucosinolates				Arylaliphatic Glucosinolates
	GIB	PRO	SIN	GAL	GRA	GNA	GBN	GIV	GER	GNL	GBS	NGBS	4HGBS	4MGBS	GST
Brassica oleracea															
White cabbage [58–61]	+	+	+	+	+	+	+	+	+	-	+	+	+	+	+
Savoy cabbage [58–61]	+	+	+	-	+	+	+	+	-	-	+	+	-	+	+
Red cabbage [58,60,61]	+	+	+	-	+	+	-	+	-	-	+	+	-	-	-
Kale [58,60–62]	+	+	+	-	+	+	-	+	-	-	+	+	+	+	+
Collard [62]	+	+	+	-	-	-	-	+	+	-	+	-	-	-	-
Tronchuda cabbage [60,63]	+	+	+	+	+	+	+	+	-	-	+	+	+	+	+
Broccoli [62,64]	+	+	+	+	+	+	+	-	+	+	+	+	+	+	+
Brussel sprouts [61,62,64]	+	+	+	-	+	+	-	+	-	-	+	+	-	-	-
Cauliflower [61,64]	+	+	+	-	+	-	-	+	-	-	+	+	-	-	-
Kohlrabi [61]	+	+	+	+	+	+	-	+	-	-	+	+	+	+	-
Brassica rapa															
Turnip [59]	+	+	-	-	-	+	+	-	+	+	+	+	+	+	+
Turnip greens [65]	+	+	-	+	+	+	+	+	-	+	+	+	+	-	+
Turnip tops [66]	+	+	-	-	-	+	+	+	-	-	+	+	+	-	+
Chinese cabbage [59]	+	+	-	-	-	+	+	-	-	+	+	+	[illegible]	+	+
Brassica napus															
Swede [59]	[illegible]	+	-	-	+	-	+	-	-	+	+	+	+	+	+
Leaf rape [67]	-	+	-	+	-	+	+	+	-	+	+	+	-	+	+

GIB: glucoiberin (3-methylsulfinylpropyl GSL); PRO: progoitrin ((R)−2-hydroxybut−3-enyl GSL); SIN: sinigrin (prop−2-enyl GSL); GAL: glucoalysiin (5-methylsulphinylpentyl GSL); GRA: glucoraphanin (4-methylsulphinylbutyl GSL); GNA: gluconapin (but−3-enyl GSL); GBN: glucobrassicanapin (pent−4-enyl GSL); GIV: glucoiberverin (3-methylsulfanylpropyl GSL); GER: glucoerucin (4-methylsulfanylbutyl GSL); GNL: gluconapoleiferin (2-hydroxypent−4-enyl GSL); GBS: glucobrassicin (indol−3-lylmethyl GSL); NGBS: neoglucobrassicin (1-methoxyindol−3-ylmethyl GSL); 4HGBS: 4-hydroxyglucobrassicin (4-hydroxyindol−3-ylmethyl GSL); 4MGBS: 4-methoxyglucobrassicin (4-methoxyindol−3-ylmethyl GSL); GST: gluconasturtiin (2-phenylethyl GSL).

GLS have no direct functions to human health: the health effects are exerted by their hydrolysis breakdown products, the isothiocyanates (ITCs). These are aromatic volatile compounds containing sulphur, derived from the hydrolytic action of the enzyme myrosinase on GLS. The plant myrosinase acts in the human gut and hydrolyses GLS in ITCs during human ingestion. However, during the cooking of the vegetables, the exposure to heat treatment can inactivate the plant myrosinase, so the ITCs are obtained thanks to the action of myrosinase produced by the human gut flora. Unfortunately, its activity and efficiency are lower than plant myrosinase [68,69]. It is possible to obtain many ITCs, and their production strictly depends on the original GLS, the substrate, the pH conditions, the availability of ferrous ions, and the level of activity of the ESP (epithiospecifier protein), a specific protein factor [52,70].

ITCs are mainly responsible for the bitterness, and spicy and typical aroma and smell of *Brassica* vegetables [63]. They possess protective and preventing effects against several kinds of cancer e.g., prostate, intestinal, liver, lung, breast, and bladder, chronic inflammation and neurodegeneration, acting on the apoptotic phase of cell developmental cycle; they are also effective in the reduction in cholesterol [19,71,72].

The most studied ITCs in medical research is sulforaphane [73,74], mainly represented in broccoli and Brussel sprout. It is the most important ITCs considering its health benefits, it derives from the glucoraphanin [75]. Sulforaphane is an indirect antioxidant, because it acts as a catalyst in the stimulation of cellular antioxidant system. In particular, sulforaphane stimulates some enzymes active against tumoral cell proliferation [65,66].

3. Sensorial Quality

The quality of vegetables for the consumer not only concerns the nutritional aspects, but also includes the sensorial parameters that can be defined by several indicators.

The principal sensorial parameters are:

- *Firmness*, which indicates the resistance of vegetables to mechanical damages; it assumes a great importance during the post-harvest management.
- *Colour*, which indicates the freshness of the product and the quality of the storage conditions; it visually attracts the consumers [76].
- *Sweetness*, which is linked to the presence of glucose, fructose, and sucrose, and provides the sweet sensation to the consumer.
- *Acidity*, which indicates the acid sensation that the product stimulates in the consumer.

Considering *Brassica* vegetable quality, fundamental sensorial parameters are those related to aroma and taste. All of these parameters can be investigated through analytical measurements or the implementation of a panel test.

3.1. *Brassica Aroma*

The typical aroma is one of the main reasons for the consumers' rejection against *Brassica* vegetables [77]. Raw vegetables are rich in aroma compounds, which are usually produced because of enzymatic reactions. The typical sulphurous and pungent odour of *Brassicaceae* crops are often attributed to GSL/ITC content. These traits predominantly stem from sulphur-compound degradation products, such as from S-methyl-L-cysteine sulfoxide (SMCSO) [78], and formation can be facilitated by factors, such as bacterial metabolism, plant senescence, cooking, and enzymatic breakdown because of tissue damage (e.g., cutting) [79–81]. Sulphides are generally undesirable odour attributes [82], and compounds such as methanethiol, dimethyl sulphide (DMS), dimethyl trisulphide (DMTS), and dimethyl disulphide (DMDS) are regularly linked with sulphurous aromas and overcooked off-flavours. The main responsible of the fresh cabbage odour is the allyl isothiocyanate, a hydrolysis product of sinigrin thanks to the action of myrosinase [83,84]. Additionally, green note is a particularly important characteristic to recognize in *Brassica* and is conferred by alcohols and aldehydes formed by the enzymatic degradation of free fatty acids [83].

Cooking is the main adopted form to eat *Brassica*, because makes those vegetables more easily digestible and causes a flavour change in them, increasing the consumers' acceptance [85]. Reductions in alcohols, aldehydes and nitriles concentration were reported in cooked *Brassica*, as well as of the sulphides amount (except in broccoli) [85]. The concentration of isothiocyanate was found to increase after cooking [85]. Additionally, the storage of vegetables in frozen form could impact their volatile profile, in particular, influencing the alcohol, aldehydes, and isothiocyanates content [85].

3.2. Brassica Taste

As previously stated, *Brassica* vegetables contain health-related compounds that possess undesirable sensory characteristics. Bitterness is particularly accentuated in *Brassica*; this sensation is caused by ITCs that derive from sinigrin, gluconapin, progoitrin, glucobrassicin, neoglucobrassicin at different intensities [86,87]. Many studies identified the relation between the bitter taste and sinigrin and goitrin in cooked Brussels sprout, and between bitterness and sinigrin and neoglucobrassicin in cooked cauliflower [88]. Several studies affirm that the GLS and their breakdown products are not the only ones responsible for the bitter taste and *Brassica* aroma, but these resulted from a synergistic activity of various phytochemicals (indole hydrolysis products, flavonoids, etc.) [65,89].

The overall taste of *Brassica* vegetables is not only linked to bitter compounds but derives from the interaction between the bitter and the sweet tastes [88]. Some evidence demonstrated how the taste is the main driver of liking a food product [90,91], and that there is an innate preference for sweet taste in respect to bitter and sour taste [92,93]. This explains why a bitter taste in vegetables could deter most consumers from buying them. Some studies demonstrated that consumers prefer *Brassica* with low amounts of bitter GLS and higher concentrations of sucrose and, more generally, that the sweet taste is a favourable characteristic for the consumer's appreciation of *Brassica* [87,91].

4. Factors Influencing the Phytochemical Compounds of *Brassica* Vegetables

The quality of the final product can be influenced by several factors such as genetic, environmental, and agricultural (Figure 1).

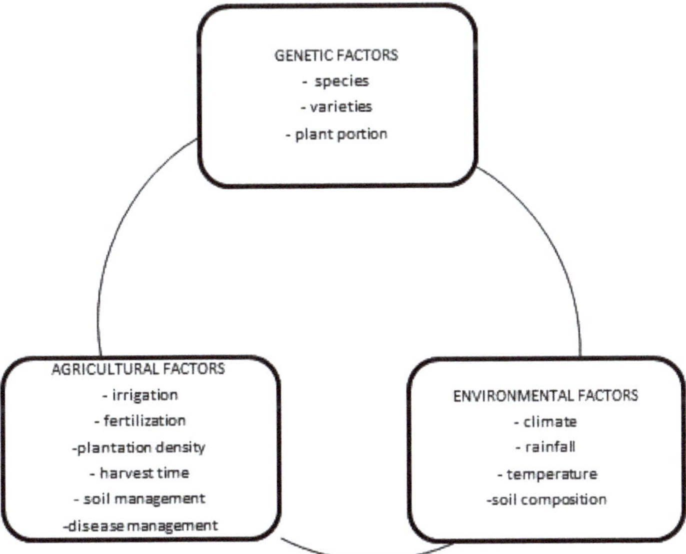

Figure 1. Factors that influence the quality of the final product.

4.1. Genetic Factors

As for all other crops, *Brassica* quality is influenced by several factors, but the principal is represented by the genotype characteristics. Many breeding programs are working towards the creation and selection of new better productive and qualitative genotypes. These programs are particularly implemented in the Mediterranean Area, where many spontaneous and wild *Brassica* provide genetic diversity and variability, allowing for the development of new pre-breeding and advanced breeding materials. Particularly in Italy and Spain, it is possible to find countless ecotypes and populations of *Brassica oleracea* and *Brassica rapa* species, handed down by generations of farmers [49].

These two species have been widely studied and showed a wide diversity in terms of nutritional quality.

4.1.1. Brassica Oleracea Species

Kale could be considered the ancestor of several *B. oleracea* vegetable crops because it has been found to be very similar to the *B. oleracea* wild type and to several wild *Brassica* species ($n = 9$) [94].

Several differences among varieties within this species were reported, e.g., the highest content of total phenolic was found in curly kale that showed a concentration 10-times higher than cauliflower and white cabbage [95]. Although the methodologies of analysis used in many studies were different, all of them agree on the lower content of phytochemicals in white cabbage, in respect to broccoli, Brussel sprouts, curly kale and red cabbage. There are controversial results regarding cauliflower as it showed high activity in liposomal phospholipid suspension system, but low activity in oxygen radical absorption capacity (ORAC method) [14,15].

As mentioned above, the variability is also expressed among genotypes of the same species in broccoli [96], cauliflower [97], cabbage [26] depending on their characteristics; in general, the higher content of antioxidants is detected in the varieties with red or purple pigmentation. Broccoli is important for its cancer-protective compounds; in particular, for its content of glucoraphanin, and its active form sulforaphane. Sicilian landraces of violet cauliflower could be considered an environmentally friendly crop, being characterized by high plant rusticity and adaptability to the Mediterranean climatic condition that allows one to limit the use of pesticides and fertilizers for its cultivation [98].

4.1.2. *Brassica rapa* Species and Other Cruciferous Crops

Brassica rapa species include turnip tops and leaves ("cima di rapa" and "friariello"), turnip, pak choi, Chinese cabbage, choy-sum and mizuna, evidencing a wide variability among close species [99] and varieties of the same species [100]. Phenolic compounds are mainly affected by the interaction between environment and genotype; this means that their variability strictly depends on the environmental conditions, hence they possess low heritability.

Choy sum, a *Brassica rapa* variety, showed the highest antioxidant potential compared to broccoli, cabbage, and cauliflower [101]. Some studies found that watercress showed a higher antioxidant potential in comparison to salad rocket; however, these two varieties, together with wild rocket and mizuna, are good sources of antioxidants [102,103]. In *B. rapa,* the aliphatic glucosinolates (GLS) is the predominant form, with gluconapin as the most abundant, followed by glucobrassicanapin [104]. *B. rapa* varieties have shown a high concentration of isorhamnetin, irrespective of the plant organs considered [105].

4.1.3. Plant Portion and Plant Developmental Stage

The variation in nutritional and phytochemical content does not differ only among species and varieties of the same species, but also can change during the growth period [6] and based on plant portion [50], as reported below.

For most of these crops, the more interesting parts of the plant for the nutritional quality are not consumed. For example, seeds seem to possess the highest content of phytochemical compounds but are not usually consumed and appreciated by consumers; this aspect is confirmed in kale, where seeds possess higher antioxidant capacity than leaves [22]. In turnip, flower buds registered the highest antioxidant content, in respect to leaves, stems and roots [106].

GLS content also differs based on plant portion. Seeds possess the highest concentration of GLS, followed by inflorescences, siliques, leaves, roots, stems and petioles [107]. Indeed, the concentration of aliphatic GLS in kale (*B. oleracea acephala*) leaves increases over time, from seedling to early flowering stages. At that stage, the aliphatic GLS content in leaves of *B. oleracea* declined drastically over time as the content in the flower buds increased [50].

A comparison study on turnip tops and turnip greens also reported several and appreciated differences in phytochemicals compounds. Turnip tops gave a higher GLS value (26.02 μmol/g dw) than turnip greens (17.78 μmol/g dw). The opposite trend was reported for total phenolic, whereby turnip greens showed a higher content (43.81 μmol/g dw) than turnip tops (37.53 μmol/g dw) [104].

Several studies confirmed the possibility to detect differences within the same portion of plant. In tronchuda cabbage, the mainly consumed portion are the internal leaves, utilised for salad or cooking; nonetheless, these have an antioxidant capacity lower than the external ones, which are usually discarded [22]. The same results were observed in Chinese cabbage, whereby the variation in bioactive compounds was also evident among different layers of the same head cabbage; phenolic acids and flavonoids were higher in the outer leaves, followed by the mid- and inner leaves. This result could be explained by the higher exposure of outer leaves to sunlight, which stimulates the production of antioxidants [108].

The stage of growth can influence the content and concentration of phytochemical compounds in *Brassica*, and the knowledge of this aspect is fundamental in choosing the proper harvesting moment for obtaining products with the highest quality. Indeed, the juvenile cabbage possesses more flavonols than the mature one [26].

Total GLS content also varies in the function of the stage of growth and increases from vegetative to reproductive stages and maturity. Consequently, the highest content is found either in flower buds, or in leaves harvested at the optimum consumption stage, 180 days after the sowing of kale [50]. In broccoli heads, the highest glucoraphanin content was also observed 180 days after sowing, with a following decline during flowering [109].

Vallejo et al. [110] found an increase in ascorbic acid and phenol compounds during the development of the inflorescence in three broccoli cultivars.

Carotenoids are also affected by the plant developmental stage. In kale, the highest content of lutein was registered in 1- to 2-week-old leaves, and the highest content of β-carotene was found in 2- to 3-week-old leaves [111].

Some of the health-promoting factors may be present 10-times higher in sprouts than in mature vegetables. Sprouting resulted in an overall increase in the total phenolic content and antioxidant capacity and, although germination time was not a discriminating factor, longer germination times resulted in the lower antioxidant capacity of the sprouts [112].

4.2. *Environmental and Agricultural Factors*

Seasonal variation, light exposure, temperature, water availability [113], phytosanitary measures, sowing date and harvesting period [114] are all factors linked to environmental conditions that can influence the quality, in particular nutritional content and profile, of *Brassica* vegetables [6]. Different responses to seasonal variations were reported in several *Brassica* crops, such as broccoli, kale, and turnip [115]; this effect is determined mainly by temperatures and day length during the period before harvest.

Countless studies agree that spring season crops, growing at intermediate temperatures, high light intensity, during longer days and in dry conditions (or low average of rainfall) during their vegetative period, contain an increased total GLS and phytochemicals concentration [50,104,114]. For example, in canola (*Brassica napus*), it was found that GLS concentration increased when a temperature of 40 °C was maintained for 4 h on five successive days, giving a total of 15-degree days of stress (15 DD/40 °C) [116]. Some authors reported that higher and lower temperatures, rather than intermediate temperatures, brought about an increase in GLS concentration, e.g., growing temperatures between 7 °C and 13 °C brought about an increase in glucoraphanin and lutein in broccoli; furthermore, they acted as a trigger for biosynthetic pathways [117]. Moreover, broccoli sprouts grown at constant high (29–33 °C) or low (11–16 °C) temperatures had higher antioxidant content than sprouts grown at intermediate temperature (21.5 °C) [118]. The same authors confirm that the main antioxidant content is observed in sprouts that grow with a strong temperature range of 30/15 °C day/night.

Autumn/winter season crops, grown at lower temperature, lower light intensity, shorter days, and higher water availability, tend to have the lowest total GLS and other phytochemicals concentration [119,120]. An exception is represented by a turnip that produces higher flavonoids and vitamin C content in the autumn/winter season; this crop accumulates and produces the main phytochemicals with low/moderate temperature and considerable radiation, mainly in turnip tops [121]. More precisely, in *Brassica rapa*, the correlation with temperature is also bound to the plant portion; indeed, the number of days with a minimum temperature below 0 °C was negatively correlated with total GLS content in turnip greens. In turnip tops, GLS content was positively correlated with the number of days with a maximum temperature above 20 °C. In the case of phenolics, no correlation was found between climatic factors and turnip greens, while in turnip tops, total flavonoids and total phenolics content seemed to be correlated with the number of days with a minimum temperature below 0 and 10 °C, respectively [104]. In broccoli, freezing temperature can positively influence the concentration of sulforaphane [122].

The biotic and abiotic factors that characterize the surrounding environment can influence the quality of *Brassicaceae*. With respect to biotic sphere, aphid infestation brought about an increased production of primary metabolites, including amino acids, as well as some secondary metabolites, as a plant defence mechanism against these pathogens. Concerning abiotic factors, the water stress condition and metal exposure produce an initial increase in photosynthetic pigments, proteins, free amino acids and sugar content, followed by a subsequent decrease [123]. In detail, a relation between copper stress and the production of amino acids was found as free amino acid production takes part in the detoxification from excess copper [124]. In *Brassica juncea*, the accumulation of metals produces a 35% increase in oil content [123]. Moderate salinity in water or soil affects the myrosinase-GLS system in broccoli, inducing the production of GLS; also, phenolic compounds increase in this stressful condition, but in the case of strong salinity both GLS and phenolics decrease [125]. Seedlings of *Brassica oleracea* L. var. *italica* subjected to water shortage (applied by increasing the time between two irrigation events) showed a decrease in inflorescence chlorophylls, carotenoids, ascorbic acid, total phenols and total soluble carbohydrates [126].

Ragusa and co-author [127] investigated the effect of different germination temperatures (10, 20 and 30 °C) on the phytochemical content as well as on reducing and antioxidant capacity of broccoli and rocket sprouts. In both seeds and sprouts, the total GLS and ascorbic acid contents did not differ between vegetables, while broccoli exhibited exceptionally higher polyphenols and a greater reduction in antioxidant capacity compared to rocket. In both species, an increase in germination temperature positively affected the glucosinolate content. Ascorbic acid increased during germination without a difference among the three tested temperatures. The phenol content increased in broccoli sprouts when grown at 30 °C, while the reverse was true in rocket. The antioxidant capacities increased with germination, and higher indexes were detected at 10 °C, particularly in rocket.

4.2.1. Cultivation System and Soil Composition

The cultivation system influences the quality of vegetable product, in particular the concentration of primary and secondary metabolites in *Brassica* vegetables.

Some authors reported a higher antioxidant (phenolic compounds, in particular flavonoids) and GLS concentration in *Brassica* growth in organic cultivation system than in conventional systems [128], as demonstrated in early harvested tronchuda cabbage [129]. This result could be linked to the fact that, under organic cultivation, crops are subjected to more biotic and abiotic stress; these stressing conditions lead to an increase in the production of secondary metabolites as a defence mechanism, and consequently obtaining vegetables with higher nutritional and antioxidant potential than in a conventional system.

Several studies described an opposite situation and contrasting evidence about phytochemical enhancement in organic vegetables [130,131]. In fact, Conversa et al. [132] reported that the choice of cultivation systems does not modify the antioxidant properties of raw and processed products, but differences can be found in the chlorophyll and carotenoid contents of organic "cima di rapa" landraces. The lipophilic antioxidant content was improved in organic product while the hydrophilic component, which constituted 99% of the total antioxidant capacity, was not affected by the different crop management in "cima di rapa". However, the organic system influenced the quality of products during storage: after 7 days of storage at 5 °C, the organic "cima di rapa" maintained the best colour with high chlorophyll levels, probably due to a higher availability of nitrogen in organic management; on the contrary, the quality declined with a higher production of strong off-odour after 14 days of storage, in comparison to the conventional products.

Regarding the soil composition effect on *Brassica* quality, it was reported that the highest GLS and phenolic compounds content were detected in locations with the highest soil pH and available potassium; the content can be also influenced by nitrogen and sulphur applications in turnip [104]. On the contrary, in *B. rapa L. Subsp. Sylvestris,* flavonols (kaempferol and quercetin derivatives) were reduced by sulphur availability [113].

4.2.2. Water Stress

It was reported that a moderate water stress increases the concentration of bioactive compounds in *Brassica*, partly due to an increased concentration per unit of dry weight; if the stress becomes intensive, the secondary metabolite production should decrease [133]. Phenolic compounds and GLS content increase in the absence of irrigation, because of a reduction in vegetative growth, mainly in turnip, cabbage and broccoli [6]. The association between low availability of water in the soil during plant growth and postharvest cold storage brought about the best maintenance of antioxidant activity in *Brassica*. Water stress conditions also affect sugar content, as it is increased in cabbage [134].

4.2.3. Plant Density, Intercropping and Trap Cropping

Plant density seems to affect the plant morphology and phytochemical compound content: a higher density decreases the head size but increases the GLS content, because the competition for nutrients in high density conditions causes stress on plants which, in turn, stimulates the production of secondary metabolites [120].

Intercropping and trap cropping are strategies utilised for weed and pest control [135]; however, the presence of another crop can generate stressful conditions, such as plant competition for light, nutrients, and water, decreasing their availability and, hence, affecting the accumulation of phytochemicals in *Brassica* plant tissue.

4.2.4. Fertilization Practices

A correct fertilization plan is fundamental for obtaining high quality, healthy and safe vegetables. The nutritional and sensorial profile of *Brassica* is conditioned by the availability of fertilizers and nutrients as they determine the biosynthesis of secondary metabolites.

Countless studies have been conducted on the effect of sulphur fertilization on phytochemical concentration, mainly on GLS production, considering their sulphurous nature [136,137]. There is a correlation between the increase in sulphur supply and higher levels of total GLS [138], in turnip [136], kale [137] and broccoli, mainly when associated with a reduction in water, at the expense of yield [139]. Vallejo and co-authors [119] suggested that the effect of sulphur application on GLS varies with the development stage of broccoli plants and differs for each kind of GLS; in fact, they found an increase in total GLS content at the start of the inflorescence development, followed by a rapid decrease thereafter. Increasing sulphur fertilization brought about a positive impact in the synthesis of polyphenols, such as flavonols and phenolic acids, increasing the total antioxidant capacity in turnip top (*B. rapa ssp. Sylvestis*) [113], and broccoli [110]. Sulphur fertilization in pre-harvest (from 2.6 mmol/L to 6.5 mmol/L) increases the lipophilic and hydrophilic antioxidant capacity but does not affect the nitrate and chlorophyll contents in ready-to-eat "friariello" product [140]. Sulphur deficiency induced an increased vulnerability of *Brassica* crops to diseases and fungal pathogens [141]. Sulphur fertilization, besides improving the antioxidant activity, it is also associated with a genotype-dependent significant reduction in leaf nitrate content, since it enhances the incorporation of nitrogen into organic compounds and consequently reducing the leaf nitrate concentration [113].

Nitrogen is the main constituent of chlorophyll structure: for this reason, its availability influences the content of carotenoids such as lutein and β-carotene, indeed high NO_3-N:NH_4-N ratio led to a higher content of both [142]. Consequently, the colour and pigmentation of leafy vegetables are also improved [132]. Nitrogen fertilisation led to a decrease in the total GLS content [136]; nonetheless, it acts differently according to the type of GLS; in fact, abundant nitrogen applications increase progoitrin and decrease sinigrin concentration in *Brassica napus* [143]. A reduced nitrogen fertilisation generated an increase in the bioactive compound content, mainly phenolics, as nitrogen stress triggers the gene expression of flavonoid pathway enzymes [128]. Combined fertilisation with NO_3^-:NH_4^+ is the optimal solution to maintain plant growth and increase the total GLS content [144].

An optimal balance between nitrogen and sulphur fertilisation influences the biosynthesis of secondary metabolites [21]. GLS, for example, can be enhanced by the presence of low nitrogen and high sulphur fertilizers: this balance influences the quantity and the quality of GLS produced, according to the corresponding amino acids synthetized. Some authors reported the effect of different nitrogen/sulphur combinations on GLS content in *Brassica*, with an increasing amount of nitrogen (80–320 kg/ha) applications. When enough sulphur was available (60 kg/ha), there were no effects on total GLS content, but their production moved to indolics; when the combination was with a low concentration of sulphur supply (10–20 kg/ha), the arylaliphatic and aliphatic GLS decreased [138]. Increased nitrogen/sulphur ratio pushes the plants towards the vegetative growth, at the expense of GLS production [136]. Fabek et al. [145] showed that the type of fertilisation may influence mineral composition in plants: nitrogen fertilisation was negatively associated with potassium (K) and calcium (Ca) content in broccoli, while sulphur fertilisation increased manganese (Mn) and zinc (Zn), and decreased copper (Cu). Applications of sodium selenate (Na_2SeO_4) produced an increase in GLS [144].

Similarly, microelements availability can influence the phytochemical concentration in *Brassica*, with salts stress increasing GLS content. In detail, selenium seems to increase the GLS content (in particular sulphuraphane), when applied up to a certain dose; above this level, it decreases the GLS production [51].

Furthermore, some *Brassica* species are used as metal hyperaccumulator, and the type and amount of metal in the soil affects the concentration of glucosinolates in plant tissues. In particular, it was reported that glucosinolate concentrations in roots and shoots of *Thlaspi caerulescens* responded in different way to enhanced Zn accumulation: decreased glucosinolate levels were observed in leaves of plants accumulating high Zn concentrations, while increased levels were detected in roots, with Zn accumulation [146]. Similarly, the content of total glucosinolates, mostly due to indolic glucosinolates as glucobrassicin, was increased only in the roots of Chinese cabbage, when subjected to high soil copper stress [147]. In two *B. juncea* cultivars subjected to high arsenic levels, the increased levels of thiol related proteins, sulphur content and phytochemicals (phenolics and ascorbic acid) in leaves allow us to better tolerate the oxidative stress induced in the plant; different response pattern of total and individual GSLs content was observed in both cultivars under arsenic stress [148].

Besides the classical fertilizers, in the last years, new proposed products that are beneficial on crops, such as improving safety, enhancing growth and production, improving the defence against weeds and pests and nutritional quality, were developed. Among these, signalling molecules, biocontrol agents, and biostimulants are now gaining high interest for improving plant resilience and quality. Leaves and cotyledons of *B. napus*, *B. rapa* and *B. juncea* showed an up to 20-fold increase in glucobrassicin content after treatment with JA (Jasmonic Acid), or MeJA (Methyl Jasmonate) [149]. In contrast, treatment with ABA (Abscisic Acid) reduced the accumulation of indole GLS in *B. napus* [150].

In summary, the levels of hormones, such as JA, SA (Salicylic Acid) and ABA, seem to be related to the regulation of GLS and of other bioactive compound content [151]. Consequently, hormonal elicitation can be a useful tool to induce the synthesis of bioactive compounds interesting for human health.

Concerning the application of biocontrol agents, Gallo et al. [152] affirmed that the use of *Trichoderma* and its metabolites led to an increase in GLS in plants. This could probably be due to their capability of inducing resistance mechanisms, stimulating the synthesis of salicylic and jasmonic acids and the cascade of events leading to the production of various metabolites; only ascorbic acid was lower compared to control plants.

Additionally, in *Brassica* spp. cultivation is increasing the use of biocontrol agents, there is also the utilisation of seaweeds extract, mycorrhizae, nematodes [153], humic acids such as vermicompost foliar sprayed [154], and protein hydrolysates, all compounds now classified as "biostimulants", useful to increase plant yield and the accumulation of bioactive compounds [155].

Brassica species can contrast the main soil-borne agents thanks to their secondary metabolites that act as biofumigants. A study reported the effectiveness of the flour of dry plants of *Brassica juncea*, *Eruca sativa*, *Raphanus sativus* and *Brassica macrocarpa* in nematodes control (*Meloidogyne* spp.) on tomatoes. Minced flour was distributed before planting (60 and 90 g m^{-2}) and was successful for the sinigrin presence [156].

5. Conclusions

Brassica vegetables are a good source of many phytochemical compounds that exert positive effects on the final consumer's health. This study presented an investigation on the presence of these bioactive compounds, analysing how they affect the sensorial and nutritional quality, and on the factors that can modify their concentration in *Brassica* food vegetables, such as genetic, environmental and agronomic factors.

There is a large possibility to improve the nutritional and sensorial quality of *Brassica* vegetables through the implementation of appropriate agronomic practices; nevertheless, the effects of the treatments are strictly genotype-dependent, and a good selection of the genotype before the start of cultivation is required. Furthermore, the environmental factors could influence to different extents the quality of *Brassica* genotypes, and they should be considered in the evaluation of the phytochemical compounds amount.

All this information is useful for developing new fresh and processed products with increased nutritional and sensorial quality, according to the final users' needs and the final purpose of consumption. If the consumer will be informed and made conscious of the healthy potential of the phytochemical compounds present in *Brassica*, they may be willing to accept these products despite the bitter taste and the intense aroma, which are often responsible for a low consumer acceptance.

Author Contributions: Conceptualization, M.V. (Marino Visciglio), B.M., F.C., and M.V. (Massimo Vagnoni); Methodology, F.B. (Francesca Biondi), F.C., E.M., and L.M.; Investigation, F.B. (Francesca Biondi), F.B. (Francesca Balducci), and L.M.; Resources, M.V. (Marino Visciglio), and B.M.; Data Curation, F.B. (Francesca Biondi), F.C., and L.M.; Writing—Original Draft Preparation, F.B. (Francesca Biondi), and L.M.; Writing—Review & Editing, B.M., and L.M.; Visualization, L.M.; Supervision, M.V. (Marino Visciglio), F.C., and B.M.; Project Administration, M.V. (Marino Visciglio), M.V. (Massimo Vagnoni), E.M., and B.M.; Funding Acquisition, M.V. (Marino Visciglio), and B.M. All authors have read and agreed to the published version of the manuscript.

Funding: This research received no external funding.

Institutional Review Board Statement: Not applicable.

Informed Consent Statement: Not applicable.

Acknowledgments: The authors thank Cecilia Limera and Maria Teresa Ariza Fernandez for extensively revise the manuscript.

Conflicts of Interest: The authors declare no conflict of interest.

References

1. Dauchet, L.; Amouyel, P.; Hercberg, S.; Dallongeville, J. Fruit and vegetable consumption and risk of coronary heart disease: A meta-analysis of cohort studies. *J. Nutr.* **2006**, *136*, 2588–2593. [CrossRef] [PubMed]
2. He, F.J.; Nowson, C.A.; MacGregor, G.A. Fruit and vegetable consumption and stroke: Meta-analysis of cohort studies. *Lancet* **2006**, *367*, 320–326. [CrossRef]
3. Farnham, M.W.; Wilson, P.E.; Stephenson, K.K.; Fahey, J.W. Genetic and environmental effects on glucosinolate content and chemoprotective potency of broccoli. *Plant Breed.* **2004**, *123*, 60–65. [CrossRef]
4. FAOSTAT—Food and Agriculture Organization of the United Nations. Top exports of Cabbages and Other Brassicas. In *FAOSTAT Database*; FAOSTAT: Rome, Italy, 2011.
5. FAOSTAT—Food and Agriculture Organization of the United Nations. Top exports of Cabbages and Other Brassicas. In *FAOSTAT Database*; FAOSTAT: Rome, Italy, 2019.
6. Francisco, M.; Tortosa, M.; Martinez-Ballesta, M.; Velasco, P.; Garcia-Viguera, C.; Moreno, D.A. Nutritional and phytochemical value of Brassica crops from the agri-food perspective. *Ann. Appl. Biol.* **2017**, *170*, 273–285. [CrossRef]
7. Ciancaleoni, S.; Chiarenza, G.L.; Raggi, L.; Branca, F.; Negri, V. Diversity characterisation of broccoli (*Brassica oleracea* L.var. italica Plenck) landraces for their on-farm (in situ) safeguard and use in breeding programs. *Genet. Resour. Crop. Evol.* **2014**, *61*, 451–464. [CrossRef]
8. Fahey, J.W. Brassicas. In *Encyclopedia of Food Sciences and Nutrition*, 2nd ed.; Caballero, B., Ed.; Academic Press: Cambridge, MA, USA, 2003; pp. 606–615, ISBN 9780122270550.
9. Köpke, U. Organic foods: Do they have a role? *Forum Nutr.* **2005**, *57*, 62–72.
10. Ou, B.X.; Huang, D.J.; Hampsch-Woodill, M.; Flanagan, J.A.; Deemer, E.K. Analysis of antioxidant activities of common vegetables employing oxygen radical absorbance capacity (ORAC) and ferric reducing antioxidant power (FRAP) assays: A comparative study. *J. Agric. Food Chem.* **2002**, *50*, 3122–3128. [CrossRef] [PubMed]
11. Zhou, K.; Yu, L. Total phenolic contents and antioxidant properties of commonly consumed vegetables grown in Colorado. *LWT Food Sci. Techol.* **2006**, *39*, 1155–1162. [CrossRef]
12. Azuma, K.; Ippoushi, K.; Ito, H.; Higashio, H.; Terao, J. Evaluation of antioxidative activity of vegetable extracts in linoleic acid emulsion and phospholipid bilayers. *J. Sci. Food Agric.* **1999**, *79*, 2010–2016. [CrossRef]
13. Wu, X.; Beecher, G.R.; Holden, J.M.; Haytowitz, D.B.; Gebhardt, S.E.; Prior, R.L. Lipophilic and hydrophilic antioxidant capacities of common foods in the United States. *J. Agric. Food Chem.* **2004**, *52*, 4026–4037. [CrossRef] [PubMed]
14. Podsedek, A. Natural antioxidants and antioxidant capacity of Brassica vegetables: A review. *LWT Food Sci. Technol.* **2007**, *40*, 1–11. [CrossRef]
15. Soengas, P.; Sotelo, T.; Velasco, P.; Cartea, M.E. Antioxidant Properties of *Brassica* Vegetables. *Funct. Plant Sci. Biotechnol.* **2011**, *43*, 55.
16. Kurilich, A.C.; Jeffery, E.H.; Juvik, J.A.; Wallig, M.A.; Klein, B.P. Antioxidant capacity of different broccoli (*Brassica oleracea*) genotypes using the oxygen radical absorbance capacity (ORAC) assay. *J. Agric. Food Chem.* **2002**, *50*, 5053–5057. [CrossRef]
17. Jahangir, M.; Abdel-Farida, I.B.; Kima, H.K.; Choia, Y.H.; Verpoortea, R. Healthy and unhealthy plants: The effect of stress on the metabolism of Brassicaceae. *Environ. Exp. Bot.* **2009**, *67*, 23–33. [CrossRef]
18. Cory, H.; Passarelli, S.; Szeto, J.; Tamez, M.; Mattei, J. The Role of Polyphenols in Human Health and Food Systems: A Mini-Review. *Front. Nutr.* **2018**, *5*, 87. [CrossRef] [PubMed]
19. Moreno, D.A.; Carvajal, M.; Lopez-Berenguer, C.; García-Viguera, C. Chemical and biological characterisation of nutraceutical compounds of broccoli. *J. Pharm. Biomed. Anal.* **2006**, *41*, 1508–1522. [CrossRef] [PubMed]

20. Heimler, D.; Vignolini, P.; Dini, M.G.; Vincieri, F.F.; Romani, A. Antiradical activity and polyphenol composition of local *Brassicaceae* edible varieties. *Food Chem.* **2006**, *99*, 464–469. [CrossRef]
21. Cartea, M.E.; Francisco, M.; Soengas, P.; Velasco, P. Phenolic Compounds in Brassica Vegetables. *Molecules* **2011**, *16*, 251–280. [CrossRef] [PubMed]
22. Ferreres, F.; Fernandes, F.; Sousa, C.; Valentão, P.; Pereira, J.A.; Andrade, P.B. Metabolic and Bioactivity Insights into *Brassica oleracea* var. *acephala*. *J. Agric. Food Chem.* **2009**, *57*, 8884–8892. [CrossRef]
23. Haytowitz, D.B.; Wu, X.; Bhagwat, S. USDA Database for the Flavonoid Content of Selected Foods, Release 3.3. U.S. Department of Agriculture, Agricultural Research Service. Nutrient Data Laboratory Home Page. 2018. Available online: http://www.ars.usda.gov/nutrientdata/flav (accessed on 3 May 2020).
24. Sousa, C.; Lopes, G.; Pereira, D.M.; Taveira, M.; Valentao, P.; Seabra, R.M.; Pereira, J.A. Screening of antioxidant compounds during sprouting of *Brassica oleracea* L. var. *costata DC. Comb. Chem. High Throughput Screen.* **2007**, *10*, 377–386. [CrossRef]
25. Ackland, M.L.; Van De Waarsenburg, S.; Jones, R. Synergistic antiproliferative action of the flavanols quercetin and kaempferol in cultured human cancer cell lines. *In Vivo* **2005**, *19*, 69–76. [PubMed]
26. Kim, D.O.; Padilla-Zakour, O.I.; Griffiths, P.D. Flavonoids and antioxidant capacity of various cabbage genotypes at juvenile stage. *J. Food Sci.* **2004**, *69*, C685–C689. [CrossRef]
27. Calderon-Montano, J.M.; Burgos-Moron, E.; Pérez-Guerrero, C.; Lòpez-Lazaro, M. A review on the dietary flavonoid kaempferol. *Med. Chem.* **2011**, *11*, 298–344. [CrossRef] [PubMed]
28. Bahorun, T.; Luximon-Ramma, A.; Crozier, A.; Aruoma, O.I. Total phenol; flavonoid; proanthocyanidin and vitamin C levels and antioxidant activities of Mauritian vegetables. *J. Sci. Food Agric.* **2004**, *84*, 1553–1561. [CrossRef]
29. Moreno, D.A.; Pérez-Balibrea, S.; Ferreres, F.; Gil-Izquierdo, A.; García-Viguera, C. Acylated anthocyanins in broccoli sprouts. *Food Chem.* **2010**, *123*, 358–363. [CrossRef]
30. Ahmadiani, N.; Robbins, R.J.; Collins, T.M.; Giusti, M.M. Anthocyanins Contents, Profiles, and Color Characteristics of Red Cabbage Extracts from Different Cultivars and Maturity Stages. *J. Agric. Food Chem.* **2014**, *62*, 7524–7531. [CrossRef]
31. Dyrby, M.; Westergaard, N.; Stapelfeldt, H. Light and heat sensitivity of red cabbage extract in soft drink model system. *Food Chem.* **2001**, *72*, 431–437. [CrossRef]
32. Milder, I.E.J.; Arts, I.C.W.; Van De Putte, B.; Venema, D.P.; Hollman, P.C.H. Lignin contents of Dutch plant foods: A database including lariciresinol; pinoresinol; secoisolariciresinol and matairesinol. *Br. J. Nutr.* **2005**, *93*, 393–402. [CrossRef]
33. Kurilich, A.C.; Tsau, G.J.; Brown, A.; Howard, L.; Klein, B.P.; Jeffery, E.H.; Kushad, M.; Walling, M.A.; Juvik, J.A. Carotene; tocopherol; and ascorbate contents in subspecies of *Brassica oleracea*. *J. Agric. Food Chem.* **1999**, *47*, 1576–1581. [CrossRef]
34. Vallejo, F.; Tomas-Barberan, F.A.; Garcia-Viguera, C. Potential bioactive compounds in health promotion from broccoli cultivars grown in Spain. *J. Sci. Food Agric.* **2002**, *82*, 1293–1297. [CrossRef]
35. Davey, M.W.; Van Montagu, M.; Inze, D.; Sanmartin, M.; Kanellis, A.; Smirnoff, N. Plant L-ascorbic acid: Chemistry; function; metabolism; bioavailability and effects of processing. *J. Sci. Food Agric.* **2000**, *80*, 825–860. [CrossRef]
36. Sanlier, N.; Guler Saban, M. The Benefits of Brassica Vegetables on Human Health. *J. Hum. Health Res.* **2018**, *1*, 104.
37. Bailey, L.B.; Rampersaud, G.C.; Kauwell, G.P.A. Folic acid supplements and fortification affect the risk for neural tube defects; vascular disease and cancer: Evolving science. *J. Nutr.* **2003**, *133*, 1961S–1968S. [CrossRef] [PubMed]
38. Atteia, B.M.R.; El-Kak, A.E.-A.A.; Lucchesi, P.A.; Delafontane, P. Antioxidant activity of folic acid: From mechanism of action to clinical application. *FASEB J.* **2009**, *23*, 103–107.
39. Quang, D.D. Antioxidant properties of folic acid: A DFT study. *Vietnam J. Sci. Technol.* **2018**, *56*, 39. [CrossRef]
40. Rice-Evans, C.; Sampson, J.; Bramley, P.M.; Holloway, D.E. Why do we expect carotenoids to be antioxidants in vivo. *Free Radic. Res.* **1997**, *26*, 381–398. [CrossRef]
41. Muller, H. Determination of the carotenoid content in selected vegetables and fruit by HPLC and photodiode array detection. *Food Res. Technol.* **1997**, *204*, 88–94. [CrossRef]
42. Wills, R.B.H.; Rangga, A. Determination of carotenoids in Chinese vegetables. *Food Chem.* **1996**, *56*, 451–455. [CrossRef]
43. Piironen, V.; Syvaoja, E.L.; Varo, P.; Salminen, K.; Koivistoinen, P. Tocopherols and tocotrienols in Finnish foods: Vegetables; fruits; and berries. *J. Agric. Food Chem.* **1986**, *34*, 742–746. [CrossRef]
44. Stampfer, M.J.; Rimm, E.B. Epidemiologic evidence for vitamin E in prevention of cardiovascular disease. *Am. J. Clin. Nutr.* **1995**, *62*, 1365–1369. [CrossRef]
45. Ali, Z.; Waheed, H.; Gul, A.; Afzal, F.; Anwaar, K.; Imran, S. Brassicaceae plants. In *Oilseed Crops*; Ahmad, P., Ed.; John Wiley & Sons: Hoboken, NJ, USA, 2017. [CrossRef]
46. Thavarajah, D.; Lawrence, T.; Powers, S.; Jones, B.; Johnson, N.; Kay, J.; Bandaranayake, A.; Shipe, E.; Thavarajah, P. Genetic variation in the prebiotic carbohydrate and mineral composition of kale (*Brassica oleracea* L. var. acephala) adapted to an organic cropping system. *J. Food Comp. Anal.* **2021**, *96*, 103718. [CrossRef]
47. Blažević, I.; Montaut, S.; Burčul, F.; Olsen, C.-E.; Burow, M.; Rollin, P.; Agerbirk, N. Glucosinolate structural diversity, identification, chemical synthesis and metabolism in plants. *Phytochemistry* **2020**, *169*, 112100. [CrossRef] [PubMed]
48. Sisti, M.; Amagliani, G.; Brandi, G. Antifungal activity of *Brassica oleracea* var. botrytis fresh aqueous juice. *Fitoterapia* **2003**, *74*, 453–458.
49. Branca, F.; Ragusa, L.; Tribulato, A.; Velasco, P.; Cartea, M.E. Glucosinolate profile in different mediterranean brassica species (n = 9). *Acta Hortic.* **2013**, *1005*, 279–284. [CrossRef]

50. Velasco, P.; Cartea, M.E.; Gonzalez, C.; Vilar, M.; Ordàs, A. Factors affecting the glucosinolate content of kale (*Brassica oleracea acephala* group). *J. Agric. Food Chem.* **2007**, *55*, 955–962. [CrossRef] [PubMed]
51. López-Berenguer, C.; Martínez-Ballesta, M.C.; García-Viguera, C.; Carvajal, M. Leaf water balance mediated by aquaporins under salt stress and associated glucosinolate synthesis in broccoli. *Plant Sci.* **2008**, *174*, 321–328. [CrossRef]
52. Mithen, R. Glucosinolates—Biochemistry; Genetics and biological activity. *Plant Growth Regul.* **2001**, *34*, 91–103. [CrossRef]
53. Verkerk, R.; Schreiner, M.; Krumbein, A.; Ciska, E.; Holst, B.; Rowland, I.; de Schrijver, R.; Hansen, M.; Gerhäuser, C.; Mithen, R.; et al. Glucosinolates in Brassica vegetables: The influence of the food supply chain on intake, bioavailability and human health. *Mol. Nutr. Food Res.* **2009**, *53*, 219–265. [CrossRef] [PubMed]
54. Blažević, I.; Montaut, S.; Burčul, F.; Rollin, P. Glucosinolates: Novel sources and biological potential. In *Glucosinolates*; Mérillon, J.-M., Ramawat, K.G., Eds.; Springer International Publishing: Cham, Switzerland, 2017; pp. 3–60.
55. Giamoustaris, A.; Mithen, R. The effect of modifying the glucosinolate content of leaves of oilseed rape (*Brassica napus* ssp. oleifera) on its interaction with specialist and generalist pests. *Ann. Appl. Biol.* **1995**, *126*, 347–363. [CrossRef]
56. Mithen, R.; Faulkner, K.; Magrath, R.; Rose, P.; Williamson, G.; Marquez, J. Development of isothiocyanate-enriched broccoli and its enhanced ability to induce phase 2 detoxification enzymes in mammalian cells. *Theor. Appl. Gen.* **2003**, *106*, 727–734. [CrossRef]
57. Sikorska-Zimny, K.; Beneduce, L. The glucosinolates and their bioactive derivatives in Brassica: A review on classification, biosynthesis and content in plant tissues, fate during and after processing, effect on the human organism and interaction with the gut microbiota. *Crit. Rev. Food Sci. Nutr.* **2020**, *25*, 1–24. [CrossRef] [PubMed]
58. VanEtten, C.H.; Daxenbichler, M.E.; Tookey, H.L.; Kwolek, W.F.; Williams, P.H.; Yoder, O.C. Glucosinolates: Potential toxicants in cabbage cultivars. *J. Am. Soc. Hort. Sci.* **1980**, *105*, 710–714.
59. Sones, K.; Heaney, R.K.; Fenwick, G.R. The glucosinolate content of UK vegetables: Cabbage (*Brassica oleracea*); swede (*B. napus*) and turnip (*B. campestris*). *Food Addit. Contam.* **1984**, *3*, 289–296. [CrossRef]
60. Cartea, M.E.; Velasco, P.; Obregòn, S.; del Rìo, M.; Padilla, G.; de Haro, A. Seasonal variation in glucosinolate content in *Brassica oleracea* crops grown in northwestern Spain. *Phytochemistry* **2008**, *69*, 403–410. [CrossRef] [PubMed]
61. Ciska, E.; Martyniak-Przybyszewska, B.; Kozlowska, H. Content of glucosinolates in cruciferous vegetables grown at the same site for two years under different climatic conditions. *J. Agric. Food Chem.* **2000**, *48*, 2862–2867. [CrossRef] [PubMed]
62. Carlson, D.G.; Daxenbichler, M.E.; Tookey, H.L. Glucosinolates in turnip tops and roots: Cultivars grown for greens and/or roots. *J. Am. Soc. Hort. Sci.* **1987**, *112*, 179–183.
63. Rosa, E.A.S.; Heaney, R.K. Seasonal variation in protein; mineral and glucosinolate composition of Portuguese cabbage and kale. *Anim. Feed Sci. Technol.* **1996**, *57*, 111–127. [CrossRef]
64. Kushad, M.M.; Brown, A.F.; Kurilich, A.C.; Juvik, J.A.; Klein, B.; Wallig, M.A.; Jeffery, E.H. Variation of glucosinolates in vegetable subspecies of *Brassica oleracea*. *J. Food Agric. Chem.* **1999**, *47*, 1541–1548. [CrossRef]
65. Padilla, G.; Cartea, M.E.; Velasco, P.; de Haro, A.; Ordàs, A. Variation of glucosinolates in vegetable crops of *Brassica rapa*. *Phytochemistry* **2007**, *68*, 536–545. [CrossRef] [PubMed]
66. Rosa, E.A.S. Glucosinolates from flower buds of Portuguese Brassica crops. *Phytochemistry* **1997**, *44*, 1415–1419. [CrossRef]
67. Cartea, M.E.; Rodrìguez, V.M.; Velasco, P.; de Haro, A.; Ordàs, A. Variation of glucosinolates and nutritional value in nabicol (*Brassica napus pabularia* group). *Euphytica* **2008**, *159*, 111–122. [CrossRef]
68. Angelino, D.; Dosz, E.B.; Sun, J.; Hoeflinger, J.L.; Van Tassell, M.L.; Chen, P.; Harnly, J.M.; Miller, M.J.; Jeffery, E.H. Myrosinase-dependent and -independent formation and control of isothiocyanate products of glucosinolate hydrolysis. *Front. Plant. Sci.* **2015**, *6*, 831. [CrossRef]
69. Fahey, J.W.; Zalcmann, A.T.; Talalay, P. The chemical diversity and distribution of glucosinolates and isothiocyanates among plants. *Phytochemistry* **2001**, *56*, 5–51. [CrossRef]
70. Grubb, C.D.; Abel, S. Glucosinolate metabolism and its control. *Trends Plant. Sci.* **2006**, *11*, 89–100. [CrossRef]
71. Keum, Y.S.; Jeong, W.S.; Kong, A.N.T. Chemoprevention by isothiocyanates and their underlying molecular signaling mechanisms. *Mutat. Res.* **2004**, *555*, 191–202. [CrossRef]
72. Cartea, M.E.; Velasco, P. Glucosinolates in Brassica foods: Bioavailability in food and significance for human health. *Phytochem. Rev.* **2008**, *7*, 213–229. [CrossRef]
73. Fahey, J.W.; Haristoy, X.; Dolan, P.M.; Kensler, T.W.; Scholtus, I.; Stephenson, K.K.; Talalay, P.; Lozniewski, A. Sulforaphane inhibits extracellular; intracellular; and antibiotic-resistant strains of Helicobacter pylori and prevents benzo[a]pyrene-induced stomach tumors. *Proc. Natl. Acad. Sci. USA* **2002**, *99*, 7610–7615. [CrossRef] [PubMed]
74. Dinkova-Kostova, A.T.; Kostov, R.V. Glucosinolates and isothiocyanates in health and disease. *Trends Mol. Med.* **2012**, *18*, 337–347. [CrossRef] [PubMed]
75. Matusheski, N.V.; Jeffery, E.H. Comparison of the bioactivity of two glucoraphanin hydrolysis products found in broccoli; sulforaphane and sulforaphane nitrile. *J. Agric. Food Chem.* **2001**, *49*, 5743–5749. [CrossRef] [PubMed]
76. Bonasia, A.; Conversa, G.; Lazzizera, C.; Elia, A. Pre-harvest nitrogen and Azoxystrobin application enhances postharvest shelf-life in butterhead lettuce. *Postharvest Biol. Technol.* **2013**, *85*, 67–76. [CrossRef]
77. Wieczorek, M.N.; Walczak, M.; Skrzypczak-Zielinska, M.; Jelen, H.H. Bitter taste of Brassica vegetables: The role of genetic factors; receptors; isothiocyanates; glucosinolates; and flavor context. *Crit. Rev. Food Sci. Nutr.* **2018**, *58*, 3130–3140. [CrossRef] [PubMed]
78. Ghawi, S.K.; Shen, Y.; Niranjan, K.; Methven, L. Consumer acceptability and sensory profile of cooked broccoli with mustard seeds added to improve chemoprotective properties. *J. Food Sci.* **2014**, *79*, 1756–1762. [CrossRef]

79. Stoewsand, S. Bioactive organosulfur phytochemicals in Brassica oleracea vegetables—A review. *Food Chem. Toxicol.* **1995**, *33*, 537. [CrossRef]
80. Palermo, M.; Pellegrini, N.; Fogliano, V. The effect of cooking on the phytochemical content of vegetables. *J. Sci. Food Agric.* **2014**, *94*, 1057. [CrossRef] [PubMed]
81. Kubec, R.; Drhova, V.; Velisek, J. Thermal Degradation of S-Methylcysteine and Its Sulfoxide Important Flavor Precursors of Brassica and Allium Vegetables. *J. Agric. Food Chem.* **1998**, *46*, 4334. [CrossRef]
82. Hansen, M.; Laustsen, A.M.; Olsen, C.E.; Poll, L.; Sørensen, H. Chemical and sensory quality of broccoli (*Brassica oleracea* L. var italica). *J. Food Qual.* **1997**, *20*, 441–459. [CrossRef]
83. Akpolat, H.; Barringer, S.A. The Effect of pH and Temperature on Cabbage Volatiles during Storage. *J. Food Sci.* **2015**, *80*, S1878–S1884. [CrossRef]
84. Bell, L.; Oloyede, O.O.; Lignou, S.; Wagstaff, C.; Methven, L. Taste and Flavor Perceptions of Glucosinolates, Isothiocyanates, and Related Compounds. *Mol. Nutr. Food Res.* **2018**, *62*, e1700990. [CrossRef]
85. Wieczorek, M.N.; Jelen, H.H. Volatile Compounds of Selected Raw and Cooked Brassica Vegetables. *Molecules* **2019**, *24*, 391. [CrossRef]
86. Rosa, E.A.S.; Heaney, R.K.; Fenwick, G.R.; Portas, C.A.M. Glucosinolates in crop plants. *Hortic. Rev.* **1997**, *19*, 99–215.
87. Schonhof, I.; Krumbein, A.; Brückner, B. Genotypic effects on glucosinolates and sensory properties of broccoli and cauliflower. *Food Nahr.* **2004**, *48*, 25–33. [CrossRef] [PubMed]
88. Beck, T.K.; Jensen, S.; Bjoern, G.K.; Kidmose, U. The Masking Effect of Sucrose on Perception of Bitter Compounds in Brassica Vegetables. *J. Sens. Stud.* **2014**, *29*, 190–200. [CrossRef]
89. Baik, H.Y.; Juvik, J.A.; Jeffery, E.H.; Wallig, M.A.; Kushad, M.; Klein, B.P. Relating glucosinolate content and flavor of broccoli cultivars. *J. Food Sci.* **2003**, *68*, 1043–1050. [CrossRef]
90. Schatzer, M.; Rust, P.; Elmadfa, I. Fruit and vegetable intake in Austrian adults: Intake frequency; serving sizes; reasons for and barriers to consumption; and potential for increasing consumption. *Public Health Nutr.* **2009**, *13*, 480–487. [CrossRef] [PubMed]
91. Cox, D.N.; Melo, L.; Zabaras, D.; Delahunty, C.M. Acceptance of health-promoting Brassica vegetables: The influence of taste perception; information and attitudes. *Public Health Nutr.* **2012**, *15*, 1474–1482. [CrossRef]
92. Reed, D.R.; Tanaka, T.; McDaniel, A.H. Diverse tastes: Genetics of sweet and bitter perception. *Physiol. Behav.* **2006**, *88*, 215–226. [CrossRef] [PubMed]
93. Drewnowski, A. The science and complexity of bitter taste. *Nutr. Rev.* **2001**, *59*, 163–169. [CrossRef]
94. Branca, F.; Ragusa, L.; Tribulato, A. Diversity of Kale Growing in Europe as a Basis for Crop Improvement. *Acta Hortic.* **2013**, *1005*, 141–147. [CrossRef]
95. Jagdish, S.; Upadhyay, A.K.; Singh, S.; Rai, M. Total phenolics content and free radical scavenging activity of Brassica vegetables. *J. Food Sci. Tech. Mysore* **2009**, *46*, 595–597.
96. Perez-Balibrea, S.; Moreno, D.A.; Garcia-Viguera, C. Genotypic effects on the phytochemical quality of seeds and sprouts from commercial broccoli cultivar. *Food Chem.* **2011**, *125*, 348–354. [CrossRef]
97. Volden, J.; Bengtsson, G.B.; Wicklund, T. Glucosinolates; L-ascorbic acid; total phenols; anthocyanins; antioxidant capacities and colour in cauliflower (*Brassica oleracea* L. ssp. botrytis); effects of long-term freezer storage. *Food Chem.* **2009**, *112*, 967–976. [CrossRef]
98. Branca, F.; Chiarenza, G.L.; Pinio, M.; Alonso, M.; Argento, S. Characteristics and seed production of sicilian landraces of violet cauliflower. *Acta Hortic.* **2013**, *1005*, 519–524. [CrossRef]
99. Singh, J.; Upadhyay, A.K.; Prasad, K.; Bahadur, A.; Rai, M. Variability of carotenes. vitamin C. E and phenolics in Brassica vegetables. *J. Food Compos. Anal.* **2007**, *20*, 106–112. [CrossRef]
100. Francisco, M.; Moreno, D.A.; Cartea, M.E.; Ferreres, F.; Garcia-Viguera, C.; Velasco, P. Simultaneous identification of glucosinolates and phenolic compounds in a representative collection of vegetable *Brassica rapa*. *J. Chromatogr. A* **2009**, *1216*, 6611–6619. [CrossRef] [PubMed]
101. Wachtel-Galor, S.; Wong, K.W.; Benzie, I.F.F. The effect of cooking on Brassica vegetables. *Food Chem.* **2008**, *110*, 706–710. [CrossRef]
102. Martinez-Sanchez, A.; Gil-Izquierdo, A.; Gil, M.I.; Ferreres, F. A comparative study of flavonoid compounds; vitamin C, and antioxidant properties of baby leaf *Brassicaceae* species. *J. Agric. Food Chem.* **2008**, *56*, 2330–2340. [CrossRef] [PubMed]
103. Martinez-Villaluenga, C.; Peñas, E.; Ciska, E.; Piskula, M.K.; Kozlowska, H.; Vidal-Valverde, C.; Frias, J. Time dependence of bioactive compounds and antioxidant capacity during germination of different cultivars of broccoli and radish seeds. *Food Chem.* **2010**, *120*, 710–716. [CrossRef]
104. Francisco, M.; Cartea, M.E.; Soengas, P.; Velasco, P. Effect of Genotype and Environmental Conditions on Health-Promoting Compounds in *Brassica rapa*. *J. Agric. Food Chem.* **2011**, *59*, 2421–2431. [CrossRef] [PubMed]
105. Jeffery, E.H.; Brown, A.F.; Kurilich, A.C.; Keek, A.S.; Matusheski, N.; Klein, B.P.; Juvik, J.A. Variation in content of bioactive components in broccoli. Study review. *J. Food Compos. Anal.* **2003**, *16*, 323–330. [CrossRef]
106. Fernandes, F.; Valentão, P.; Sousa, C.; Pereira, J.A.; Seabra, R.M.; Andrade, P.B. Chemical and antioxidative assessment of dietary turnip (*Brassica rapa* var. rapa L.). *Food Chem.* **2007**, *105*, 1003–1010. [CrossRef]
107. Brown, P.D.; Tokuhisa, J.G.; Reichelt, M.; Gershenzon, J. Variation of glucosinolate accumulation among different organs and developmental stages of *Arabidopsis thaliana*. *Phytochemistry* **2003**, *62*, 471–481. [CrossRef]

108. Seong, G.U.; Hwang, I.W.; Chung, S.K. Antioxidant capacities and polyphenolics of Chinese cabbage (*Brassica rapa* L. ssp. pekinensis) leaves. *Food Chem.* **2016**, *199*, 612–618. [CrossRef]
109. Rybarczyk, A.; Wold, B.A.; Hansen, M.K.; Hagen, S.F. Vitamin C in broccoli (*Brassica oleracea* L. var. italica) flower buds as affected by postharvest light, UV-B irradiation and temperature. *Postharvest Biol. Technol.* **2014**, *98*, 82–89.
110. Vallejo, F.; Tomas-Barberan, F.; Garcia-Viguera, C. Changes in broccoli (*Brassica oleracea* L. var. *italica*) health-promoting compounds with inflorescence development. *J. Agric. Food Chem.* **2003**, *51*, 3776–3782. [CrossRef]
111. Lefsrud, M.; Kopsell, D.; Wenzel, A.; Sheehan, J. Changes in kale (*Brassica oleracea* L. var. acephala) carotenoid content and chlorophyll pigment concentrations during leaf ontogeny. *Sci. Hort. Amst.* **2007**, *112*, 136–141. [CrossRef]
112. Vale, A.P.; Cidade, H.; Pinto, M.; Oliveira, M.B.P.P. Effect of sprouting and light cycle on antioxidant activity of *Brassica oleracea* varieties. *Food Chem.* **2014**, *165*, 379–387. [CrossRef]
113. De Pascale, S.; Maggio, A.; Pernice, R.; Fogliano, V.; Barbieri, G. Sulphur fertilization may improve the nutritional value of *Brassica rapa* L. subsp. sylvestris. *Eur. J. Agron.* **2007**, *26*, 418–424. [CrossRef]
114. Renaud, E.N.C.; Lammerts Van Bueren, E.T.; Myers, J.R.; Paulo, M.J.; Van Eeuwijk, F.A.; Zhu, N.; Juvik, J.A. Variation in broccoli cultivar phytochemical content under organic and conventional management systems: Implications in breeding for nutrition. *PLoS ONE* **2014**, *9*, e95683. [CrossRef]
115. Rosa, E.A.S.; Rodrigues, A.S. Total and individual glucosinolate content in 11 broccoli cultivars grown in early and late season. *HortScience* **2001**, *36*, 56–59. [CrossRef]
116. Aksouh, N.M.; Jacobs, B.C.; Stoddard, F.L.; Mailer, R.J. Response of canola to different heat stresses. *Aust. J. Agric. Res.* **2001**, *52*, 817–824. [CrossRef]
117. Schonhof, I.; Klaring, H.P.; Krumbein, A.; Clauβen, W.; Schreiner, M. Effect of temperature increase under low radiation conditions on phytochemicals and ascorbic acid in greenhouse grown broccoli. *Agric. Ecosyst. Environ.* **2007**, *119*, 103–111. [CrossRef]
118. Pereira, F.M.V.; Rosa, E.; Fahey, J.W.; Stephenson, K.K.; Carvalho, R.; Aires, A. Influence of temperature and ontogeny on the levels of glucosinolates in broccoli (*Brassica oleracea* var. italica) sprouts and their effect on the induction of mammalian phase 2 enzymes. *J. Agric. Food Chem.* **2002**, *50*, 6239–6244. [CrossRef]
119. Vallejo, F.; Tomas-Barberan, F.; Garcia-Viguera, C. Healthpromoting compounds in broccoli as influenced by refrigerated transport and retail sale period. *J. Agric. Food Chem.* **2003**, *51*, 3029–3034. [CrossRef] [PubMed]
120. Bjorkman, M.; Klingen, I.; Birch, A.; Bones, A.; Bruce, T.; Johansen, T.J.; Meadow, R.; Molmann, J.; Seljasen, R.; Smart, L.E.; et al. Phyto-chemicals of Brassicaceae in plant protection and human health. Influences of climate; environment and agronomic practice. *Phytochemistry* **2011**, *72*, 538–556. [CrossRef]
121. Francisco, M.; Cartea, M.E.; Butrón, A.M.; Sotelo, T.; Velasco, P. Environmental and genetic effects on yield and secondary metabolite production in Brassica rapa crops. *J. Agric. Food Chem.* **2012**, *60*, 5507–5514. [CrossRef] [PubMed]
122. Pék, Z.; Daood, H.; Nagyné, M.G.; Neményi, A.; Helyes, L. Effect of environmental conditions and water status on the bioactive compounds of broccoli. *Cent. Eur. J. Biol.* **2013**, *8*, 777–787. [CrossRef]
123. Singh, S.; Sinha, S. Accumulation of metals and its effects in *Brassica juncea* L. Czern. (cv. Rohini) grown on various amendments of tannery waste. *Ecotox. Environ. Saf.* **2005**, *62*, 118–127. [CrossRef]
124. Xiong, Z.T.; Liu, C.; Geng, B. Phytotoxic effects of copper on nitrogen metabolism and plant growth in *Brassica pekinensis* Rupr. *Ecotox. Environ. Saf.* **2006**, *64*, 273–280. [CrossRef]
125. Rodriguez-Hernandez, M.C.; Moreno, D.A.; Carvajal, M.; Martìnez-Ballesta, M.C. Genotype influences sulfur metabolism in broccoli (*Brassica oleracea* L.) under elevated CO2 and NaCl stress. *Plant. Cell Physiol.* **2014**, *55*, 2047–2059. [CrossRef]
126. Sakr, M.Y.; Ibrahim, H.M.; ElAwady, A.E.; AboELMakarm, A.A. Growth, yield and biochemical constituents as well as post-harvest quality of water-stressed broccoli (*Brassica oleraceae* L. var. italica) as affected by certain biomodulators. *Sci. Hortic.* **2021**, *275*, 109605. [CrossRef]
127. Ragusa, L.; Picchi, V.; Tribulato, A.; Cavallaro, C.; Lo Scalzo, R.; Branca, F. The effect of the germination temperature on the phytochemical content of broccoli and rocket sprouts. *Int. J. Food. Sci. Nutr.* **2016**, *68*, 411–420. [CrossRef]
128. Sousa, C.; Pereira, D.M.; Pereira, J.A.; Bento, A.; Rodrigues, M.A.; Dopico-Garcìa, S.; Valentao, P.; Lopes, G.; Ferreres, F.; Seabra, R.M.; et al. Multivariate analysis of tronchuda cabbage (*Brassica oleracea* L. var. costata DC) phenolics: Influence of fertilizers. *J. Agric. Food Chem.* **2008**, *56*, 2231–2239. [CrossRef]
129. Vrchoska, V.; Sousa, C.; Valentao, P.; Ferreres, F.; Pereira, J.A.; Seabra, R.M.; Andrade, P.B. Antioxidative properties of tronchuda cabbage (*Brassica oleracea* L. Var. costata DC) external leaves against DPPH; superoxide radical; hydroxyl radical and hypochlorous acid. *Food Chem.* **2006**, *98*, 416–425. [CrossRef]
130. Dangour, A.D.; Dodhia, S.K.; Hayter, A.; Allen, E.; Lock, K.; Uauy, R. Nutritional quality of organic foods: A systematic review. *Am. J. Clin. Nutr.* **2009**, *90*, 680–685. [CrossRef]
131. Hoefkens, C.; Sioen, I.; Baert, K.; De Meulenaer, B.; De Henauw, S.; Vandekinderen, I.; Devlieghere, F.; Opsomer, A.; Verbeke, W.; Van Camp, J. Consuming organic versus conventional vegetables: The effect on nutrient and contaminant intakes. *Food Chem. Toxicol.* **2010**, *48*, 3058–3066. [CrossRef]
132. Conversa, G.; Bonasia, A.; Lazzizera, C.; Elia, A. Bio-physical; physiologiacal; and nutritional aspects of ready-to-use cima di rapa (*Brassica rapa* L. subsp. sylvestris L. Janch. var. esculenta Hort.) as affected by conventional and organic growing systems and storage time. *Sci. Hortic.* **2016**, *213*, 76–86. [CrossRef]

133. Khan, M.A.M.; Ulrichs, C.; Mewis, I. Drought stress—Impact on glucosinolate profile and performance of phloem feeding cruciferous insects. *Acta Hortic.* **2011**, *917*, 111–117. [CrossRef]
134. Cogo, S.L.P.; Chaves, F.C.; Schirmer, M.A.; Zambiazi, R.C.; Nora, L.; Silva, J.A.; Rombaldi, C.V. Low soil water content during growth contributes to preservation of green colour and bioactive compounds of cold-stored broccoli (*Brassica oleracea* L.) florets. *Postharvest Biol. Technol.* **2011**, *60*, 158–163. [CrossRef]
135. Hooks, C.R.R.; Johnson, M.W. Impact of agricultural diversification on the insect community of cruciferous crops. *Crop. Prot.* **2003**, *22*, 223–238. [CrossRef]
136. Kim, S.J.; Matsuo, T.; Watanabe, M.; Watanabe, Y. Effect of nitrogen and sulphur application on the glucosinolate content in vegetable turnip rape (*Brassica rapa* L.). *Soil Sci. Plant. Nutr.* **2002**, *48*, 43–49. [CrossRef]
137. Kopsell, D.E.; Kopsell, D.A.; Randle, W.A.; Coolong, T.W.; Sams, C.E.; Curran-Celentano, J. Kale carotenoids remain stable while flavor compounds respond to changes in sulfur fertility. *J. Agric. Food Chem.* **2003**, *51*, 5319–5325. [CrossRef]
138. Li, S.; Schonhof, I.; Krumbein, A.; Li, L.; Stutzel, H.; Schreiner, M. Glucosinolate concentration in turnip (*Brassica rapa* ssp. rapifera L.) roots as affected by nitrogen and sulfur supply. *J. Agric. Food Chem.* **2007**, *55*, 8452–8457. [CrossRef]
139. Schreiner, M.; Huyskens-Keil, S.; Peters, P.; Schonhof, I.; Krumbein, A.; Widell, S. Seasonal climate effects on root colour and compounds of red radish. *J. Sci. Food Agric.* **2002**, *82*, 1325–1333. [CrossRef]
140. Barbieri, G.; Bottino, A.; Orsini, F.; De Pascale, S. Sulfur fertilization and light exposure during storage are critical determinants of the nutritional value of ready-to-eat friariello campano (*Brassica rapa* L. subsp. sylvestris). *J. Sci. Food Agric.* **2009**, *89*, 2261–2266. [CrossRef]
141. Dubuis, P.H.; Marazzi, C.; Städler, E.; Mauch, F. Sulphur deficiency causes a reduction in antimicrobial potential and leads to increased disease susceptibility of oilseed rape. *J. Phytopathol.* **2005**, *153*, 27–36. [CrossRef]
142. Kopsell, D.A.; Kopsell, D.E.; Curran-Celentano, J. Carotenoid pigments in kale are influenced by nitrogen concentration and form. *J. Sci. Food Agric.* **2007**, *87*, 900–907. [CrossRef]
143. Zhao, F.; Evans, E.J.; Bilsborrow, P.E.; Syers, J.K. Influence of nitrogen and sulphur on the glucosinolate profile of rapeseed (*Brassica napus*). *J. Sci. Food Agric.* **1994**, *64*, 295–304. [CrossRef]
144. Robbins, R.J.; Keck, A.S.; Banuelos, G.; Finley, J.W. Cultivation conditions and selenium fertilization alter the phenolic profile; glucosinolate; and sulforaphane content of broccoli. *J. Med. Food* **2005**, *8*, 204–214. [CrossRef] [PubMed]
145. Fabek, S.; Toth, N.; Redovnikovic, I.R.; Custic, M.H.; Benko, B.; Žutic, I. The effect of nitrogen fertilization on nitrate accumulation; and the content of minerals and glucosinolates in broccoli cultivars. *Food Technol. Biotechnol.* **2012**, *50*, 9.
146. Tolrà, R.P.; Poschenrieder, C.; Alonso, R.; Barceló, D.; Barceló, J. Influence of zinc hyperaccumulation on glucosinolates in Thlaspi caerulescens. *New Phytol.* **2001**, *151*, 621–626. [CrossRef]
147. Aghajanzadeh, T.A.; Prajapati, D.H.; Burow, M. Copper toxicity affects indolic glucosinolates and gene expression of key enzymes for their biosynthesis in Chinese cabbage. *Arch. Agron. Soil Sci.* **2020**, *66*, 1288–1301. [CrossRef]
148. Pandey, C.; Augustine, R.; Panthri, M.; Zia, I.; Bisht, N.C.; Gupta, M. Arsenic affects the production of glucosinolate, thiol and phytochemical compounds: A comparison of two Brassica cultivars. *Plant Physiol. Biochem.* **2017**, *111*, 144–154. [CrossRef] [PubMed]
149. Bodnaryk, R.P. Potent effect of jasmonates on indole glucosinolates in oilseed rape and mustard. *Phytochemistry* **1994**, *35*, 301–305. [CrossRef]
150. Mollers, C.; Nehlin, L.; Glimelius, K.; Iqbal, M.C.M. Influence of in vitro culture conditions on glucosinolate composition of microspore-derived embryos of *Brassica napus*. *Physiol. Plant.* **1999**, *107*, 441–446. [CrossRef]
151. Guo, R.; Shen, W.; Qian, H.; Zhang, M.; Liu, L.; Wang, Q. Jasmonic acid and glucose synergistically modulate the accumulation of glucosinolates in *Arabidopsis thaliana*. *J. Exp. Bot.* **2013**, *64*, 5707–5719. [CrossRef]
152. Gallo, M.; Esposito, G.; Ferracane, R.; Vinale, F.; Naviglio, D. Beneficial effects of *Trichoderma* genus microbes on qualitative parameters of *Brassica rapa* L. subsp. sylvestris L. Janch. Var. esculenta Hort. *Eur. Food Res. Technol.* **2013**, *236*, 1063–1071. [CrossRef]
153. Muelchen, A.M.; Rand, R.E.; Parke, J.L. Evaluation of crucifer manures for controlling aphanomyces root rot of peas. *Plant Dis.* **1990**, *74*, 651–654. [CrossRef]
154. Pant, A.P.; Radovich, T.J.K.; Hue, N.V.; Talcott, S.T.; Krenek, K.A. Vermicompost extracts influence growth; mineral nutrients; phytonutrients and antioxidant activity in pak choi (*Brassica rapa* cv. *bonsai; Chinensis* group) grown under vermicompost and chemical fertiliser. *J. Sci. Food Agric.* **2009**, *89*, 2383–2392. [CrossRef]
155. Lola-Luz, T.; Hennequart, F.; Gaffney, M. Enhancement of phenolic and flavonoid compounds in cabbage (*Brassica oleraceae*) following application of commercial seaweed extracts of the brown seaweed; (*Ascophyllum nodosum*). *J. Agric. Food Sci.* **2013**, *22*, 288–295. [CrossRef]
156. Argento, S.; Raccuia, S.A.; Melilli, M.G.; Toscano, V.; Branca, F. Brassicas and their glucosinolate content for the biological control of root-knot nematodes in protected cultivation. *Acta Hortic.* **2013**, *1005*, 539–544. [CrossRef]

Article

Influence of Post-Flowering Climate Conditions on Anthocyanin Profile of Strawberry Cultivars Grown from North to South Europe

Erika Krüger [1,*], Frank Will [2], Keshav Kumar [2], Karolina Celejewska [3], Philippe Chartier [4], Agnieszka Masny [3], Daniela Mott [5], Aurélie Petit [4], Gianluca Savini [5] and Anita Sønsteby [6]

1 Institute of Pomology, Hochschule Geisenheim University, 65366 Geisenheim, Germany
2 Institute of Beverage, Hochschule Geisenheim University, 65366 Geisenheim, Germany; Frank.Will@hs-gm.de (F.W.); Keshav.Kumar@hs-gm.de (K.K.)
3 The National Institute of Horticultural Research (INHORT), Konstytucji 3 Maja 1/3, 96-100 Skierniewice, Poland; karolina.celejewska@inhort.pl (K.C.); agnieszka.masny@inhort.pl (A.M.)
4 INVENIO, Maison Jeannette, 24140 Douville, France; p.chartier@invenio-fl.fr (P.C.); a.petit@invenio-fl.fr (A.P.)
5 Sant'Orsola Società Cooperativa Agricola, Via Lagorai, 127, 38057 Pergine Valsugana, Italy; Daniela.Mott@santorsola.com (D.M.); Gianluca.Savini@santorsola.com (G.S.)
6 NIBIO, Norwegian Institute of Bioeconomy Research, 1431 Ås, Norway; Anita.Sonsteby@Nibio.no
* Correspondence: Erika.Krueger@hs-gm.de

Abstract: The effect of cultivar and environmental variations and their interaction on anthocyanin components of strawberry were assessed for six cultivars grown in five locations from North to South of Europe in two different years. To evaluate the impact of latitude- and altitude-related factors, daily mean (T_{mean}), maximum (T_{max}) and minimum (T_{min}) temperature and global radiation accumulated for 3, 5, 10 and 15 days before fruit sampling, was analyzed. In general, fruits grown in the south were more enriched in total anthocyanin and pelargonidin-3-glucoside (pel-3-glc), the most abundant anthocyanin in strawberry. Principal component analysis (PCA) provided a separation of the growing locations within a cultivar due to latitudinal climatic differences, temporary weather changes before fruit collection and cultivation technique. PCA also depicted different patterns for anthocyanin distribution indicating a cultivar specific reaction on the environmental factors. The linear regression analysis showed that pel-3-glc was relatively less affected by these factors, while the minor anthocyanins cyanidin-3-glucoside, cyanidin-3-(6-O-malonyl)-glucoside, pelargonidin-3-rutinoside and pelargonidin-3-(6-O-malonoyl)-glucoside were sensitive to T_{max}. The global radiation strongly increased cya-3-mal-glc in 'Frida' and pel-3-rut in 'Frida' and 'Florence'. 'Candonga' accumulated less pel-3-glc and total anthocyanin with increased global radiation. The anthocyanin profiles of 'Gariguette' and 'Clery' were unaffected by environmental conditions.

Keywords: anthocyanins; *Fragaria* × *ananassa*; latitude; temperature; global radiation; cultivar × environmental interaction

Citation: Krüger, E.; Will, F.; Kumar, K.; Celejewska, K.; Chartier, P.; Masny, A.; Mott, D.; Petit, A.; Savini, G.; Sønsteby, A. Influence of Post Flowering Climate Conditions on Anthocyanin Profile of Strawberry Cultivars Grown from North to South Europe. *Appl. Sci.* **2021**, *11*, 1326. https://doi.org/10.3390/app11031326

Received: 29 December 2020
Accepted: 28 January 2021
Published: 2 February 2021

1. Introduction

Strawberry (*Fragaria* × *ananassa* Duch.) is the most important berry crop being cultivated from North to South of Europe. Beside its unique color, taste and aroma, strawberry fruits are enriched with several nutritious and bioactive compounds providing health benefits by reducing risk of diseases such as inflammation disorders and oxidative stress, obesity-related disorders and heart disease, and protection against various types of cancer [1–4]. Anthocyanins are a type of flavonoids that are commonly found in strawberries. The functional properties and the sensory qualities of the anthocyanins could easily be explained based on their chemical reactivity [5]. The antioxidative activity of anthocyanins could mainly be attributed to the presence of the flavylium cation moiety. Despite their low bioavailability [6], anthocyanins have been shown to exhibit a range of biological effects, including antioxidant activity, photoprotection, anti-carcinogenesis, induction of apoptosis, and prevention of DNA damage [3,7]. Anthocyanins also serve as visual attractants for pollinators and seed dispensers and play a crucial role in plant protection against biotic and abiotic stress, and hence, in adaptability to environmental conditions at site [8].

Thus, the total anthocyanin contents in strawberry are both qualitatively and quantitatively known to be strongly influenced by the genotype (among others [9–14]) and likewise, by external factors such as high or low temperature and light (photoperiod, quantity and wavelength including UV-light). Recent review articles have highlighted the influence of temperature and light on the synthesis and accumulation of plant secondary metabolites including anthocyanins [15–17]. For example, some studies reported positive correlation between anthocyanin contents and temperature in strawberries grown in controlled environment [18–20], as well as in ambient conditions [14,21]. The studies describing the influence of incident light are mostly related to protected cultivation systems and includes shading [22], UV-B radiation [23–26] and blue and red LED-light [27,28].

The synthesis and accumulation of anthocyanins in strawberries are primarily known to be influenced by the genotype, however, little is known about latitudinal effects on the anthocyanin content of strawberry causing quantitative or qualitative changes in the content of these compounds [21,23]. The ambient conditions, including temperature and photoperiod, varies with latitude, and may also influence the different anthocyanin components. In future, due to the climate changes, the northern regions of Europe will become more suitable for berry production [25]. Moreover, it is desirable that quality traits of a cultivar, including bioactive compounds such as anthocyanins, remain invariant to the changes caused by environmental conditions during the cultivation season in different years at all growing locations. Cultivars with such stable fruit qualities could serve as useful parents in breeding programs with regard to climate warming.

The objective of the present work was to study the cultivar and environmental impact and their interaction on the anthocyanin content of six strawberry cultivars: 'Candonga', 'Clery', 'Frida', 'Florence', 'Gariguette' and 'Sonata' grown at five geographical distant locations throughout Europe from South-East Norway to South-West France, covering a distance of more than 15.5 degree of latitudes or more than 2000 km. These cultivars are mainly selected because of their diversity and popularity in Europe. The present study would help in identifying the appropriate cultivar with desirable anthocyanin traits that could be cultivated in a given environmental condition.

2. Materials and Methods

2.1. Experimental Sites, Plant Material and Cultivation

The experiments were conducted during two consecutive cropping seasons (2017 and 2018) at five locations from North to South Europe at NIBIO (Norwegian Institute of Bioeconomy Research, Bergen, Norway) (NO, 60° N), INHORT (PL, 51° N), HGU (DE, 49° N), Sant'Orsola (IT, 46° N) and Invenio (FR, 44° N), hereinafter named as Norway, Poland, Germany, Italy and France. As common for the different regions, experiments in Norway, Poland and Germany were carried out in open field, whereas in Italy and France they were performed in polytunnels that were open-sided after anthesis. Details of the respective latitude, altitude, yearly mean temperature, soil type, soil pH and cultivation type, as well as start of flowering, harvest season and day length at start of harvest are given in Table 1. Air temperature (T_{mean}, T_{max} and T_{min}) and global radiation were recorded at each location. To describe the environmental conditions of the respective harvest season, weekly mean temperature and global radiation were calculated starting 2 weeks prior to start of harvest of the earliest cultivar 'Clery' (Figure 1).

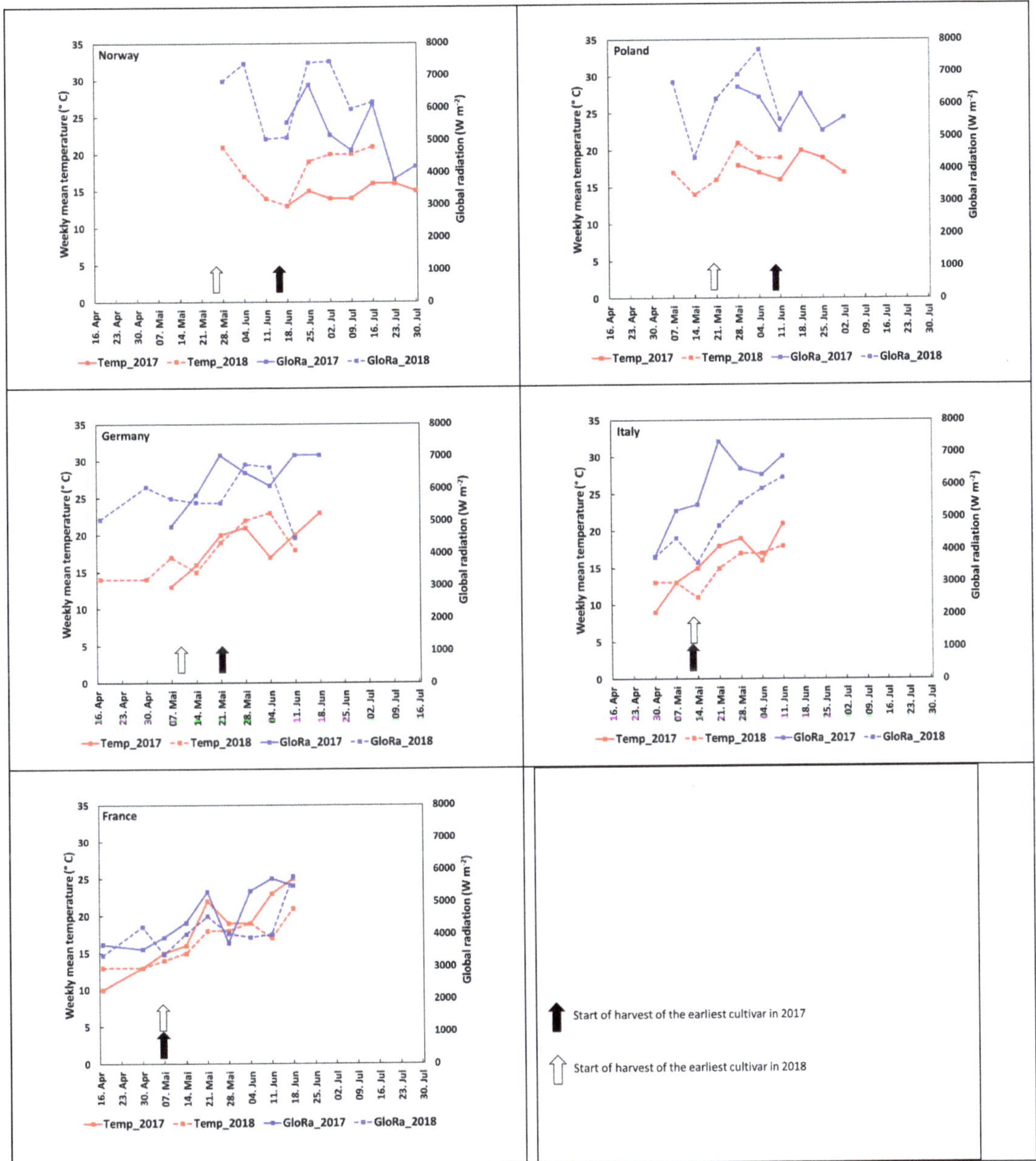

Figure 1. Weekly mean temperature and global radiation courses during the harvest season for the five experimental locations in 2017 and 2018, starting two weeks before the first picking of the earliest cultivar 'Clery'.

Table 1. Geographical location, soil and climatic conditions, cultivation type and harvest season for the five experimental locations in Europe.

	NIBIO Norway		INHORT Poland		HGU Germany		Sant'Orsola Italy		INVENIO France	
Latitude	60°40′ N		51°95′ N		49°59′ N		46°4′ N		44°85′ N	
Altitude (m a.s.l.)	262		252		95		925		145	
Yearly mean temperature (°C) [(a)]	5.0		7.9		9.9		11.3		12.9	
Soil type	Loam		Pseudopodsol with light clay		Sandy loam		Soilless culture		Soilless culture	
pH of the soil/substrate	5.7–6.2		6.5–7.0		6.5–7.0		5.5–6.0		6.1	
Cultivation type	open field		open field		open field		tunnel		tunnel	
	2017	2018	2017	2018	2017	2018	2017	2018	2017	2018
Yearly mean temperature (°C) [(a)]	5.0	5.7	9.1	9.8	11.3	12.4	15.6	15.1	13.2	13.8
Start of flowering [(b)]	06.06.	21.05.	16.05	30.04.	04.04. [(d)]	16.04.	04.04.	11.04.	04.04.	07.04.
Start of harvest [(b)]	07.07.	17.06.	13.06.	22.05.	29.05.	14.05.	19.05.	17.05.	09.05.	09.05.
End of harvest [(c)]	16.08.	21.07.	07.07.	19.06.	16.06.	18.06.	12.06.	13.06.	22.06.	22.06.
Day length at start of harvest (h) [(b)]	18:40	19:09	16:43	16:04	15:56	15:20	14:30	14:30	14.36	14:36

[(a)] For the period 1981–2010. [(b)] Earliest cultivar. [(c)] Latest cultivar. [(d)] Warm temperature in March enhanced start of flowering followed by extremely cold temperature delaying fruit development.

The six short-day strawberry genotypes used in this study were selected for their diverse genetic background and adaptability to different environments: 'Candonga'® (E), 'Clery' (IT), 'Florence' (UK), 'Frida' (NO), 'Gariguette' (FR) and 'Sonata' (NL). All cultivars were cultivated in Norway, Poland, Germany and France, while Italy only grew 'Clery', 'Frida', 'Gariguette' and 'Sonata'. Each season, all locations propagated their own plants from mother-plants purchased from the same nursery to avoid epigenetic variation within a cultivar due to different origin. Thus, each year, plug-plants were planted in week 32 in Norway, Poland and Germany for the harvest season the next year. At all sites, three randomized replicated plots, each with 20 plants per genotype, were trialed. In Italy and France, where strawberry are commonly grown in tunnels, cold-stored plants were set in peat bags, as usual, in week 9 in 2017 and 2018 for harvesting in the respective year. Furthermore, in Italy and France, the experimental design consisted of three plots, and of three replications per genotype with 22–24 plants per plot for the harvest in the same year. Plant protection, fertilization and irrigation in open-field sites, and fertigation of the plants in peat bags were performed according to local guidelines.

To evaluate the impact of latitude-related factors on the anthocyanin synthesis and accumulation, daily temperature (T_{mean}, T_{max} and T_{min}) and global radiation were accumulated for 3, 5, 10 and 15 days (3d, 5d, 10d and 15d) before sampling of the berries to be analyzed for each cultivar.

2.2. Sample Preparation for Anthocyanin Analysis

Three independent biological replications were collected twice in an approximately weekly interval at mid-harvest at each location per cultivar (Figure 1 and Table 1). Thus, the samples (500 g) consisted mainly of secondary and tertiary fruits, being fully ripe and undamaged. Each biological replicate of freshly harvested strawberries was promptly prepared for analytical purposes by first cutting the calyx and then slicing the fruits into 2–4 pieces. Slices were frozen in a liquid nitrogen bath for at least 10 s. For anthocyanin analysis, the frozen slices were ground either in a lab mill (e.g., Retsch GM 200) or crushed under liquid nitrogen with a precooled mortar and pistil to a fine powder. The frozen powders were filled into 50 mL plastic tubes with screw caps, and stored at −20° C. Anthocyanin analysis was carried out at Geisenheim. Suitable shipping conditions were chosen to prevent thawing of the samples during transportation until solvent extraction. The frozen powders (5.0 g) were weighed into a 50 mL plastic tube and extracted twice in an ultrasonic bath (30 min, 120 W, Bandelin Sonorex RK 106, VWR, Darmstadt, Germany) and intermediate centrifugation (4500 upm 15 min, Hettich Rotanta 460, Tuttlingen, Germany) with a total of 2 × 10 mL acidified methanol/water/formic acid (80/20/1 *v*/*v*/*v*). Sonication and centrifugation were performed under cool conditions. After final centrifugation, the pooled supernatants were made up to 25 mL in a volumetric flask; aliquots were 0.45 µm filtered prior to HPLC analysis.

2.3. HPLC Analysis of Anthocyanins

HPLC analysis of the methanolic extracts was performed on an Accela HPLC system coupled with a PDA detector (Thermo Fisher, Dreieich, Germany) using a 125 × 2 mm i.d., 3 µm ODS-3 column (Dr. Maisch, Ammerbuch, Germany) protected with a guard column of the same material. Injection volume was 4 µL, elution conditions were: 250 µL/min flow rate at 40 °C; solvent A was 5% formic acid (ULC/MS grade, Promochem, Wesel, Germany); solvent B, methanol (gradient grade, Roth, Karlsruhe, Germany); 1 min isocratic conditions with 10% B, linear gradient from 10% to 40% B in 12 min, followed by washing with 100% B and re-equilibrating the column. Quantitation was carried out using peak areas (500 nm trace for pelargonidins, 515 nm for cyanidins) from external calibration via the reference substances pelargonidin-3-glucoside and cyanidin-3-glucoside, respectively. Anthocyanin analysis was carried out in duplicate.

2.4. Electrospray Ionization (ESI)-MS Identification of Anthocyanins

For mass detection, the Accela HPLC system was coupled to a ThermoFinnigan LXQ mass spectrometer (Thermo Fisher, Dreieich, Germany) equipped with an ESI source and an ion trap mass analyzer. The whole system was controlled by Xcalibur software. For anthocyanins, the mass spectrometer was operated in the positive mode under the following conditions: source voltage 4.5 kV; capillary voltage 32 V; capillary temperature 275 °C; collision energy 30% (MS^2) and 33% (MS^3).

2.5. Software Used for Statistical Analysis

All statistical analysis was conducted on the MATLAB (2016b) platform. Significant differences were calculated using post hoc analysis with Tukey's honestly significant difference criteria on the ANOVA (analysis of variance) results. Principal component analysis (PCA) [29] was performed using the PLS-Toolbox (Eigenvector Research, Manson, WA, USA). Linear regression analysis was used to study the correlation between different cultivar-specific anthocyanin components and the mean, minimum, maximum temperatures, and global radiation summarized 3, 5, 10 and 15 days before start of harvest. Coefficient of determinations (R^2) ≥ 0.2 at $p \leq 0.05$ are presented.

2.6. Data Arrangement for PCA

The data for 'Clery', 'Frida', 'Gariguette' and 'Sonata' grown at the locations in France, Italy, Germany, Poland and Norway were arranged in a matrix of dimension 60 × 22; 60 represents the number of samples (five locations, two cropping seasons, two picking dates and three biological replicates) and 22, the number of variables (cya-3-glc, pel-3-glc, pel-3-rut, cya-3-mal-glc, pel-3-mal-glc, total anthocyanin as well as global radiation, mean, maximum and minimum temperatures summarized 3, 5, 10 and 15 days prior to harvest). The data for 'Candonga' and 'Florence' were arranged in matrices of dimension 48 × 22, where 48 is the number of samples (four location × two seasons × two picking dates × three biological replicates) and 22 is the same variables as specified above. The PCA analysis carried out on all the cultivars together clearly indicated (given in the Supplementary Figure S1) that the anthocyanin profiles of the strawberries are mainly influenced by the location. Thus it was important that each of the six cultivars were analyzed separately.

To ensure that each component had equal variance and comparable impact on the PCA modeling, the specific data for each cultivar were auto-scaled prior to PCA. Autosscaling [29] is a common pre-processing method that uses mean-centering followed by dividing each variable by the corresponding standard deviation. Each variable upon auto-scaling has a mean of zero and unit variance.

3. Results and Discussion

3.1. Harvest Season and Environmental Characterization of the Growing Locations

This study investigated the adaption of six strawberry genotypes to different environments related to anthocyanin accumulation in the fruits. As expected, the harvest period varied along the North-South axis (Table 1), due to the longitudinal difference between the growing sites (>15.5° of latitude). In addition, yearly differences in temperature and global radiation affected the harvest period.

For instance, the harvest period was much earlier in 2018 compared to 2017 in Norway, Poland and Germany, while there were no variations in Italy and France. To highlight the environmental differences between the five growing sites, ambient weekly mean temperature and global radiation during the harvest seasons are shown in Figure 1, starting two weeks prior to start of harvest of the earliest cultivar 'Clery'. Higher year-on-year variations were observed in Norway. Here, weekly mean temperature was, on average, 3 °C lower, and global radiation 1200 W m^{-2} less in 2017 compared to 2018 (18 °C and 6412 W m^{-2}). In this year, weekly mean temperature and global radiation in Norway were similar to those in both years in Poland and Germany, and for temperature in France, only. Due to its high altitude (Table 1), Italy exhibited lower weekly mean temperature in both years, thus being similar to Norway in 2017. In Italy, large yearly variations occurred for global radiation in 2017 (5917 W m^{-2}) with similar values as for Norway, Poland and Germany, while in 2018, Italy had the same low values as for both years in France. The low values for global radiation at the southern locations are probably due to the earlier cropping season and shorter photoperiod compared to Norway, as the other extreme.

The temperature and global radiation may vary within the harvest season, therefore, the sum of temperature and global radiation 3, 5, 10 and 15 days before the two fruit samplings were calculated for each cultivar and location, and used for principal component and regression analyses. The complete list is provided in the Supplementary Materials (Table S1).

3.2. Effect of Genotype on Total and Individual Anthocyanins

Cyanidin-3-glucoside (cya-3-glc, m/z [M^+] 449, MS^2 287), pelargonidin-3-glucoside (pel-3-glc, m/z [M^+] 433, MS^2 271), pelargonidin-3-rutinoside (pel-3-rut, m/z [M^+] 579, MS^2 433, 271), cyanidin-3-(6-O-malonyl)-glucoside (cya-3-mal-glc, m/z [M^+] 535, MS^2 287), and pelargonidin-3-(6-O-malonoyl)-glucoside (pel-3-mal-glc, m/z [M^+] 519, MS^2 271) were assigned by their mass spectra as the major strawberry anthocyanins (Tables 2–7). Over all, the pel-3-glc showed the highest concentrations (Table 4).

Table 2. Effect of growing location and year on total anthocyanin content as the mean of two picking dates in fruits of six strawberry cultivars.

		Total Anthocyanins HPLC (mg kg^{-1} Fresh Weight)						
		Cultivar						
Location	**Year**	**Can**	**Cle**	**Flo**	**Fri**	**Gar**	**Son**	***Yearly Mean per Location***
Norway	**2017**	212.6 ± 25.6	284.6 ± 30.5	374.5 ± 30.4	453.4 ± 42.4	263.1 ± 58.2	191.6 ± 16.2	*296.6 B*
	2018	133.8 ± 33.8	195.2 ± 48.7	405.5 ± 43.8	451.6 ± 68.8	198.9 ± 25.8	145.6 ± 30.4	*255.1 B*
	mean	*173.2 ab*	*239.9 c*	*390.0 d*	*452.5 e*	*231.0 bc*	*168.6 a*	
Poland	**2017**	245.9 ± 53.0	310.1 ± 34.3	359.1 ± 89.4	438.0 ± 58.8	235.0 ± 34.9	210.2 ± 27.2	*299.7 B*
	2018	225.0 ± 40.8	249.6 ± 22.4	448.5 ± 68.5	397.9 ± 74.6	217.6 ± 40.1	222.6 ± 35.4	*293.5 B*
	mean	*235.5 a*	*279.9 a*	*403.8 b*	*417.9 b*	*226.3 a*	*216.4 a*	

Table 2. *Cont.*

		Total Anthocyanins HPLC (mg kg^{-1} Fresh Weight)						
		Cultivar						
Location	Year	Can	Cle	Flo	Fri	Gar	Son	*Yearly Mean per Location*
Germany	2017	196.5 ± 22.3	285.5 ± 46.5	437.3 ± 57.7	429.2 ± 21.1	225.5 ± 44.4	199.4 ± 44.2	*295.6 B*
	2018	223.6 ± 31.3	307.7 ± 21.1	507.9 ± 26.5	385.9 ± 27.9	273.5 ± 19.9	217.7 ± 51.1	*319.4 B*
	mean	*211.0 a*	*296.6 b*	*472.6 d*	*407.6 c*	*249.5 ab*	*208.6 a*	
Italy	2017	-	363.0 ± 27.2	-	432.4 ± 52.7	326.4 ± 19.4	209.2 ± 17.7	*332.8 B*
	2018	-	276.6 ± 44.9	-	302.2 ± 56.2	166.9 ± 22.3	124.7 ± 34.4	*217.6 A*
	mean	-	*319.8 bc*	-	*367.4 c*	*246.6 b*	*167.0 a*	
France	2017	315.5 ± 23.2	218.9 ± 63.8	455.2 ± 81.3	446.7 ± 44.3	261.5 ± 39.3	212.0 ± 48.3	*318.3 B*
	2018	306.9 ± 45.2	365.6 ± 61.3	420.9 ± 66.3	342.8 ± 64.3	221.2 ± 34.4	198.6 ± 49.2	*309.2 B*
	mean	*311.2 b*	*292.3 b*	*438.0 c*	*394.4 c*	*241.3 ab*	*205.3 a*	
Cultivar mean over all locations		*232.5 b*	*285.7 c*	*426.1 d*	*408.0 d*	*239.0 b*	*193.1 a*	
Significance		*Cultivar*	*Location*	*Year*	*Cultivar x location*	*Cultivar x year*	*Location x year*	
		***	*ns*	*	*ns*	*ns*	*	

Data are expressed as means ± SD (standard deviation) of two sampling dates per year. Before performing the statistical analysis, the homogeneity of the data were ensured using Bartlett's test. Mean values (n = 3) of different cultivars grown at a particular location followed by lower-case letters represent significant differences ($p \leq 0.05$) between cultivars. Mean values of all the cultivars grown in a particular location followed by different upper-case letters represent significant difference between the two years 2017 and 2018 ($p \leq 0.05$). Mean values of all cultivars grown at all the locations followed by different lower-case letters represent significant differences ($p \leq 0.05$). * = 0.05; *** = 0.001; ns = not significant. Can = 'Candonga'; Cle = 'Clery'; Flo = 'Florence'; Fri = 'Frida'; Gar = 'Gariguette'; Son = 'Sonata'.

Table 3. Effect of growing location and year on cyanidin 3-glucoside content as the mean of two picking dates in fruits of six strawberry cultivars.

		Cyanidin 3-Glucoside (mg kg^{-1} Fresh Weight)						
		Cultivar						
Location	Year	Can	Cle	Flo	Fri	Gar	Son	*Yearly Mean per Location*
Norway	2017	9.6 ± 2.6	3.6 ± 1.5	9.4 ± 2.8	14.1 ± 2.4	3.2 ± 0.7	5.7 ± 1.7	*7.6 A*
	2018	12.7 ± 3	7.4 ± 5.2	14.3 ± 2.2	13.2 ± 1.1	4.4 ± 1.2	6.0 ± 2.3	*9.6 A*
	mean	*11.2 b*	*5.5 a*	*11.8 b*	*13.7 b*	*3.8 a*	*5.8 a*	
Poland	2017	9.7 ± 1.7	3.2 ± 0.6	12.7 ± 4.5	15.0 ± 3.3	3.3 ± 0.5	6.3 ± 2.6	*8.4 A*
	2018	29.2 ± 2.6	8.8 ± 1.9	47.8 ± 6.7	27.8 ± 4.1	9.2 ± 2.1	13.4 ± 5.9	*22.7 B*
	mean	*19.4 bc*	*6.0 a*	*30.3 c*	*21.4 bc*	*6.2 a*	*9.9 ab*	
Germany	2017	10.7 ± 2.4	2.5 ± 0.6	17.4 ± 3.6	12.8 ± 1.2	2.5 ± 1.3	4.4 ± 1.6	*8.4 A*
	2018	12.9 ± 4.8	5.0 ± 4.8	26.4 ± 4.4	16.9 ± 2.0	4.0 ± 0.5	6.0 ± 2.4	*11.9 A*
	mean	*11.8 b*	*3.7 a*	*21.9 c*	*14.8 b*	*3.3 a*	*5.2 a*	
Italy	2017	-	1.6 ± 0.3	-	11.9 ± 4.6	2.5 ± 0.5	2.8 ± 0.5	*4.7 B*
	2018	-	1.6 ± 0.7	-	2.4 ± 1.5	1.5 ± 0.7	2.7 ± 1.8	*2.0 A*
	mean	-	*1.6 a*	-	*7.0 b*	*2.0 a*	*2.7 a*	
France	2017	11.3 ± 1.5	1.3 ± 0.1	19.0 ± 14.5	12.5 ± 1.7	1.7 ± 0.3	2.6 ± 0.7	*8.0 A*
	2018	10.7 ± 3.0	1.8 ± 0.4	16.9 ± 5.8	9.0 ± 2.1	2.7 ± 0.6	3.1 ± 1.3	*7.4 A*
	mean	*11.0 b*	*1.6 a*	*18.0 c*	*10.7 b*	*2.2 a*	*2.8 a*	
Cultivar mean over all locations		*13.3 b*	*3.7 a*	*20.5 c*	*13.5 b*	*3.5 a*	*5.3 a*	

Table 3. *Cont.*

		Cyanidin 3-Glucoside (mg kg^{-1} Fresh Weight)						
		Cultivar						
Location	**Year**	**Can**	**Cle**	**Flo**	**Fri**	**Gar**	**Son**	***Yearly Mean per Location***
Significance		*Cultivar* ***	*Location r* ***	*Year* ***	*Cultivar x location* *ns*	*Cultivar x year* *ns*	*Location x year* ***	

Data are expressed as means ± SD (standard deviation) of two sampling dates per year. Before performing the statistical analysis, the homogeneity of the data were ensured using Bartlett's test. Mean values (n =3) of different cultivars grown at a particular location followed by lower-case letters represent significant differences ($p \leq 0.05$) between cultivars. Mean values of all the cultivars grown in a particular location followed by different upper-case letters represent significant difference between the two years 2017 and 2018 ($p \leq 0.05$). Mean values of all cultivars grown at all the locations followed by different lower-case letters represent significant differences ($p \leq 0.05$). *** = 0.001; ns = not significant. Can = 'Candonga'; Cle = 'Clery'; Flo = 'Florence'; Fri = 'Frida'; Gar = 'Gariguette'; Son = 'Sonata'.

Table 4. Effect of growing location and year on pelargonidin-3-glucoside content as the mean of two picking dates in fruits of six strawberry cultivars.

-		Pelargonidin-3-glucoside (mg kg^{-1} Fresh Weight)						
		Cultivar						
Location	**Year**	**Can**	**Cle**	**Flo**	**Fri**	**Gar**	**Son**	***Yearly Mean per Location***
Norway	**2017**	184.3 ± 23.9	228.4 ± 23.1	330.3 ± 26.7	328.1 ± 36.7	178.97 ± 39.8	145.4 ± 13.9	*232.6 A*
	2018	108.8 ± 27.7	146.5 ± 41.2	346.0 ± 35.6	340.1 ± 39.2	133.67 ± 16.2	95.4 ± 48.9	*195.1 A*
	mean	*146.6 ab*	*187.4 b*	*338.1 c*	*334.1 c*	*156.3 ab*	*120.4 a*	
Poland	**2017**	209.9 ± 48.0	246.2 ± 24.0	306.0 ± 85	306.5 ± 37.4	160.4 ± 24.1	154.3 ± 17.2	*230.6 A*
	2018	162.2 ± 15.2	177.9 ± 18.5	333.4 ± 54.6	253.2 ± 50.8	131.6 ± 25.5	146.7 ± 21.3	*200.8 A*
	mean	*186.0 ab*	*212.1 b*	*319.7 c*	*279.9 c*	*146.0 a*	*150.5 a*	
Germany	**2017**	164.5 ± 17.6	216.4 ± 35.3	382.3 ± 49.8	292.4 ± 12.7	149.8 ± 24.3	140.7 ± 26.6	*224.3 A*
	2018	189.5 ± 21.6	233.2 ± 15.7	420.8 ± 16.8	268.1 ± 18.8	183.8 ± 9.0	158.4 ± 37.6	*242.1 A*
	mean	*177.0 a*	*224.8 b*	*401.5 d*	*280.3 c*	*166.8 a*	*149.5 a*	
Italy	**2017**	-	288.5 ± 22.4	-	306.2 ± 41.7	222.2 ± 19.9	160.1 ± 15.8	*244.3 B*
	2018	-	200.8 ± 33.4	-	221.0 ± 39.5	102.1 ± 17.5	91.4 ± 24.0	*153.8 A*
	mean	-	*244.7 b*	-	*263.8 b*	*162.1 a*	*125.8 a*	
France	**2017**	279.5 ± 21.6	171.1 ± 46.4	395.0 ± 50.6	321.8 ± 33.1	176.0 ± 27.8	159.9 ± 36.1	*250.5 B*
	2018	271.0 ± 39.5	274.0 ± 53.1	345.7 ± 55.5	225.2 ± 42.7	125.5 ± 31.6	140.7 ± 33.6	*230.4 A*
	mean	*275.2 b*	*222.6 b*	*370.3 c*	*273.5 b*	*150.8 a*	*150.3 a*	
Cultivar mean over all locations		*196.2 b*	*218.3 b*	*357.4 d*	*286.3 c*	*156.4 a*	*139.3 a*	
Significance		*Cultivar* ***	*Location* *ns*	*Year* ***	*Cultivar x location* *ns*	*Cultivar x year* *ns*	*Location x Year* *	

Data are expressed as means ± SD (standard deviation) of two sampling dates per year. Before performing the statistical analysis, the homogeneity of the data were ensured using Bartlett's test. Mean values (n = 3) of different cultivars grown at a particular location followed by lower-case letters represent significant differences ($p \leq 0.05$) between cultivars. Mean values of all the cultivars grown in a particular location followed by different upper-case letters represent significant difference between the two years 2017 and 2018 ($p \leq 0.05$). Mean values of all cultivars grown at all the locations followed by different lower-case letters represent significant differences ($p \leq 0.05$). * = 0.05; *** = 0.001; ns = not significant. Can = 'Candonga'; Cle = 'Clery'; Flo = 'Florence'; Fri = 'Frida'; Gar = 'Gariguette'; Son = 'Sonata'.

Table 5. Effect of growing location and year on pelargonidin-3-rutinoside content as the mean of two picking dates in fruits of six strawberry cultivars.

		Pelargonidin-3-rutinoside (mg kg^{-1} Fresh Weight)						
		Cultivar						
Location	**Year**	**Can**	**Cle**	**Flo**	**Fri**	**Gar**	**Son**	*Yearly Mean per Location*
Norway	2017	18.2 ± 2.5	5.3 ± 1.1	1.4 ± 0.2	27.0 ± 3.7	0.9 ± 0.5	7.3 ± 1.1	*10.0 A*
	2018	11.9 ± 3.4	2.5 ± 1.5	1.4 ± 0.2	23.3 ± 11.9	0.5 ± 0.1	19.2 ± 3.2	*9.8 A*
	mean	*15.1 cd*	*3.9 abc*	*1.4 ab*	*25.2 d*	*0.7 a*	*13.2 bcd*	
Poland	2017	26.2 ± 4.4	7.1 ± 0.4	2.1 ± 1.7	25.2 ± 3.2	0.9 ± 0.3	5.9 ± 0.9	*11.2 A*
	2018	28.3 ± 5.7	9.1 ± 1.1	3.7 ± 0.3	25.9 ± 4.8	3.0 ± 0.2	8.5 ± 0.6	*13.1 A*
	mean	*27.3 c*	*8.1 b*	*2.9 a*	*25.5 c*	*2.0 a*	*7.2 b*	
Germany	2017	21.0 ± 3.6	6.7 ± 0.7	1.4 ± 0.3	25.4 ± 1.8	0.9 ± 0.5	6.4 ± 1.9	*10.3 A*
	2018	21.2 ± 9.6	9.6 ± 6.8	6.2 ± 1.5	20.9 ± 3.6	2.3 ± 0.7	6.8 ± 2.5	*11.2 A*
	mean	*21.1 c*	*8.2 b*	*3.8 ab*	*23.1 c*	*1.6 a*	*6.6 b*	
Italy	2017	-	8.1 ± 0.9	-	24.0 ± 3.8	2.0 ± 0.2	5.5 ± 0.5	*9.9 B*
	2018	-	7.1 ± 1.2	-	7.1 ± 2.5	1.3 ± 0.1	2.6 ± 1.2	*4.5 A*
	mean	-	*7.6 b*	-	*15.6 c*	*1.6 a*	*4.1 ab*	
France	2017	23.6 ± 3.2	4.5 ± 1.6	2.0 ± 0.1	20.4 ± 3.7	1.2 ± 0.4	4.8 ± 1.8	*9.4 A*
	2018	24.3 ± 3.1	7.2 ± 1.3	1.8 ± 0.4	14.1 ± 3.8	1.8 ± 0.9	3.6 ± 1.5	*8.8 A*
	mean	*23.9 d*	*5.8 b*	*1.9 a*	*17.2 c*	*1.5 a*	*4.2 ab*	
Cultivar mean over all locations		*21.8 c*	*6.7 b*	*2.5 a*	*21.3 c*	*1.5 a*	*7.0 b*	
Significance		*Cultivar*	*Location*	*Year*	*Cultivar x location*	*Cultivar x year*	*Location x year*	
		***	*ns*	*ns*	**	*	*ns*	

Data are expressed as means ± SD (standard deviation) of two sampling dates per year. Before performing the statistical analysis, the homogeneity of the data were ensured using Bartlett's test. Mean values (n = 3) of different cultivars grown at a particular location followed by lower-case letters represent significant differences ($p \leq 0.05$) between cultivars. Mean values of all the cultivars grown in a particular location followed by different upper-case letters represent significant difference between the two years 2017 and 2018 ($p \leq 0.05$). Mean values of all cultivars grown at all the locations followed by different lower-case letters represent significant differences ($p \leq 0.05$). * = 0.05; ** = 0.01; *** = 0.001; ns = not significant. Can = 'Candonga'; Cle = 'Clery'; Flo = 'Florence'; Fri = 'Frida'; Gar = 'Gariguette'; Son = 'Sonata'.

Table 6. Effect of growing location and year on cyanidin-3-(6-O-malonyl)-glucoside content as the mean of two picking dates in fruits of six strawberry cultivars.

		Cyanidin-3-(6-O-malonyl)-glucoside (mg kg^{-1} Fresh Weight)						
		Cultivar						
Location	**Year**	**Can**	**Cle**	**Flo**	**Fri**	**Gar**	**Son**	*Yearly Mean per Location*
Norway	2017	0.1 ± 0.2	0.7 ± 0.2	0.8 ± 0.2	3.3 ± 1.5	1.4 ± 0.6	1.2 ± 0.5	*1.3 A*
	2018	0.2 ± 0.1	1.5 ± 0.9	2.6 ± 0.5	2.8 ± 0.4	2.3 ± 0.5	0.8 ± 0.7	*1.7 A*
	mean	*0.2 a*	*1.1 b*	*1.7 b*	*3.0 c*	*1.8 b*	*1.0 ab*	
Poland	2017	0.0 ± 0.0	0.7 ± 0.2	1.6 ± 0.8	3.5 ± 1.7	1.2 ± 0.5	1.4 ± 0.9	*1.4 A*
	2018	2.5 ± 0.0	4.4 ± 0.4	11.0 ± 1.6	8.6 ± 1.2	5.6 ± 1.2	5.6 ± 1.0	*6.3 B*
	mean	*1.2 a*	*2.5 a*	*6.3 b*	*6.1 b*	*3.4 ab*	*3.5 ab*	
Germany	2017	0.0 ± 0.0	0.7 ± 0.2	2.3 ± 0.6	3.8 ± 0.3	1.2 ± 1.1	1.3 ± 0.7	*1.6 A*
	2018	0.0 ± 0.0	0.8 ± 0.4	3.5 ± 0.7	4.1 ± 0.9	1.7 ± 0.2	1.6 ± 0.8	*2.0 A*
	mean	*0.0 a*	*0.8 b*	*2.9 c*	*4.0 d*	*1.5 b*	*1.5 b*	
Italy	2017	-	0.7 ± 0.1	-	2.6 ± 1.4	0.9 ± 0.4	0.8 ± 0.1	*1.2 B*
	2018	-	0.7 ± 0.2	-	0.7 ± 0.4	0.7 ± 0.3	0.7 ± 0.4	*0.7 A*
	mean	-	*0.7 a*	-	*1.7 b*	*0.8 a*	*0.8 a*	

Table 6. *Cont.*

Location	Year	Can	Cle	Flo	Fri	Gar	Son	*Yearly Mean per Location*
		Cyanidin-3-(6-O-malonyl)-glucoside (mg kg^{-1} Fresh Weight)						
		Cultivar						
France	**2017**	0.1 ± 0.2	0.2 ± 0.1	1.6 ± 0.9	3.1 ± 0.3	0.5 ± 0.2	0.7 ± 0.2	*1.0 A*
	2018	0.3 ± 0.1	0.6 ± 0.2	3.3 ± 1.4	3.1 ± 0.6	1.7 ± 0.4	1.0 ± 0.4	*1.7 B*
	mean	*0.2 a*	*0.4 ab*	2.5 *c*	3.1 *c*	1.1 *b*	0.8 *ab*	
Cultivar mean over all locations		*0.4 a*	*1.1 ab*	*3.4 c*	*3.6 c*	*1.7 b*	*1.5 b*	
Significance		*Cultivar* ***	*Location* ***	*Year* ***	*Cultivar x location ns*	*Cultivar x year* *	*Location x year* ***	

Data are expressed as means ± SD (standard deviation) of two sampling dates per year. Before performing the statistical analysis, the homogeneity of the data were insured using Bartlett's test. Mean values (n =3) of different cultivars grown at a particular location followed by lower-case letters represent significant differences ($p \leq 0.05$) between cultivars. Mean values of all the cultivars grown in a particular location followed by different upper-case letters represent significant difference between the two years 2017 and 2018 ($p \leq 0.05$). Mean values of all cultivars grown at all the locations followed by different lower-case letters represent significant differences ($p \leq 0.05$). * = 0.05; *** = 0.001; ns = not significant. Can = 'Candonga'; Cle = 'Clery'; Flo = 'Florence'; Fri = 'Frida'; Gar = 'Gariguette'; Son = 'Sonata'.

Table 7. Effect of growing location and year on pelargonidin-3-(6-O-malonyl)-glucoside content as the mean of two picking dates in fruits of six strawberry cultivars.

location	Year	Can	Cle	Flo	Fri	Gar	Son	*Yearly Mean per Location*
		Pelargonidin-3-(6-O-malonyl)-glucoside (mg kg^{-1} Fresh Weight)						
		Cultivar						
Norway	**2017**	0.4 ± 0.6	46.7 ± 7.7	32.6 ± 5.9	80.9 ± 10.3	78.7 ± 19.1	32.0 ± 5.3	*45.2 A*
	2018	0.2 ± 0.1	37.3 ± 8.6	41.3 ± 7.9	72.2 ± 24.5	58.0 ± 9.3	24.2 ± 5.5	*38.9 A*
	mean	*0.3 a*	*42.0 b*	*36.9 b*	*76.5 c*	*68.3 c*	*28.1 b*	
Poland	**2017**	0.2 ± 0.4	52.9 ± 10.4	36.6 ± 8.0	87.8 ± 17.9	69.2 ± 11.8	42.4 ± 7.5	*48.2 A*
	2018	2.8 ± 0.3	49.4 ± 3	52.5 ± 9.9	82.4 ± 16.1	68.1 ± 12.7	48.4 ± 8.2	*50.6 A*
	mean	*1.5 a*	*51.2 b*	*44.5 b*	*85.1 d*	*68.6 c*	*45.4 b*	
Germany	**2017**	0.4 ± 0.3	59.2 ± 12.8	33.9 ± 16.7	94.8 ± 10.5	71.1 ± 18.8	46.7 ± 14.4	*51.0 A*
	2018	0.0 ± 0.0	59.2 ± 6.4	51.1 ± 8.5	75.9 ± 12.8	81.6 ± 10.6	44.9 ± 11.8	*52.1 A*
	mean	*0.2 a*	*59.2 c*	*42.5 b*	*85.4 d*	*76.4 d*	*45.8 b*	
Italy	**2017**	-	64.1 ± 11.0	-	87.8 ± 13.0	98.8 ± 7.7	40.0 ± 5.9	*72.6 B*
	2018	-	66.3 ± 13	-	71.5 ± 13.3	61.3 ± 8.8	27.3 ± 7.4	*56.6 A*
	mean	-	*65.2 b*	-	*79.6 b*	*80.1 b*	*33.6 a*	
France	**2017**	1.1 ± 1.5	41.8 ± 16.0	37.7 ± 29.4	88.8 ± 14.3	82.1 ± 13.2	44.2 ± 11.8	*49.3 A*
	2018	0.7 ± 0.3	82.0 ± 14.8	53.1 ± 10.1	90.9 ± 17.8	89.4 ± 24.4	50.3 ± 15.6	*61.1 A*
	mean	0.9 a	61.9 b	45.4 b	89.9 c	85.8 c	47.2 b	
Cultivar mean over all locations		*0.7 a*	*55.9 c*	*42.3 b*	*83.3 d*	*75.8 d*	*40.2 b*	
significance		*Cultivar* ***	*Location* ***	*Year ns*	*Cultivar x location ns*	*Cultivar x year ns*	*Location x Year* *	

Data are expressed as means ± SD (standard deviation) of two sampling dates per year. Before performing the statistical analysis, the homogeneity of the data were ensured using Bartlett's test. Mean values (n =3) of different cultivars grown at a particular location followed by lower-case letters represent significant differences ($p \leq 0.05$) between cultivars. Mean values of all the cultivars grown in a particular location followed by different upper-case letters represent significant difference between the two years 2017 and 2018 ($p \leq 0.05$). Mean values of all cultivars grown at all the locations followed by different lower-case letters represent significant differences ($p \leq 0.05$). * = 0.05; *** = 0.001; ns = not significant. Can = 'Candonga'; Cle = 'Clery'; Flo = 'Florence'; Fri = 'Frida'; Gar = 'Gariguette'; Son = 'Sonata'.

The genotype influenced significantly ($p \leq 0.05$) the abundance of total and individual anthocyanins in the fruits (Tables 2–7). When considering the cultivar means for all locations, 'Florence' and 'Frida' showed the highest total anthocyanin content (426.1 and 408.0 mg kg^{-1} fresh weight (FW), respectively) whereas 'Sonata' had the lowest (193.2 mg kg^{-1} FW) (Table 2). As expected, total anthocyanin content was strongly related to the predominant anthocyanin pel-3-glc (Table 4), resulting in the highest content for 'Florence' (357.4 mg kg^{-1} FW) and, however, being significantly different from 'Florence', for 'Frida' (286.3 mg kg^{-1} FW) and the lowest for 'Sonata' (139.3 mg kg^{-1} FW) and also for 'Gariguette' (156.4 mg kg^{-1} FW). Thereby, pel-3-glc contribution to the total anthocyanin content in the cultivars was in the range of 66–84%. 'Florence' (20.5 mg kg^{-1} FW) also had, on average, a higher content of cya-3-glc compared to the other cultivars (Table 3). Pel-3-rut was highest in 'Candonga' and 'Frida' (21.8 and 21.3 mg kg^{-1} FW) (Table 5), whereas cya-3-mal-glc was enriched in 'Florence' and 'Frida' (20.5 and 13.5 mg kg^{-1} FW) (Table 6). The level of pel-3-mal-glc, the second abundant anthocyanin in strawberry, was again highest in 'Frida' (83.3 mg kg^{-1} FW) and in 'Gariguette' (75.8 mg kg^{-1} FW) (Table 7). The contribution of pel-3-mal-glc varied widely in a range of 9.9–31.7% of the total anthocyanin. In contrast to the other investigated cultivars, 'Candonga' contained only cya-3-glc, pel-3-glc, and pel-3-rut in large quantities, while cya-3-mal-glc and pel-3-mal-glc were only found in very small amounts in some locations and years (on average <0.4 and 0.7 mg kg^{-1} FW). Each cultivar had an individual anthocyanin profile that will be discussed later (see Section 3.3). In general, the values for total and individual anthocyanin are in similar ranges as previously reported for these cultivars [11,12,14,21,30–32].

3.3. Anthocyanins are Affected by Location

To better characterize the influence of location and thus mainly latitude and climate, as well as yearly weather parameters on the anthocyanin profile of six strawberry cultivars, a cultivar-specific PCA was conducted. The score and loading plots comprised by the first two principal components (PC1 and PC2) explained ~70% of the total variance of the data set for each of the four cultivars ('Clery', 'Frida', 'Gariguette' and 'Sonata') grown in five locations (Figures 2 and 3), and of the two cultivars ('Candonga' and ''Florence') grown in four locations (Figure 4).

In addition, the PCA models also captured the effect of the two years and the two picking dates within each year. PC2 described 15.7–25.3% of the data variation and was mainly responsible for the separation of each cultivar by location, and thus by latitude and climatic factors, as well as local pre-harvest weather conditions on the anthocyanin synthesis and accumulation.

In general, the loading plots (Figures 2–4) of each cultivar indicates that pel-3-gluc are closely related to total anthocyanin, indicating a similar reaction on the different locations and their environmental conditions. Likewise, cya-3-glc and cya-3-mal-glc were grouped closely together. Thus, they were similar, but in an opposite way as pel-3-glc and total anthocyanins, influenced by locations and yearly weather variations. An exception was observed for 'Frida' where the total anthocyanin and pel-3-glc were clustered together with both cya-derivates, and hence, underlying the same location and environmental effects. Even though all pel-derivates have the same synthesis pathway, pel-3-rut and pel-3-mal-glc showed a cultivar and location-dependent distribution pattern. In the case of 'Clery', 'Gariguette' and 'Candonga', pel-3-rut was clustered more or less separately between both cya-derivates and total anthocyanin and pel-3-glc, while in 'Sonata', it was more related to the cya-derivates, in 'Frida', to cya-derivates, pel-3-glc, and total anthocyanin, while in 'Florence', pel-3-rut was linked to total anthocyanin and the other two pel-derivates. Pel-3-mal-glc was more or less related to pel-3-glc and total anthocyanin in the case of 'Clery', 'Gariguette' and 'Sonata'. In contrast, pel-3-mal-glc was clustered separately for 'Frida' and together with all other pel-derivates and total anthocyanins in the case of 'Florence'.

The score plots (Figures 2–4) indicated a similar clustering of location for each of the cultivars, showing a typical latitudinal division by the North-South axis. However, modification also occurred by cultivar, yearly variations and weather conditions during the harvest period.

The high amount of cya-3-glc and cya-3-mal-glc in 'Clery' samples from Poland and Norway (Tables 3 and 6), and the enriched levels of total anthocyanins, pel-3-glc, and pel-3-mal-glc and in fruits grown in France (Tables 2, 4 and 7), mainly contributed to the clustering of the different locations (Figure 2). The pel-3-rut content was high in 'Clery' fruits from Germany (Table 5), however, this was less important for the separation by location. Interesting to note is the effect of yearly variations combined with harvest period on the synthesis and accumulation of total anthocyanins and their individual compounds at the different locations. For example, 17A samples from France are overlapping with those from Italy in 2018. Another example is the clear separation of the Polish 18B samples from their other ones.

The loading plot for 'Frida' (Figure 2) separated the locations more clearly than for 'Clery'. However, the effects of yearly variations and harvest period are less pronounced. The main difference between locations were due to the high abundance of pel-3-glc in samples from Norway in both years (Table 4), and total anthocyanin both in Norway and Poland (Table 2). In addition, clustering was due to low levels of cya-3-glc in fruits from Italy, especially in 2017 (Table 3), and low values of pel-3-rut (Table 5) in both Italy and France in 2018. Moreover, pel-3-mal-glc (Table 7) was found to be relatively less abundant in samples from Norway.

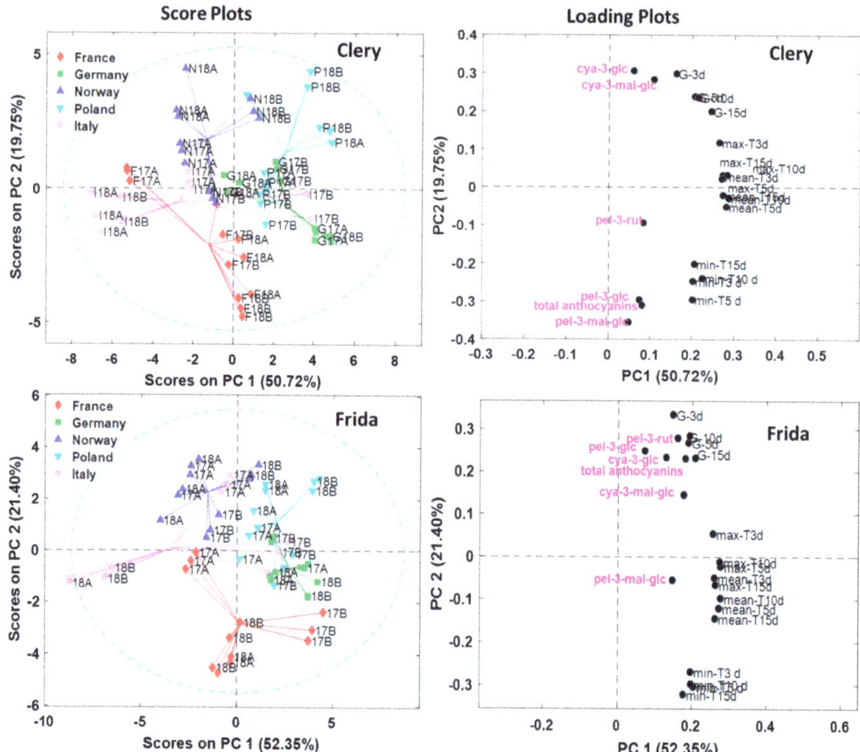

Figure 2. Score and loading plots of PCA for 'Clery', and 'Frida' grown at five locations (F: France; G: Germany; N: Norway; P: Poland and I: Italy) characterized by total and individual anthocyanin and climatic data. The number and capital letters in the different panel are referring to the year (2017 and 2018, labeled as 17 and 18) and A or B to the harvest date, being one week apart.

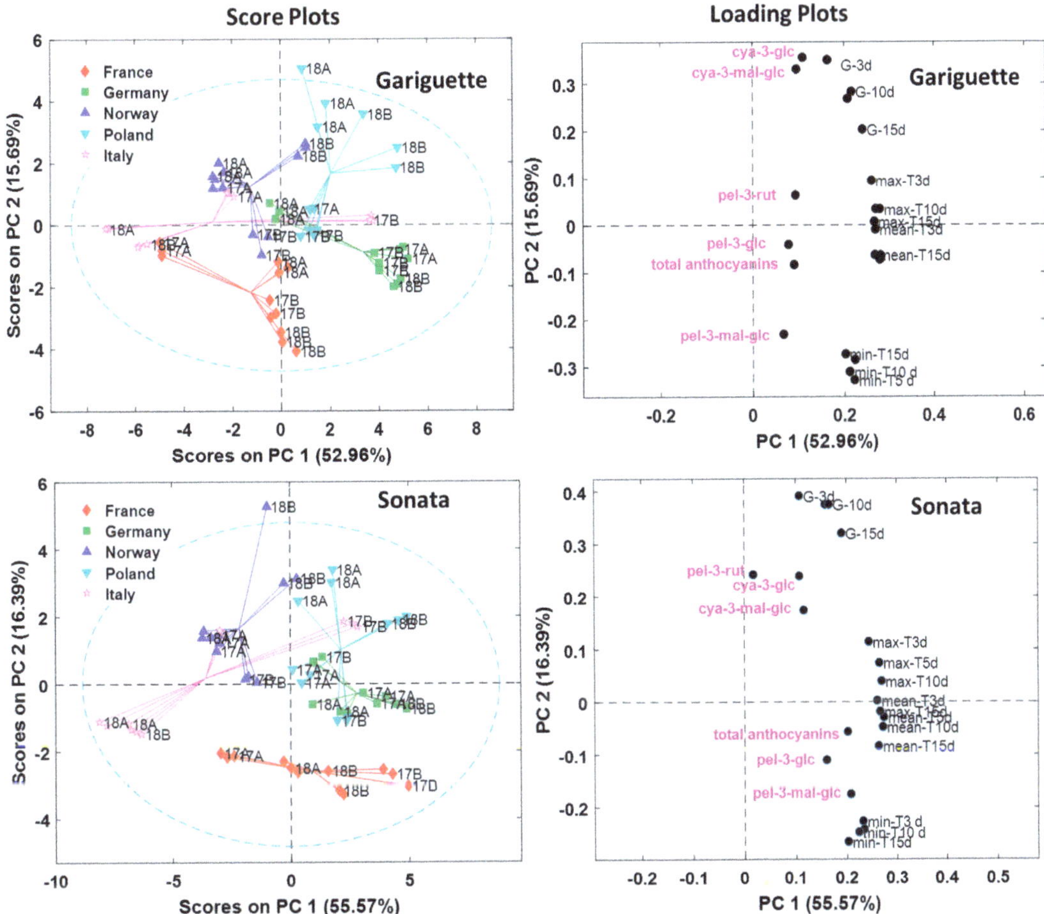

Figure 3. Score and loading plots of PCA for 'Gariguette', and 'Sonata' grown at five locations (F: France; G: Germany; N: Norway; P: Poland and I: Italy) characterized by total and individual anthocyanin and climatic data. The number and capital letters in the different panel are referring to the year (2017 and 2018, labeled as 17 and 18) and A or B to the harvest date, being one week apart.

Although the mean levels of total and individual anthocyanins in fruits of 'Gariguette' did not vary much between locations, as was the case for the other cultivars, the different locations were separated as well (see score plot in Figure 3). This was mainly due to the high level of cya-derivates in Poland in 2018 (Tables 3 and 6), and high amounts of pel-3-mal-glc in samples from France in both years, from Norway and Italy in 2017 and from Germany in 2018 (Table 7). Interestingly, total anthocyanin (Table 2) and its main component pel-3-glc (Table 4) did not contribute much to the separation of the locations because of contrasting year-on-year effects mainly in Norway, and even more pronounced, in Italy. In general, the anthocyanin profile of 'Gariguette' seemed to be less affected by the growing locations.

In the case of 'Sonata', total and individual anthocyanins did not vary much between the locations (Tables 2–7). Samples from France were clearly separated from the other locations (score plot in Figure 3), properly due to their relative low level of cya-derivates (Tables 3 and 6). Interestingly, Italian samples contained similar levels of these anthocyanins, but also less pel-3-glc (Table 4), and thus total anthocyanin (Table 2) in 2018, and this distinguished these samples from those of France and partly from the other locations. Moreover, in 2018, Norwegian samples were enriched in pel-3-rut (Table 5), but contained, at the same time, low levels of pel-3-glc (Table 4), and total anthocyanin (Table 2) like samples from Italy. Both variables were of the factors responsible for the clustering of the Norwegian samples.

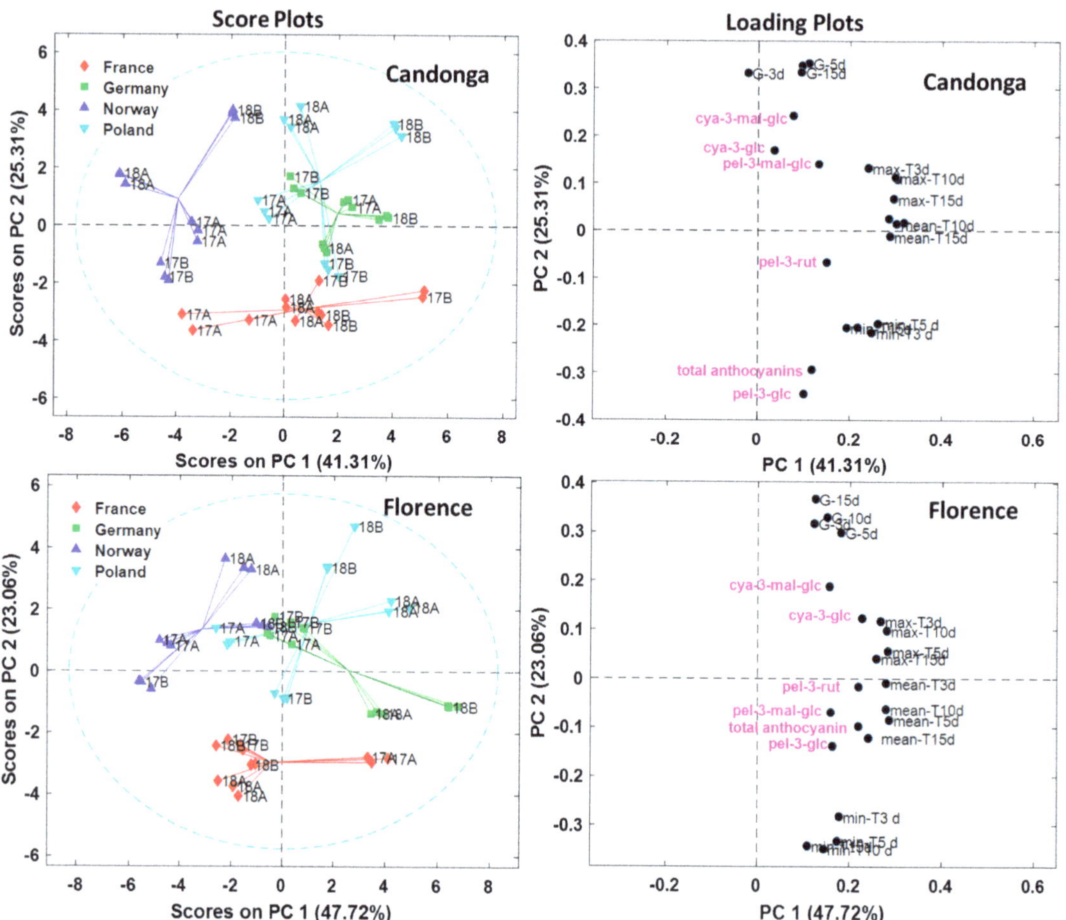

Figure 4. Score and loading plots of PCA for 'Candonga', and 'Florence' grown at four locations (F: France; G: Germany; N: Norway and P: Poland) characterized by total and individual anthocyanin and climatic data. The number and capital letters in the different panel are referring to the year (2017 and 2018, labeled as 17 and 18) and A or B to the harvest date, being one week apart.

Total anthocyanin and pel-3-glc were abundant in fruits of 'Candonga' grown in France but less enriched in those grown in Norway in 2018 (Tables 2 and 4). Thus, both compounds contribute to the separation of France from Norway, and both of these from the two other locations (Figure 4). Moreover, pel-3-rut (Table 5) was low in samples grown in Norway whereas samples grown in Poland were enriched in cya-3-glc (Table 3) and pel-3-mal-glc (Table 7) in 2018, and thus showing clear seasonal effects.

Fruits of 'Florence' grown in France were clearly separated from those grown in Norway (Figure 4). Moreover, PCA segregated both these locations from Germany and Poland. In both seasons, the level of pel-3-glc and anthocyanins were high in the French samples, whereas seasonal effects only enriched the levels of total anthocyanins in Poland and Germany, of pel-3-glc in Germany and of pel-3-mal-glc in all locations in 2018 (Tables 2, 4 and 7). In addition, fruits grown in Poland and Germany were enriched in cya-derivates in 2018, too (Tables 3 and 6).

In general, the PCA separated samples of all cultivars from France from those from the other locations. One reason might be the low values of global radiation for France in both years (Figure 1). However, more reasonable is the fact, that here, as in Italy, strawberries were cultivated in an open-sided tunnel covered with a standard plastic film that is well known to be non-transparent for UV radiation. Among others, flavonoid syntheses in plants is strongly induced by light and UV-B wavelength (280–315 nm). They are effective scavengers of reactive oxygen species (ROS) and absorb selectively UV radiation [17]. One of these flavonoids are anthocyanins being synthesized in higher amounts by excess UV-light. Previous studies reported a retarded coloring of the ripening fruit resulting in a decreased level of total anthocyanin in strawberries grown under UV opaque film compared to UV transparent film [26]. Moreover, no effect of UV radiation on total anthocyanin were observed for strawberry, raspberry and blueberry when grown under films varying from UV blocking to highly transparent [24]. In a study by Josuttis et al. [23], it was shown that the anthocyanin cya-3-glc decreased in strawberries when grown under a UV- blocking plastic film. Cya-3-glc is a minor anthocyanin in strawberry cultivars but abundant, for example, in red *Lettuce sativa* types [24,33]. Additionally, enhanced levels of the derivate cya-3-galactoside were found in the skin of apple [34,35] when exposed to low night temperature and light including UV wavelength. For peach, a different genetic background-dependent cya-3-glc level was detected in two peach cultivars after postharvest treatments with UVA or UVB light [36]. Accompanying transcriptomic studies identified different cultivar-specific expressed genes related to anthocyanin synthesis. In the current study, a cultivar-dependent reaction to UV exclusion was observed in the way that only fruits of 'Clery' and 'Sonata' showed a decreased cya-3-glc content under the UV-blocking tunnel production in France and Italy compared to those from open-field production in Norway, Poland and Germany.

The fact that the Italian samples of each cultivar were not clustered like the French samples but located between France and the other locations, may be due to the lower temperature during the harvest period (Table 1) in Italy, because of latitude and altitude. Thus, temperature effects probably modified the UV-reducing tunnel effect in samples from Italy.

3.4. Impact of Temperature and Global Radiation on Cultivar-Specific Anthocyanin Profiles

As shown by the PC analysis, the synthesis and accumulation of total and individual anthocyanin were influenced by temperature (T_{mean}, T_{max} and T_{min}) and global radiation and its interactions, being altogether affected by the latitude of the growing location. Environmental factors changing with latitude are mainly photoperiod, quantity and spectral composition of the solar radiation [17], as well as air temperature being indirectly dependent on solar radiation. However, weather conditions may vary between seasons at site and during the harvest period. Consequently, linear regression analyses were assessed to better explain the dependency of the cultivar-specific anthocyanin synthesis and accumulation on T_{mean}, T_{max}, T_{min} and global radiation. Moreover, to evaluate the effective time span of these factors, they were summarized 3, 5, 10 and 15 days prior to harvest. Overall, the percentage of variations (Figure 5), explained by these environmental factors, were rather low with some exceptions (highest R^2 = 0.62), highlighting again, the genotype specific-based anthocyanin syntheses. For example 'Gariguette' and 'Clery', with the exception of cya-3-mal-glc, were not affected by the environmental factors tested. For 'Clery', this result supports previous studies showing also no or little environmental effects on anthocyanin accumulation in fruits from this cultivar, when grown at three locations from North to Central Europe [21], and at different altitude in Switzerland [31]. In contrast, 'Florence',

'Frida' and 'Sonata' reacted most sensitively, but differently, to the pre-harvest temperature and global radiation conditions. For these cultivars, in general, (T_{max}) had a higher impact on the syntheses of the less abundant anthocyanins cya-3-glc, pel-3-rut, cya-3-mal-glc, pel-3-mal-glc and on the total anthocyanin, than the minimum temperature. Thereby, the influence of T_{max} often seemed to be stronger than the related T_{mean} itself (for example, for the cya-derivates in 'Florence' and 'Frida'). Only the synthesis of pel-3-glc, the main anthocyanin of strawberry, was slightly more influenced by T_{min} than by T_{max}. In the current study, global radiation was positive correlated with cyl-3-glc and pel-3-rut only in fruits of 'Frida'.

Noticeable is the contrasting behavior of 'Candonga' shown as the only cultivar with a negative relationship between global radiation and the main anthocyanin pel-3-glc ($R^2 = 0.38 - 0.44$), and thus also to the related total anthocyanin, and between cya-3-mal-glc and T_{min}. 'Candonga' was bred and selected for tunnel production in Spain, which is the common production system for that area. Therefore, it is assumed that protection against UV-radiation was not an important growing factor for this genotype. In a Spanish study performed in tunnels, no correlation was found for 'Candonga' between total anthocyanin and T_{mean}, T_{max} and solar radiation and, in contrast to this study, a positive relationship to T_{min} [14]. In the same study, however, other cultivars reacted also on T_{mean} or T_{max}.

Noteworthy also is the positive relationship between global radiation and the content of cya-derivates in the Scandinavian cultivar 'Frida'. Moreover, 'Frida', together with 'Florence'—bred in the United Kingdom—were both sensitive to T_{max} with high R^2 values for cya-derivates and pel-3-rut, and in the case of 'Frida', also of pel-3-mal-glc. Thus, both cultivars showed a high adaption to their breeding locations to benefit best from T_{max} in areas with, in general, lower temperature, and in case of 'Frida', also from the latitudinal long photoperiod in Norway.

Numerous studies evaluated the effect of temperature on the anthocyanin content of strawberry. In general, the content of total anthocyanin was enhanced with increasing temperature when plants were grown under controlled conditions [9,19,20]). However, the major anthocyanin pel-3-glc was less affected than the minor abundant pel-3-mal-glc [19]. Moreover, increased night temperature induced higher anthocyanin levels [9]. A contrasting result was obtained in a study with Japanese cultivars where high growing temperature (30/15 °C day/night) decreased the anthocyanin content compared to the control (20/15 °C), due to differently expressed genes involved in the anthocyanin synthesis [37].

There are only few studies evaluating latitudinal or altitudinal effects on the anthocyanin content of strawberry. In general, although pre-harvest weather conditions modified the data, fruits grown at higher latitudes had less anthocyanins than the southern ones [13,22,38]. In addition, quantitative changes in the anthocyanin profiles were found [20], giving northern fruits a higher percentage of minor anthocyanins like cya-3-glc and cya-3-mal-glc. These latitudinal depending results confirm our findings. In a study comparing different altitudes (differences ~600 m) with increased temperature but lower radiation in the period 10 days before harvest at the higher altitude, no influence on total anthocyanin and pel-3-glc were found, but cultivar specifically increased levels of cya-3-gluc and minor pel-derivates [31]. In contrast, Guerrero-Chavez et al. [39] found a decrease of total anthocyanin, mainly pel-3-glc and an unnamed pel-derivate, with increasing altitude (~600 m) in fruits of 'Elsanta'. Furthermore, in a recent study, temperature, UV radiation and sunshine duration were found to affect bioactive compounds of strawberry stronger than locations differing nearly by 800 m altitude. However, anthocyanin was the compound class that showed significant differences between locations in one cultivar only [32].

In this experiment, 'Florence' and 'Frida' showed a high adaptation to their breeding place, and in case of 'Candonga', to the tunnel cultivation technique used during selection. High adaption ability is well known for wild species. For example, wild populations of different *Vaccinium* species, grown from South to North Scandinavia, showed significant variations in the anthocyanin profile and total anthocyanin content giving northern populations a higher anthocyanin content in their berries [40–42]. It was explained by the long photoperiod at northern sites and its intense radiation of UV, visible and far-red wavelength [42], but also by its lower mean temperature or by the interaction of both. The adaption was under strong genetic control and remained when cloned plants of the Nordic populations were grown in South Scandinavia [42].

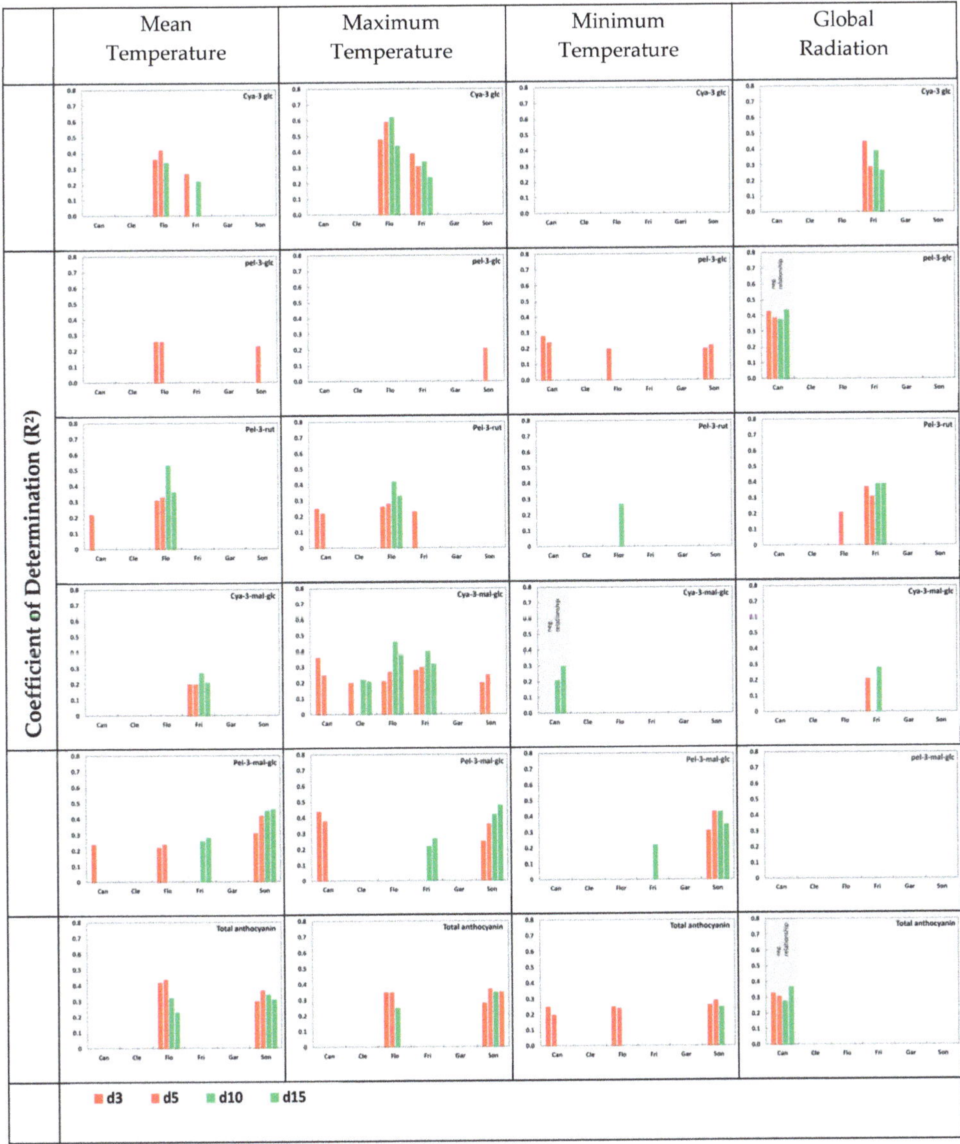

Figure 5. Coefficient of determinations (R^2) between cya-3-glc, pel-3-glc, pel-3-rut, cya-3-mal-glc, ple-3-mal-glc and total anthocyanin (shown top to bottom) and the environmental factors' mean temperature (T_{mean}), maximum temperature (T_{max}), minimum temperature (T_{min}) and global radiation accumulated for 3, 5, 10 and 15 days (3d, 5d, 10d and 15d) before sampling of the berries to be analyzed for each cultivar 'Candonga' (Can), 'Clery' (Cle), 'Florence' (Flo), 'Frida' (Fri), 'Gariguette' (Gar) and 'Sonata' (Son), shown left to right. Cultivars with no data (blank) indicate the observed correlation between different anthocyanin components, and mean, maximum, minimum temperatures and global radiation were statistically insignificant.

The pigmentation of strawberry occurs relatively rapid at the end of the fruit development. Metabolomic studies indicate the first appearance of anthocyanins with rapidly increasing amounts until ripeness around 5–10 days after the fruit's white stage [38,43–45]. These previous studies have focused on fruit developmental stages and total anthocyanins only [38,44], or cya-hexose, pel-hexose and pel-rutinose [45]. The current study considered different time intervals up to 15 days prior to harvest and took into account not only the total anthocyanin but all individual anthocyanins evaluated. As expected, pel-3-glc were cultivar-specific, and presented only the last five days before harvest. However, in our study, it was surprising to find environmental effects occurring at an earlier fruit developmental stage affecting the different anthocyanins. For example, in 'Candonga' fruits, global radiation inhibited its synthesis up to 15 days pre-harvest. Furthermore, 'Candonga' was the exclusive cultivar where pel-3-rut, cya-3-mal-glc and pel-3-glc were not found for T_{max} at d10 and d15 before harvest whereas they were partly present in 'Florence', 'Frida' and 'Sonata'. The synthesis of the other individual anthocyanins was affected by temperature and less by global radiation when taking the whole period into account. It is assumed that at this early stage, photosynthesis was enhanced by favorable temperature and radiation conditions and, in that way, precursors of the anthocyanins like carbon skeletons were accumulated. When evaluating primary and secondary metabolites during strawberry development [45], a decrease of diverse sugars was found over time, while anthocyanins increased at the late fruit developmental stages.

Herein, the impact of location on anthocyanin profiles of certain cultivars was systematically studied. The obtained results showed that the present work could serve as a useful starting point towards evaluating the latitudinal effects on plant performance and internal fruit quality. However, some further specific studies are required to validate this. In addition, certain optimizations are still required for accounting the seasonal variations in the global radiation and temperatures. These issues will be addressed in detail in our near future research work.

4. Conclusions

Our study indicated that the anthocyanin content of strawberry cultivars are, beside the well-known genetic origin, partly affected at site by the local environmental factors, namely temperature, global radiation and cultivation technique. While 'Clery' and 'Gariguette' displayed a very high stability in their anthocyanin content regardless of the growing location, the other cultivars partly reacted on the local conditions of the growing sites. Thus, a high cultivar x environment interaction was observed for the evaluated cultivars. T_{max}, T_{min} and global radiation, relatively less affected pel-3-glc. Cya-3-glc, cya-3-mal-glc, pel-3-rut, pel-3-mal-glc were found to be highly sensitive to T_{max}. In addition, global radiation strongly increased cya-3-mal-glc and pel-3-rut in case of 'Frida', while in case of 'Candonga', the abundance of pel-3-glc decreased with global radiation. The anthocyanin profiles of 'Gariguette' and 'Clery' were unaffected by environmental conditions.

The minor strawberry anthocyanins cya-3-glc, pel-3-rut, cya-3-mal-glc and pel-3-mal-glc seemed to be cultivar-specific more sensitive to such environmental variations than the abundant pel-3-glc. Thereby, the minor anthocyanins might be useful to breed cultivars which are able to accumulate anthocyanins even under sub-optimal conditions, for instance, like 'Frida' which produced high content of anthocyanins in Norway being sensitive to T_{max} and global radiation. In contrast, cultivars with high stability in their anthocyanin content like 'Clery' and 'Gariguette' may be valuable parents as well in breeding programs focusing on the challenge of increased temperature due to climate change and on weather instability between and within harvest periods. Thus, a better understanding of the cultivar x environmental interactions will be necessary. However, the cultivar-specific relationship between fruit anthocyanin content and the evaluated environmental factors in our study were rather low, indicating that other factors not considered were involved.

Supplementary Materials: The following are available online at https://www.mdpi.com/2076-3417/11/3/1326/s1, Table S1: Sum of mean, maximum and minimum temperature as well as sum of global radiation, 3, 5, 10, and 15 days prior to sampling of fruits for analyses of 6 cultivars from North to South of Europe, Figure S1: PCA score plots for (a) the 6 cultivars (Clery (Cle), Candonga (Can), Frida (Fri), Florence (Flo), Gariguette (Gar) and Sonata (Son)) in 4 locations and (b) the 4 cultivars (Clery (Cle), Frida (Fri), Gariguette (Gar) and Sonata (Son)) in 5 locations clearly indicate that the location has major impact than the genetic variation. The '17' and '18' indicate years 2017 and 2018, respectively. The 'A' and 'B' indicate the two picking dates. In order to have better understanding, it is essential to analyze each cultivar separately.

Author Contributions: Conceptualization, E.K.; data curation, K.K.; investigation, E.K., F.W., P.C., A.M., D.M., A.P., G.S., K.C. and A.S.; writing—original draft, E.K., F.W. and K.K.; writing—review and editing, P.C., A.M., D.M., A.P., G.S., K.C. and A.S. All authors have read and agreed to the published version of the manuscript.

Funding: This research was funded by European Union's Horizon 2020 research and innovation programme, grant number 679303.

Institutional Review Board Statement: Not applicable.

Informed Consent Statement: Not applicable.

Data Availability Statement: Not applicable.

Acknowledgments: The authors would like to thank all farm workers and laboratory stuff for skillful field management, sample collection and analyses, respectively.

Conflicts of Interest: The authors declare no conflict of interest. The funders had no role in the design of the study; in the collection, analyses, or interpretation of data; in the writing of the manuscript, or in the decision to publish the results.

References

1. Nile, S.H.; Park, S.W. Edible berries: Bioactive components and their effect on human health. *Nutrition* **2014**, *30*, 134–144. [CrossRef]
2. Giampieri, F.; Tulipani, S.; Alvarez-Suarez, J.M.; Quiles, J.L., Mezzetti, B.; Battino, M. The strawberry: Composition, nutritional quality, and impact on human health. *Nutrition* **2012**, *28*, 9–19. [CrossRef]
3. Giampieri, F.; Forbes-Hernandez, T.Y.; Gasparrini, M.; Alvarez-Suarez, J.M.; Afrin, S.; Bompadre, S.; Quiles, J.L.; Mezzetti, B.; Battino, M. Strawberry as a health promoter: An evidence based review. *Food Funct.* **2015**, *6*, 1386–1398. [CrossRef]
4. Afrin, S.; Gasparrini, M.; Forbes-Hernandez, T.Y.; Reboredo-Rodriguez, P.; Mezzetti, B.; Varela-López, A.; Giampieri, F.; Battino, M. Promising health benefits of the strawberry: A focus on clinical studies. *J. Agric. Food Chem.* **2016**, *64*, 4435–4449. [CrossRef] [PubMed]
5. Tena, N.; Martín, J.; Asuero, A.G. State of the art of anthocyanins: Antioxidant activity, sources, bioavailability, and therapeutic effect in human health. *Antioxidants* **2020**, *9*, 451. [CrossRef] [PubMed]
6. Giampieri, F.; Alvarez-Suarez, J.M.; Mazzoni, L.; Forbes-Hernandez, T.Y.; Gasparrini, M.; Gonzàlez-Paramàs, A.M.; Santos-Buelga, C.; Quiles, J.L.; Bompadre, S.; Mezzetti, B.; et al. An anthocyanin-rich strawberry extract protects against oxidative stress damage and improves mitochondrial functionality in human dermal fibroblasts exposed to an oxidizing agent. *Food Funct.* **2014**, *5*, 1939–1948. [CrossRef] [PubMed]
7. Kuntz, S.; Rudloff, S.; Asseburg, H.; Borsch, C.M.; Fröhling, B.; Unger, F.; Dold, S.; Spengler, B.; Römpp, A.; Kunz, C. Uptake and bioavailability of anthocyanins and phenolic acids from grape/blueberry juice and smoothie in vitro and in vivo. *Br. J. Nutr.* **2015**, *113*, 1044–1055. [CrossRef] [PubMed]
8. Treutter, D. Significance of flavonoids in plant resistance and enhancement of their biosynthesis. *Plant. Biol.* **2005**, *7*, 581–591. [CrossRef]
9. Wang, S.Y.; Zheng, W.; Galletta, G.J. Cultural system affects fruit quality and antioxidant capacity in strawberries. *J. Agric. Food Chem.* **2002**, *50*, 6534–6542. [CrossRef]
10. Tulipani, S.; Mezzetti, B.; Capocasa, F.; Bompadre, S.; Beekwilder, J.; De Vos, C.H.R.; Capanoglu, E.; Bovy, A.; Battino, M. Antioxidants, phenolic compounds, and nutritional quality of different strawberry genotypes. *J. Agric. Food Chem.* **2008**, *56*, 696–704. [CrossRef]
11. Aaby, K.; Mazur, S.; Nes, A.; Skrede, G. Phenolic compounds in strawberry (*Fragaria* × *ananassa* Duch.) fruits: Composition in 27 cultivars and changes during ripening. *Food Chem.* **2012**, *132*, 86–97. [CrossRef]
12. Ariza, M.; Martínez-Ferri, E.; Domínguez, P.; Medina, J.J.; Miranda, L.; Soria, C. Effects of harvest time on functional compounds and fruit antioxidant capacity in ten strawberry cultivars. *J. Berry Res.* **2015**, *5*, 71–80. [CrossRef]

13. Cocco, C.; Magnani, S.; Maltoni, M.L.; Quacquarelli, I.; Cacchi, M.; Antunes, M.L.C.; D'Antuono, F.; Faedi, W.; Baruzzi, G. Effects of site and genotype on strawberry fruits quality traits and bioactive compounds. *J. Berry Res.* **2015**, *5*, 145–155. [CrossRef]
14. Cervantes, L.; Ariza, M.T.; Miranda, L.; Lozano, D.; Medina, J.J.; Soria, C.; Martínez-Ferri, E. Stability of fruit quality traits of different strawberry varieties under variable environmental conditions. *Agronomy* **2020**, *10*, 1242. [CrossRef]
15. Jaakola, L.; Hohtola, A. Effect of latitude on flavonoid biosynthesis in plants. *Plant Cell Environ.* **2010**, *33*, 1239–1247. [CrossRef] [PubMed]
16. Jaakola, L. New insights into the regulation of anthocyanin biosynthesis in fruits. *Trends Plant. Sci.* **2013**, *18*, 477–483. [CrossRef] [PubMed]
17. Zoratti, L.; Karpinnen, K.; Escobar, A.L.; Häggman, H.; Jaakola, L. Light-controlled flavonoid biosynthesis in fruits. *Front. Plant. Sci.* **2014**, *5*, 1–16. [CrossRef] [PubMed]
18. Wang, S.Y.; Zheng, W. Effect of plant growth temperature on antioxidant capacity in strawberry. *J. Agric. Food Chem.* **2001**, *49*, 4977–4982. [CrossRef]
19. Josuttis, M.; Dietrich, H.; Patz, C.-D.; Krüger, E. Effects of air and soil temperatures on the chemical composition of fruit and agronomic performance in strawberry (*Fragaria* × *ananassa* Duch.). *J. Hortic. Sci. Biotechnol.* **2011**, *86*, 415–421. [CrossRef]
20. Balasooriya, B.L.H.N.; Dassanayake, K.; Ajlouni, S. High temperature effects on strawberry fruit quality and antioxidant contents. *Acta Hortic.* **2020**, *1278*, 225–234. [CrossRef]
21. Josuttis, M.; Carlen, C.; Crespo, P.; Nestby, R.; Dietrich, H.; Krüger, E. A comparison of bioactive compounds of strawberry fruit from Europe affected by genotype and latitude. *J. Berry Res.* **2012**, *2*, 73–95. [CrossRef]
22. Anttonnen, M.J.; Hoppula, K.J.; Nestby, R.; Verheul, M.J.; Karjalainen, R.O. Influence of fertilization, mulch color, early forcing, fruit order, planting date, shading, growing environment and genotype, on the content of selected phenolics in strawberry (*Fragria* × *ananassa* Duch.) fruits. *J. Agric. Food Chem.* **2006**, *54*, 2614–2620. [CrossRef] [PubMed]
23. Josuttis, M.; Dietrich, H.; Treuter, D.; Will, F.; Linnemannstöns, L.; Krüger, E. Solar UVB response of bioactives in strawberry (*Fragaria* × *ananassa* Duch.): A comparison of protected and open-field cultivation. *J. Agric. Food Chem.* **2010**, *58*, 12692–12702. [CrossRef] [PubMed]
24. Ordidge, M.; García-Macías, P.; Battey, N.; Gordon, M.H.; Hadley, P.; John, P.; Lovegrove, J.A.; Vysini, E.; Wagstaffe, A. Phenolic contents of lettuce, strawberry, raspberry, and blueberry crops cultivated under plastic films varying in ultraviolet transparency. *Food Chem.* **2010**, *119*, 1224–1227. [CrossRef]
25. Heide, O.M.; Sønsteby, A. Climate-photothermographs, a tool for ecophysiological assessment of effects of climate warming in crop plants: Examples with three berry crops. *J. Berry Res.* **2020**, 411–418. [CrossRef]
26. Tsormpatsidis, E.; Ordrige, M.; Henbest, R.G.C.; Wagstaffe, A.; Battey, N.H.; Hadley, P. Harvesting fruit of equivalent chronological age and fruit position shows individual effect of UV radiation on aspects of the strawberry ripening process. *Environ. Exp. Botany.* **2011**, *74*, 178–185. [CrossRef]
27. Zhang, Y.; Leiyu, J.; Li, Y.; Chen, Q.; Ye, Y.; Zhang, Y.; Luo, Y.; Sun, B.; Wang, X.; Tang, H. Effect of red and blue light on anthocyanin accumulation and differential gene expression in strawberry (*Fragaria* × *ananassa*). *Molecules* **2018**, *23*, 820. [CrossRef]
28. Zhang, Y.; Hu, W.; Peng, X.; Sun, B.; Wang, X.; Tang, H. Characterization of anthocyanin and proanthocyanidin biosynthesis in two strawberry genotypes during fruit development in response to different light qualities. *J. Photochem. Photobiol. B* **2018**, *186*, 225–231. [CrossRef]
29. Bro, R.; Smilde, A.K. Principal component analysis. *Anal Methods* **2014**, *6*, 2812–2831. [CrossRef]
30. Bueniída, B.; Gil, M.I.; Tudela, J.A.; Gady, A.L.; Medina, J.J.; Soria, C.; López, J.M.; Tomás-Barberán, F. HPLC-MS analysis of proanthocyanidin oligomers and other phenolics in 15 strawberry cultivars. *J. Agric. Food Chem.* **2010**, *58*, 3916–3926. [CrossRef]
31. Crespo, P.; Bordonaba, J.G.; Terry, L.A.; Carlen, C. Characterisation of major taste and health related compounds of four strawberry genotypes grown at different Swiss production sites. *Food Chem.* **2010**, *122*, 16–24. [CrossRef]
32. Palmieri, L.; Masuero, D.; Martinatti, P.; Baratto, G.; Martens, S.; Vrhovsek, U. Genotype-by-environment effect on bioactive compounds in strawberry (*Fragaria* × *ananassa* Duch.). *J. Sci. Food Agric.* **2017**, *97*, 4180–4189. [CrossRef] [PubMed]
33. Behn, H.; Schurr, U.; Ulbrich, A.; Noga, G. Development-dependent UV-B responses in red oak leaf lettuce (*Lactuca sativa* L.): Physiological mechanisms and significance for hardening. *Eur. J. Hortic. Sci.* **2011**, *76*, 33–44.
34. Awad, M.A.; Wagenmakers, P.; de Jager, A. Effects of light on flavonoid and chlorogenic acid levels in skin of 'Jonagold' apples. *Sci. Hortic.* **2001**, *88*, 289–298. [CrossRef]
35. Ubi, B.E.; Honda, C.; Bessho, H.; Kondo, S.; Wada, M.; Kobayashi, S.; Moriguchi, T. Expression analysis of anthocyanin biosynthetic genes in apple skin: Effect of UV-B and temperature. *Plant. Sci.* **2006**, *170*, 571–578. [CrossRef]
36. Zhao, Y.; Dong, W.; Wang, K.; Zhang, B.; Allan, A.C.; Lin-Wang, K.; Chen, K.; Xu, C. Differential sensitivity of fruit pigmentation to ultraviolet light between two peach cultivar. *Front. Plant. Sci.* **2017**, *8*, 1552. [CrossRef]
37. Matsushita, K.; Sakayori, T.; Ikeda, S. The effect of high air temperature on anthocyanin and the expression of its biosynthetic genes in strawberry 'Sachinoka'. *Environ. Control. Biol.* **2016**, *54*, 101–107. [CrossRef]
38. Carbonne, F.; Preuss, A.; de Vos, R.C.H.; D'Amico, E.; Perrotta, G.; Bovy, A.G.; Martens, S.; Rosati, C. Developmental, genetic and environmental factors affect the expression of flavonoid genes, enzymes and metabolites in strawberry fruits. *Plant. Cell Environ.* **2009**, *32*, 1117–1131. [CrossRef]
39. Guerrero-Chavez, G.; Scampicchio, M.; Andreotti, C. Influence of the site altitude on strawberry phenolic composition and quality. *Sci. Hortic.* **2015**, *192*, 21–28. [CrossRef]

40. Lätti, A.K.; Riihinen, K.; Kainulainen, P.S. Analysis of anthocyanin variation in wild populations of bilberry (*Vaccinium myrtillus* L.) in Finland. *J. Agric. Food Chem.* **2008**, *56*, 190–196. [CrossRef]
41. Lätti, A.K.; Jaakola, L.; Riihinen, K.; Kainulainen, P.S. Anthocyanin and flavanol variation in bog bilberries (*Vaccinium uliginosum* L.) in Finland. *J. Agric. Food Chem.* **2010**, *58*, 427–433. [CrossRef] [PubMed]
42. Åkerström, A.; Jaakola, L.; Båång, U.; Jäderlund, A. Effects of latitude-related factors and geographical origin on anthocyanidin concentrations in fruits of *Vaccinium myrtillus* L. (Bilberries). *J. Agric. Food Chem.* **2010**, *58*, 11939–11945. [CrossRef] [PubMed]
43. Ferreyra, R.M.; Viña, S.Z.; Mudridge, A.; Chaves, A.R. Growth and ripening season effects on antioxidant capacity of strawberry cultivar Selva. *Sci. Hortic.* **2007**, *112*, 27–32. [CrossRef]
44. Halbwirth, H.; Puhl, I.; Haas, U.; Jezik, K.; Treutter, D.; Stich, K. Two-phase flavonoid formation in developing strawberry (*Fragaria* × *ananassa* Duch.) fruit. *J. Agric. Food. Chem.* **2006**, *54*, 1479–1485. [CrossRef]
45. Fait, A.; Hanhineva, K.; Beleggia, R.; Dai, N.; Rogachey, I.; Nikiforova, V.J.; Fernie, A.R.; Aharoni, A. Reconfiguration of the achene and receptacle metabolic networks during strawberry fruit development. *Plant. Physiol.* **2008**, *148*, 730–750. [CrossRef]

Article

Comparative Effects of Organic and Conventional Cropping Systems on Trace Elements Contents in Vegetable Brassicaceae: Risk Assessment

Fernando Cámara-Martos [1], Jesús Sevillano-Morales [1], Luis Rubio-Pedraza [2], Jesús Bonilla-Herrera [2] and Antonio de Haro-Bailón [3,*]

1 Departamento de Bromatología y Tecnología de los Alimentos, Campus Universitario de Rabanales, Universidad de Córdoba, Edificio C-1, 14014 Córdoba, Spain; bt2camaf@uco.es (F.C.-M.); oo3semoj@uco.es (J.S.-M.)

2 Departamento de Agricultura, Ayuntamiento de Alcalá la Real, 23680 Alcalá la Real (Jaén), Spain; luisrubio@alcalalareal.es (L.R.-P.); jebohe@gmail.com (J.B.-H.)

3 Departamento de Mejora Genética Vegetal, Instituto de Agricultura Sostenible (IAS–CSIC), Alameda del Obispo s/n, 14004 Córdoba, Spain

* Correspondence: adeharobailon@ias.csic.es; Tel.: +34-957-499235; Fax: +34-957-499252

Abstract: Genotypes selected from 3 plant species (*Brassica rapa, Eruca vesicaria* and *Sinapis alba*) belonging to the *Brassicaceae* family were chosen to compare the concentrations of 9 inorganic elements (Cd, Co, Cr, Cu, Fe, Ni, Mn, Pb and Zn) in these varieties, that were grown under both conventional and organic conditions during two agricultural seasons (2018/2019 and 2019/2020) on two different experimental farms (Farm I and Farm II). We found that, together with agriculture practices, the inorganic element concentrations in Brassicas depended on many other factors, including soil characteristics. However, there were no conclusive results indicating a lower heavy metal content or a higher nutritionally beneficial trace elements content in vegetables grown under organic agriculture. Finally, a probabilistic assessment (@Risk) derived from the consumption of 150–200 g of these vegetables showed that organic Brassicas fulfill in comparison with the conventional ones, similar Dietary Reference Intakes (DRI) percentages for Co, Cr, Cu, Fe, Mn and Zn. Regarding heavy metals (Cd, Ni and Pb), we only found slight differences (mainly in the case of Pb) in the Tolerable Intakes (TI) between both cropping systems.

Keywords: organic farming; conventional farming; trace elements; heavy metals; risk assessment

Citation: Cámara-Martos, F.; Sevillano-Morales, J.; Rubio-Pedraza, L.; Bonilla-Herrera, J.; de Haro-Bailón, A. Comparative Effects of Organic and Conventional Cropping Systems on Trace Elements Contents in Vegetable Brassicaceae: Risk Assessment. *Appl. Sci.* **2021**, *11*, 707. https://doi.org/10.3390/app11020707

Received: 21 December 2020
Accepted: 10 January 2021
Published: 13 January 2021

1. Introduction

Nowadays, the maintenance of a good health status via appropriate dietary habits has become of great social concern, but we also have to bear in mind what has been called the "health trilemma", which tells us that food, health and the environment are closely linked, should also be borne in mind, in order to establish a balance between them for a healthier life on a more sustainable planet.

Currently, the consumption of plant-based foods with nutraceutical properties is one of the crucial factors for the welfare and promotion of health, preventing various pathologies like cancer, cardiovascular and neurodegenerative diseases [1,2]. Vegetable species belonging to the *Brassicaceae* (formerly *Cruciferae*) family are considered as being one of the first cultivated and domesticated plant groups and is appreciated for constituting a good source of minerals and trace elements [3] and for their health-promoting phytochemicals such as glucosinolates [4]. Plant species from this family include nutritionally important human and animal foodstuffs such as broccoli, turnip, cabbage, cauliflower, rapeseed, mustard, rocket, and is one of the ten most economically important plant families in the world [5,6].

In addition, in recent years, developed countries have been showing a greater interest in organic agriculture, with an increase of around 250% in the last 10 years [7]. This type of agriculture is based on the non-use of synthetic fertilizers or pesticides, and, instead, fertilization of the land with composted material, rich in organic matter, derived from the biodegradation of plant and animal sources [8].

Some authors have pointed out that organic foods contain higher concentrations of nutritionally beneficial trace elements and lower concentrations of harmful heavy metals [9]. In fact, vegetables can uptake and retain these inorganic elements from the surrounding environment through their roots and leaves [10]. However, the data existing in the bibliography on this topic are inconclusive, and it is difficult to make a valid comparison between both vegetable groups due to the limited availability of well-controlled or paired studies [11].

Therefore, for all the above reasons, the objectives of this research were (a) to compare the concentrations of nine inorganic elements (Co, Cr, Cu, Cd, Fe, Ni, Mn, Pb and Zn), well known both for their nutritional and toxicologic role in 3 species of *Brassicaceae (Brassica rapa, Eruca vesicaria* and *Sinapis alba)* grown under both conventional and organic conditions during two agricultural seasons (2018/2019 and 2019/2020) on two different experimental farms; (b) to make a probabilistic estimation with computer software (@Risk) of the contributions of the inorganic elements present in these vegetables to the recommended intakes or to the toxicologic limits established for them. This was to find out whether organically-grown *Brassicas* have a greater nutritional value than the conventionally-grown ones.

The novelty of this work lies in the fact that few long-term studies have been made comparing vegetables grown under organic and conventional conditions during two agriculture seasons.

2. Material and Methods

2.1. Plant Material

Genotypes selected from 3 plant species belonging to the *Brassicaceae* family were chosen based on previous studies showing their differences in their glucosinolates profile and trace elements concentration: *Brassica rapa* L. (turnip greens and top greens), *Eruca vesicaria* L. (rocket) and *Sinapis alba* L. (white mustard). Turnip greens are the young leaves harvested in the vegetative growth period, and turnip tops are the fructiferous stems with flower buds and the surrounding leaves that are consumed before opening and while still green. These plant species are well adapted to Mediterranean environmental conditions and have been obtained by the Plant Breeding Group at the Institute for Sustainable Agriculture (IAS-CSIC) after several generations of breeding for seed yield and glucosinolate content.

This material was sown and cultivated during two seasons, 2018–2019 and 2019–2020, in two Farms (I and II) in Southern Spain (see Figure 1).

The Farm I land (37°51′ N, 4°48′ W) is located in Córdoba, next to the Guadalquivir River, in a position of the first terrace (altitude of 106 m), with a deep soil (Typic Xerofluvent) of sandy-loam texture with high pH (around 8), intermediate organic matter content (1.6%), and high carbonate content (17%). The experimental plot size for conventional cultivation of *Brassicaceae* species on Farm I was 25 × 25 m. The climate is typically continental Mediterranean (Csa in Köppen's climate classification), with relatively cold winters, intensely hot dry summers and mean annual precipitations of 650 mm. On this Farm, the three Brassicaceae species were only grown under conventional conditions with herbicides and mineral fertilization being applied. In pre-sowing, an herbicide with triflularin as its active matter was used at a dose of 1.5 L/ha. Moreover, before sowing, a basic dressing with 8-15-15 bottom fertilizer was applied at a rate of 600 kg/ha. A top dressing (cover fertilization) with 300 kg/ha of Ammonium Nitrate was applied after the winter stop at the resumption of vegetative growth.

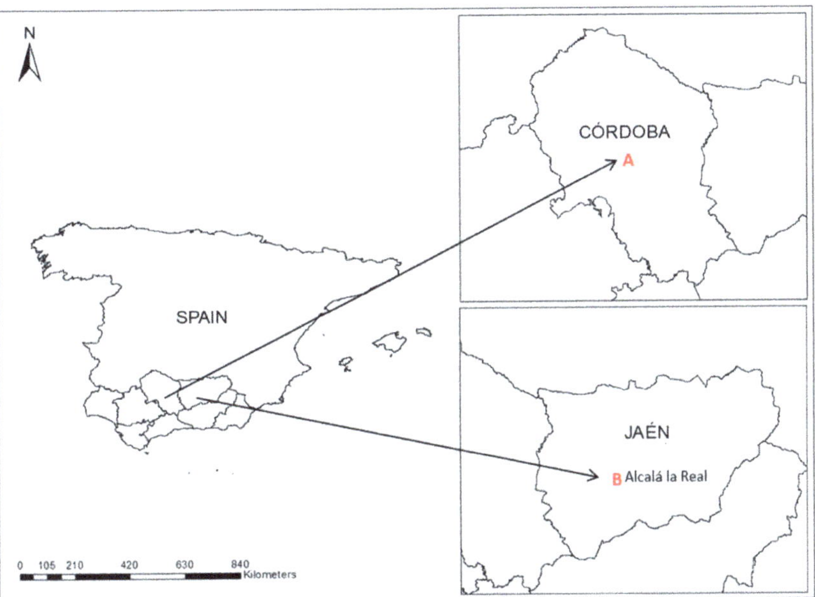

Figure 1. Two Farms in the Southern Spain (A: Farm I. Alameda del Obispo (Córdoba). *Brassicaceae* grown in Conventional Cropping System; B: Farm II. Ribera Alta. Alcalá la Real (Jaén). *Brassicaceae* grown in Conventional and Organic Cropping System. The distance between both Farms is 121 km.

Farm II is located in the municipal district of Alcalá la Real (Jaén) (37°27′ N 3°55′ W, Spain) in the Sub-Baetic zone, next to the Velillos River (altitude 920 m) with a moderately stony structure and clay loam texture (Xerofluvent-Fluvisol calcareous) with high pH (8.2), high organic matter content (3%), and high carbonate content (16%). The experimental plot size for conventional and ecological cultivation of *Brassicaceae* species in Farm II were 25 × 25 m each. Both experimental plots were close together and separated only by a 2-m-wide border. The climate is typically continental Mediterranean (Csa in Köppen's climate classification), with short summers, very hot, arid and mostly cloudless, winters are long, very cold and partially cloudy and mean annual precipitations of 650 mm. In this Farm, the three *Brassicaceae* species were grown both under conventional and organic conditions.

The conventional cultivation conditions on Farm II were similar to those of the Farm I. In organic cultivation, neither herbicides nor mineral fertilizers were applied. Instead, only treatment with a mixture of goat and sheep manure was applied at a rate of 3 kg/m^2.

When plants from the different species reached their optimal moment of consumption (from 3 to 5 months after sowing), leaf samples from individual plants of each species were harvested, pooled, and processed for chemical analysis. The number of analyzed plants throughout the two-years duration of the study were turnip greens (*Brassica rapa*) (n = 60), turnip tops (*Brassica rapa*) (n = 85), *Eruca vesicaria* (n = 18), and *Sinapis alba* (n = 12). The higher number of harvested *Brassica rapa* samples is due to turnip greens and turnip tops having good commercial prospects, and their consumption, both fresh and processed, has considerably increased in the last years. Furthermore, *Sinapis alba* and *Eruca vesicaria* are currently minority crops consumed only in salads, although their consumption may increase in the future due to their special composition in glucosinolates with medicinal properties (Sinalbine in *Sinapis alba* and Glucorafanine in *Eruca vesicaria*).

Plants were thoroughly washed with tap water to remove dirt and dust, and they were finally rinsed with deionized water. Then, they were stored at −80 °C until freeze-drying, which was done in Telstar® model Cryodos-50 equipment (Telstar, Terrasa, Spain).

The freeze-dried samples were ground in a Janke and Kunkel Model A10 mill (IKA-Labortechnik, Staufen, Germany) for about 20 s, and stored in a desiccator until their analysis.

2.2. Materials and Reagents

All the reagents were of an analytical-reagent grade. Ultrapure water (18 MΩ/SCF) prepared with a Milli-Q Reference Water Purification (Millipore, Madrid, Spain) was used throughout the experiments. All the glassware and plastic containers were soaked in 50% nitric acid overnight, then in 20% hydrochloric acid for an additional night and rinsed three times with de-ionized water prior to use. Hyperpure nitric acid (65%) and hydrochloric acid (35%) were obtained from Panreac (Barcelona, Spain). Hydrogen peroxide (33%) was acquired from Sigma Aldrich (St. Loius, MO, USA).

Standard solutions for measuring the elements Cd, Co, Cr, Cu, Fe, Mn, Ni, Pb and Zn were prepared immediately before use by dilution with distilled deionized water of 1000 mg/L standard solutions (Certipur-Merck, Darmstad, Germany).

2.3. Trace Element Determination

To determine the trace element content (Co, Cr, Cu, Fe, Mn, Ni, Pb and Zn) of *Brassicaceae* species, 0.5 g of freeze-dried sample was weighed in a porcelain crucible. Samples were incinerated in a muffle furnace at 460 °C for 15 h. The ash was bleached after cooling by adding 200 µL of hyperpure HNO_3 and 1 mL of deionized water, drying this on thermostatic hotplates, and maintaining it in a muffle furnace at 460 °C for 1 h more. Ash recovery was performed with 100 µL of hyperpure HNO_3, making up to 10 mL with deionized water. To Cd analysis, in order to avoid Cd volatilization, 0.5 g of freeze-dried sample was placed in a Teflon vessel. Then, 3 mL of of hyperpure HNO_3 and 2 mL of H_2O_2 were added to each vessel and kept for 10 min at room temperature. After sealing the vessels hermetically, they were placed in a microwave oven (Multiwave GO, Anton Paar, Germany) and digested following the instrumental parameters indicated by the manufacturer. Every sample was diluted up to a volume of 20 mL with ultrapure water

Elemental analyses for Fe (λ = 248.3 nm; Slit width = 0.2 nm), Mn (λ = 279.5 nm; Slit width = 0.2 nm) and Zn (λ = 213.9 nm; Slit width = 0.7 nm) were performed by flame absorption atomic spectroscopy (FAAS) with a Varian Spectra AA-50B model, equipped with standard air-acetylene flame, and single- element hollow cathode lamps. Finally, electrothermal atomic absorption spectroscopy (ET-AAS) was used for the determination of Cd, Co, Cr, Cu, Ni and Pb (Agilent Technologies model 240Z AA with a graphite furnace and autosampler). This equipment was certified by an equipment qualification report (EQR; Agilent Technologies). In this latter, analytical methodology was developed following the instrumental parameters indicated by the manufacturer with slight modifications (Table 1). For Cd and Pb analysis, a chemical modifier (200 mL solution) was prepared containing a mixture of 0.1% Paladium matrix modifier 10 g/L (Merck, Spain) plus 0.06% Magnesium nitrate hexahydrate in 10 mL HNO_3 hyperpure solution (69%). For each measurement, 15 µL of sample and 5 µL of modifier solution were injected. The accuracy and precision of the different analytical techniques used in determining trace element concentrations were validated by recovery experiments using Certified Reference Materials (Table 2).

Table 1. Instrumental conditions for Cd, Co, Cr, Ni and Pb analysis by ET-AAS in *Brassicaceae* samples.

	Cd (λ = 228.8 nm)		Co (λ = 240.7 nm)		Cr (λ = 357.9 nm)		Cu (λ = 324.8 nm)		Ni (λ = 232.0 nm)		Pb (λ = 283.3 nm)		
Step	T (°C)	t (s)	T (°C)	t (s)	T (°C)	t (s)	T (°C)	t (s)	T (°C)	t (s)	T (°C)	t (s)	**Argon Flow (L/min)**
Drying	85	5	85	5	85	5	85	5	85	5	85	5	0.3
	95	40	95	40	95	40	95	40	95	40	95	40	0.3
	120	10	120	8	120	20	120	10	120	10	120	10	0.3
Pyrolysis	300	5	750	5	1000	5	800	5	800	5	500	10	0.3
	300	3	750	3	1000	3	800	3	800	3	500	7	0
Atomization	1800	2.8	2300	2.8	2600	2.8	2300	2.8	2400	2.8	2300	5	0
Cleaning	1800	2	2300	2	2600	2	2300	2	2400	2	2400	4	0.3

Table 2. Analysis of certified references materials (mean ± standard deviation), limit of detection and limit of quantification.

Element	Certified References Material (mg kg^{-1})							
	White Cabbage BCR-679					Peach Leaves NIST-1547		
	LOD (μg/g)	**LOQ (μg/g)**	**Certified**	**Found**	**Recovery (%)**	**Certified**	**Found**	**Recovery (%)**
Cd	0.005	0.016	1.66 ± 0.07	1.57 ± 0.01	95			
Co	0.023	0.077				0.07 *	0.063 ± 0.020	90
Cr	0.016	0.052	0.6 ± 0.1 *	0.61 ± 0.03	102			
Cu	0.078	0.260	2.89 ± 0.12	3.13 ± 0.25	108			
Fe	1.09	3.62	55.0 ± 2.5	56.2 ± 3.0	102			
Mn	0.138	0.459	13.3 ± 0.5	13.2 ± 0.4	99			
Ni	0.056	0.187	27.0 ± 0.8	25.6 ± 0.9	95			
Pb	0.048	0.150	-	-	-	0.869 ± 0.018	0.883 ± 0.073	102
Zn	1.20	4.84	79.7 ± 2.7	78.2 ± 3.0	98			

* Indicative value.

2.4. Statistical Analyses and Risk Assessment

The IBM SPSS 25 statistical software package was used for statistical analysis. The data were expressed as mean and standard deviation. Data were analyzed using ANOVA tests. Significant differences were considered when $p < 0.05$.

A probabilistic model was developed to estimate the intake level for Cd, Co, Cr, Cu, Fe, Mn, Ni, Pb and Zn derived from feeding with Brassica vegetables. It should be pointed out that in developing the model, only the concentrations of trace elements for the *Brassica rapa* species were considered as it is the one most consumed in Spain, Portugal and Southern Italy, being part of very traditional recipes. Nowadays, turnip greens and turnip tops have good commercial prospects, and their consumption, both fresh and processed, has considerably increased in the last years. *Eruca vesicaria* and *Sinapis alba* have only eaten in small amounts in some salads, and they have a slightly spicy flavor like mustard greens.

This model followed a probabilistic approach in which variables were described by probability distributions, and they were fitted to concentration data obtained in this study for each element. Additionally, to estimate the intake level, serving size was considered assuming 15–20 g per day (around 150–200 g of fresh matter) of *Brassica rapa* (turnip greens or turnip tops). Daily intake was defined by a uniform distribution in the probabilistic model, meaning that all values in that range had the same probability to occur.

The probability distributions describing the trace element concentration data were fitted using @Risk v7.5 (Palisade, Newfield, NY, USA). The simulation ran with 10,000 iterations per element. The goodness of fit assessed how well the fitted distribution described the data; in this section, the Akaike Information Criterion (AIC) and Chi-square statistical tests were used. Additionally, the visual analysis was considered to assess the fit of the probability distributions to intake data. Data obtained through this probabilistic model was compared to the Spanish DRI for adult population [12]. In the case of Cd, Ni and Pb [13–15] considering that they are heavy metals, tolerable intakes (μg/day) were considered.

3. Results and Discussion

3.1. Trace Element Contents in Brassicaceae Species: Conventional Versus Organic

The concentrations of Cu, Mn, and Zn for organic turnip greens (*Brassica rapa*) grown on Farm II were 7.4; 100.4 and 24.4 μg/g d.w (see Table 3). The concentrations of these trace elements in ones conventionally grown on the same farm were practically the same, with values of 7.5; 110.8 and 22.9 μg/g d.w. Conversely, there were statistically significant differences for Zn ($p < 0.05$) and Cu, Mn ($p < 0.01$) between the turnip greens grown on Farm I (conventional system) and those grown on Farm II (both conventional and organic systems). The highest concentrations of Cu and Zn (11.3; 38.9 μg/g d.w) and the lowest one Mn (56.0 μg/g d.w) found in turnip greens grown on Farm I compared to those analyzed on Farm II, demonstrate that, together with farming systems, the trace element concentrations in foodstuffs depend on many other factors, including soil characteristics, pollution from anthropogenic sources, genetic factors, seasonal influences and interactions between the elements [16].

Table 3. Total trace elements concentration (dry matter) in *Brassicaceae* species analyzed (mean ± standard deviation).

	Co (μg/g)	Cr (μg/g)	Cu (μg/g)	Fe (μg/g)	Mn (μg/g)	Zn (μg/g)
Conventional						
Farm I						
Brassica rapa (turnip greens)	0.23 ± 0.10	2.41 ± 1.50	11.3 ± 3.8	223 ± 133	56.0 ± 20.1	38.9 ± 23.1
Brassica rapa (turnip tops)	0.14 ± 0.04	0.87 ± 0.59	8.06 ± 3.80	71 ± 22	24.2 ± 9.9	29.9 ± 9.1
Eruca vesicaria	0.39 ± 0.08	7.17 ± 0.40	8.19 ± 1.73	539 ± 324	34.8 ± 8.6	37.1 ± 5.0
Farm II						
Brassica rapa (turnip greens)	0.28 ± 0.12	2.06 ± 0.98	7.54 ± 1.55	194 ± 75	110.8 ± 60.6	22.9 ± 4.7
Brassica rapa (turnip tops)	0.16 ± 0.08	0.52 ± 0,21	4.56 ± 2.27	60 ± 14	25.5 ± 13.3	25.1 ± 6.2
Eruca vesicaria	0.26 ± 0.09	3.29 ± 1.72	9.33 ± 1.30	413 ± 175	33.2 ± 8.0	34.2 ± 4.5
Sinapis alba	<LOQ	0.51 ± 0.03	5.57 ± 0.24	153 ± 32	18.7 ± 0.7	25.9 ± 3.5
Organic						
Brassica rapa (turnip greens)	0.22 ± 0.06	1.20 ± 0.66	7.39 ± 1.79	105 ± 63	100.4 ± 38.9	24.4 ± 3.3
Brassica rapa (turnip tops)	0.19 ± 0.06	0.48 ± 0.24	9.04 ± 6.33	72 ± 43	24.8 ± 6.6	31.2 ± 5.9
Eruca vesicaria	0.44 ± 0.11	5.17 ± 2.25	10.2 ± 1.1	700 ± 207	48.1 ± 9.8	33.0 ± 3.0
Sinapis alba	<LOQ	0.26 ± 0.08	5.97 ± 1.56	68 ± 11	17.8 ± 2.4	23.8 ± 3.2

The Cu, Mn and Zn in turnip tops grown under the three experimental conditions were Farm I, conventional, 8.1; 24.2 and 29.9 μg/g d.w. respectively; Farm II, conventional 4.6; 25.5 and 25.1 μg/g d.w.; Farm II, organic, 9.0; 24.8 and 31.2 μg/g d.w. The concentrations in the turnip tops of the three trace elements were lower than those analyzed in the corresponding turnip greens. This could be explained by considering that turnip greens are the vegetative *Brassica rapa* leaves, whereas the turnip tops are the fructiferous stems with flower buds and the surrounding leaves. Previous studies have indicated that metals tend to accumulate preferentially in roots rather than in storage organs or fruits [17,18]. Moreover, unlike what happens to the turnip greens, statistically significant differences were found for Cu ($p < 0.05$) and Zn ($p < 0.01$) between the conventional and organic turnip tops grown on Farm II, the highest concentrations being found in the organic ones. These results are in agreement with those reported by Kelly and Bateman [9] for similar studies made with other vegetables species (tomatoes and lettuces)

Cu, Mn and Zn concentrations in *Sinapis alba* were 6.0; 17.8 and 23.8 μg/g d.w for the organic cropping system and 5.6; 18.7 and 25.9 μg/g d.w for the conventional one. Similarly, concentrations for these trace elements in *Eruca vesicaria* were 10.2; 48.1 and 33.0 μg/g d.w for the organic system and 9.3; 33.2 and 34.2 μg/g d.w for the conventional one. Zn concentrations found in these *Brasicaceae* species are in agreement with those reported by Cámara–Martos et al. [3] in a previous study (*Sinapis alba* 20.8 μg/g d.w; *Eruca vesicaria* 23.5 μg/g d.w). However, our work again failed to find statistically significant differences in the concentrations of these trace elements between organic and conventional agriculture.

The results appearing in the bibliography do not show any clear trend for these trace elements either. Krejcova et al. [19], in conventionally-grown carrots, have shown a higher content of Mn and Cu but a lower content of Zn than in organic ones. Hadayat et al. [18], for organic lettuce, potato and carrot reported higher Cu contents than those in the same conventional vegetables. However, for conventional lettuce and carrot, higher Zn concentrations than those in the organic ones, were also found.

On the other hand, although in our study no differences were found in Mn concentrations between conventionally-grown *Brassicas* and organic ones, some authors have demonstrated that a lower mean concentration of Mn in organic crops is a common pattern. This could be due to the high concentrations of *arbuscular mycorrhizal fungi* in organic soils [9,20]. Although, this aspect has not been completely clarified. Other studies have shown that Mn is used as an additive to livestock feed supplements and, in turn, that this trace element would be present in the manure used in organic farming [9].

Co concentrations in turnip greens (*Brassica rapa*) were very similar in the three studied conditions (Farm I conventional 0.23 μg/g d.w; Farm II conventional 0.28 μg/g d.w and Farm II organic 0.22 μg/g d.w, with no significant differences between them (Table 3). Regarding Cr contents in turnip greens, in organic plants harvested on Farm II, lower concentrations (1.20 μg/g d.w) ($p < 0.01$) than in conventional ones of Farm I (2.41 μg/g d.w) and Farm II (2.06 μg/g d.w) were found. These results are in agreement with those reported by Krejcova et al. [19], who also found higher Cr concentration in conventional carrots (0.059 μg/g) than in organic carrots (0.046 μg/g). We have presumed that the main chemical form in which Cr is found in *Brassicaceae* vegetables, would be Cr (III). This chemical form is considered as being a beneficial element for human health, and according to several previous research works [21,22], it is the main chemical form (unlike Cr(VI) in which Cr is found in waters and foods. Therefore, according to their Co and Cr content, organic turnip greens would not have a higher nutritional value than conventional grown ones.

For the same reason as that already mentioned for the previous trace elements, Co and Cr concentrations in turnip tops (*Brassica rapa*) also decreased with respect to the corresponding turnip greens (*Brassica rapa*) (Farm I, conventional, 0.14 and 0.87 μg/g d.w.); (Farm II, conventional, 0.16 and 0.52 μg/g d.w.); (Farm II, organic, 0.19 and 0.48 μg/g d.w.) Again we find differences in Cr concentrations between turnip tops grown on the two Farms but not between the cropping systems (conventional versus organic). Thus, there were significant statistical differences for Cr contents ($p < 0.05$) between conventional turnip tops grown on Farm I and conventional turnip tops grown on Farm II. We also found statistically significant differences for Cr contents ($p < 0.01$) between conventional turnip tops grown on Farm I and organic ones grown on Farm II. Therefore, this again indicates that the total inorganic element content in vegetables does not only depend on the farming system [16].

Sinapis alba showed higher Cr concentrations ($p < 0.01$) for conventionally produced vegetables (0.51 μg/g d.w.) than organically ones (0.26 μg/g d.w.). Co contents for this *Brassicaceae* species were below the quantification limit (LOQ < 0.07 μg/g). On the other hand, Co and Cr contents in *Eruca vesicaria* were Farm I, conventional, 0.39 and 7.17 μg/g d.w; Farm II, conventional, 0.26 and 3.29 μg/g d.w, and, Farm II, organic 0.44 and 5.17 μg/g d.w. Cr contents in *Eruca vesicaria* are slightly higher with those reported in a previous study (2.59 μg/g) [3]. In addition, there were no statistically significant differences for both elements between organic and conventional *Eruca vesicaria*. Previous studies have reported that there is no clear trend in this matter. Thus, Hadayat et al. [18] gave higher Co contents in conventional potato, onion, tomato and carrot versus organic ones, whereas Cr contents were higher for organic onion, carrot and potato.

Regarding Fe, we found statistically significant differences ($p < 0.01$) between organic turnip greens (*Brassica rapa*) (0.10 mg/g d.w.) and conventional ones (0.19 µg/g d.w.) from Farm II; and between organic *Sinapis alba* (0.07 mg/g d.w.) and conventional one (0.15 mg/g d.w.) from Farm II (Table 3). Similarly, Krejcova et al. [19] have also reported higher Fe content for conventional carrots (5.24 µg/g) versus organic ones (4.96 µg/g). Nevertheless, for the rest of the vegetable *Brassicaceae studied*, the differences between conventional and organic cultivation were scant. In relation to this latter aspect, Kelly and Bateman [9] observed only minor variations in Fe concentrations between tomatoes and lettuces cultivated on organic and conventional farms.

While the above trace elements have a clear nutritional role, Pb and Cd are considered to be heavy metals that have harmful effects on the environment and human health. With respect to Ni, although the nutritional and/or toxicologic role of Ni in humans is unclear, in animal models, severe Ni deficiency can affect vision, Fe metabolism, and Na homeostasis [23]. However, high concentrations of this element can also affect vital processes in plants and induce toxic effects at morphologic, physiologic and biochemical levels [24].

For this latter element, we found statistically significant differences ($p < 0.05$) for Ni content in turnip tops (*Brassica rapa*) grown by organic agriculture (0.85 µg/g d.w.) on Farm II and turnip tops grown under conventional agriculture (1.02 µg/g d.w.) on the same Farm, with the lowest concentrations in organic ones (see Table 4). Similarly, Krejcova et al. [19] have also reported a lower Ni content in organic carrots (0.79 µg/g) than in conventional ones (1.58 µg/g). Other vegetables such as tomato and onion have also shown lower Ni contents when they are grown by organic systems. Conversely, other vegetable foodstuffs such as organic wheat (semolina samples) have shown higher Ni content than conventional wheat.

Table 4. Total heavy metals concentration (dry matter) in *Brassicaceae* species analyzed (mean ± standard deviation).

	Cd (µg/g)	Ni (µg/g)	Pb (µg/g)
Conventional			
Farm I			
Brassica rapa (turnip greens)	0.28 ± 0.16	1.56 ± 1.11	0.99 ± 0.80
Brassica rapa (turnip tops)	0.12 ± 0.03	0.81 ± 0.32	0.90 ± 0.88
Eruca vesicaria	0.72 ± 0.10	3.99 ± 1.47	3.40 ± 0.61
Farm II			
Brassica rapa (turnip greens)	0.19 ± 0.06	1.39 ± 0.91	0.62 ± 0.37
Brassica rapa (turnip tops)	0.13 ± 0.03	1.02 ± 0.21	<LOQ
Eruca vesicaria	0.48 ± 0.05	1.76 ± 0.48	0.55 ± 0.07
Sinapis alba	0.23 ± 0.01	0.21 ± 0.03	0.16 ± 0.02
Organic			
Brassica rapa (turnip greens)	0.18 ± 0.05	1.02 ± 0.46	0.33 ± 0.23
Brassica rapa (turnip tops)	0.12 ± 0.05	0.85 ± 0.22	<LOQ
Eruca vesicaria	0.54 ± 0.04	2.55 ± 0.71	0.91 ± 0.39
Sinapis alba	0.07 ± 0.01	<LOQ	0.45 ± 0.31

Ni values for conventional turnip tops grown on Farm I were (0.81 µg/g d.w.) and Farm II (1.02 µg/g d.w.) with statistically significant differences ($p < 0.05$) between them. Trace element content in organic *Sinapis alba* was below the quantification limit (LOQ < 0.18 µg/g) whereas Ni values for *Eruca vesicaria* ranged between 1.76–3.99 µg/g d.w without significant differences between conventional and organic system. These results are in agreement with those found in a previous study [3] for this latter *Brassica* specie (1.12 µg/g).

Regarding Pb, there were statistically significant differences ($p < 0.05$) between organic turnip greens (*Brassica rapa*) (0.33 µg/g d.w.) and conventional ones (0.62 µg/g d.w.) from Farm II (Table 4). Nevertheless, for the rest of the *Brassicaceae* studied, the differences between conventional and organic cultivation were scant. The influence of the soil in which the plants are grown has also been demonstrated. Thus, we found statistically significant differences ($p < 0.01$) for *Eruca vesicaria* between conventional plants grown on Farm I (3.40 µg/g d.w.) and plants grown on Farm II (0.55 µg/g d.w.). Furthermore, while for organic and conventional turnip tops grown on Farm II, Pb concentrations were below the quantification limit (LOQ < 0.160 µg/g), concentrations in conventional turnip tops grown on Farm I reached mean values of 0.90 µg/g d.w.

There are no conclusive results regarding a lower Pb content in vegetables grown through organic agriculture. Thus, Hadayat et al. [18] found lower Pb concentrations in organic tomato, lettuce, onion and carrot but not in potato. Krejcova et al. [19] reported higher Pb contents in conventional carrots (0.064 µg/g) versus organic ones (0.043 µg/g). However, Zaccone et al. [16] found higher contents of this heavy metal in organic wheat (94 µg/g) versus that in conventionally grown wheat (82 µg/g). Finally, Karavoltsos et al. [25] have indicated that, although the majority of organic vegetables may have lower Pb content, organic agriculture as such does not necessarily reduce the content of this heavy metal in organically cultivated products.

Another heavy metal whose consumption is aimed to reduce with the development of organic agriculture is Cd. Nevertheless, in the present study, we only found statistically significant differences for Cd levels ($p < 0.01$) between organic (0.07 µg/g d.w.) and conventional (0.23 µg/g d.w.) (Table 4) *Sinapis alba* grown on Farm II. For the rest of the *Brassicaceae* cultivated on Farm II, Cd values were very similar both in organic and conventional plants. Comparing our results with those in a previous study, Hadayat et al. (2018) reported lower Cd concentrations in organic tomato, onion, carrot and potato but not in lettuce. Cámara–Martos et al. [26] for infant foods, such as weaning jars, made with organic vegetable ingredients, it was also found that Cd concentrations were considerably lower than those reported in weaning formulas which were not categorized as organic. Krejcova et al. [19] showed slightly higher mean Cd concentrations in conventional carrots (0.066 µg/g) versus organic ones (0.060 µg/g), and Hoefkens et al. [11] indicated significant higher or lower concentrations and even non-significant differences in Cd concentrations, depending on the food matrix.

A factor that again influenced the Cd concentrations in *Brassicaceae* was the soil in which they were grown. Thus, we found statistically significant differences in Cd levels ($p < 0.05$) for conventional *Eruca vesicaria* grown on Farm I (0.72 µg/g d.w.) and those conventionally grown on Farm II (0.48 µg/g d.w.)

According to Karavoltsos et al. [25], organic agriculture could eventually lead to the production of foodstuffs with a lower heavy metal content, although organic agriculture as such is not able to secure low metal contents in its products. Our results show that this final content is also influenced by other factors such as soil, vegetable variety or even the presence of these elements in the air and in irrigation waters.

3.2. *Probabilistic Assessment: Conventional Versus Organic*

As already mentioned in the Material and methods section, a probabilistic model approach was developed to estimate the intake level of trace elements, which were derived from the consumption of 15–20 g (around 150–200 g of fresh matter) of these *Brassicas*. It should be pointed out that in developing the model, only the concentrations of trace elements for the *Brassica rapa* species (turnip greens and turnip top) were considered as they are the one most consumed.

Dietary reference intakes (DRI) for the Spanish population were considered [12]. There is not DRI for Co; however, this element represents approximately a 4.3% of vitamin B12. Considering a DRI for vitamin B12 between 2–2.4 µg, this corresponds to around 0.10 µg/day [27]. In the case of Ni [14] and Cd [13], considering that they are heavy metals, the tolerable intake (TI) of 2.8 μg/kg body weight·day and 2.5 μg/kg body weight·week (0.36 μg/kg body weight·day) was used. For Pb a benchmark dose ($BMDL_{01}$) for cardiovascular effects (1.50 μg/kg body weight·day) was considered [15]. It should also be noted that the present statistical tool was completed using the variability of inorganic elements present in plants as well as the variability of the *Brassica* vegetable ingested. Both aspects determine the total intake of the inorganic elements.

Thus, the results obtained from the simulation of the probabilistic model with conventional *Brassicas,* indicated values for Co, Cr, Cu, Fe, Mn and Zn of 2.58 μg, 17.41 μg, 0.14 mg, 1.74 mg, 0.55 mg and 0.56 mg, respectively, for 50th percentile (Figure 2). This shows that the intake of trace elements of at least half of the population consuming these conventional *Brassicas* will not be lower than these values. That intake fulfills Co DRI and complies with the following percentages of DRI for studied elements: Cr 69.6%, Cu 12.7%, Fe: 9.7%, Mn 30.6% and Zn 8.0%. When the same simulation of the probabilistic model is developed with concentrations belonging to organic *Brassicas*, the values obtained for 50th percentile were 3.61 μg, 9.60 μg, 0.12 mg, 1.20 mg, 0.65 mg and 0.49 mg for Co, Cr, Cu, Fe, Mn and Zn, respectively (Figure 3), which satisfy similarly to the conventional *Brassicas* the DRI percentages for these trace elements.

Regarding heavy metals, we have considered intakes for 95th percentile as being the most unfavorable situation. Thus, Ni, Pb and Cd intakes for 95th percentile with conventional *Brassicas* were 39.00, 47.34 and 6.82 μg, respectively (Figure 4), whereas intakes with organic *Brassicas* were 25.70, 15.55 and 4.56 μg for Ni, Pb and Cd, respectively (Figure 5). According to these values, organic *Brassicas* led to a decrease in the intake of these three elements metals. However, when these results were expressed as percentages of TI for these elements, we only found slight differences (mainly in the case of Pb) between both agriculture systems. The TI percentages for a mean body weight of 70 kg per person were 19.9, 45.1 and 27.1% for Ni, Pb and Cd, respectively, with conventional *Brassicas,* and 13.1, 14.8 and 18.1 % for organic *Brassicas*. These results indicate that Cd, Ni and Pb contents in vegetable *Brassica rapa* harvested under both conventional and organic farming conditions are below the accepted safety limits and do not represent any toxicologic risk.

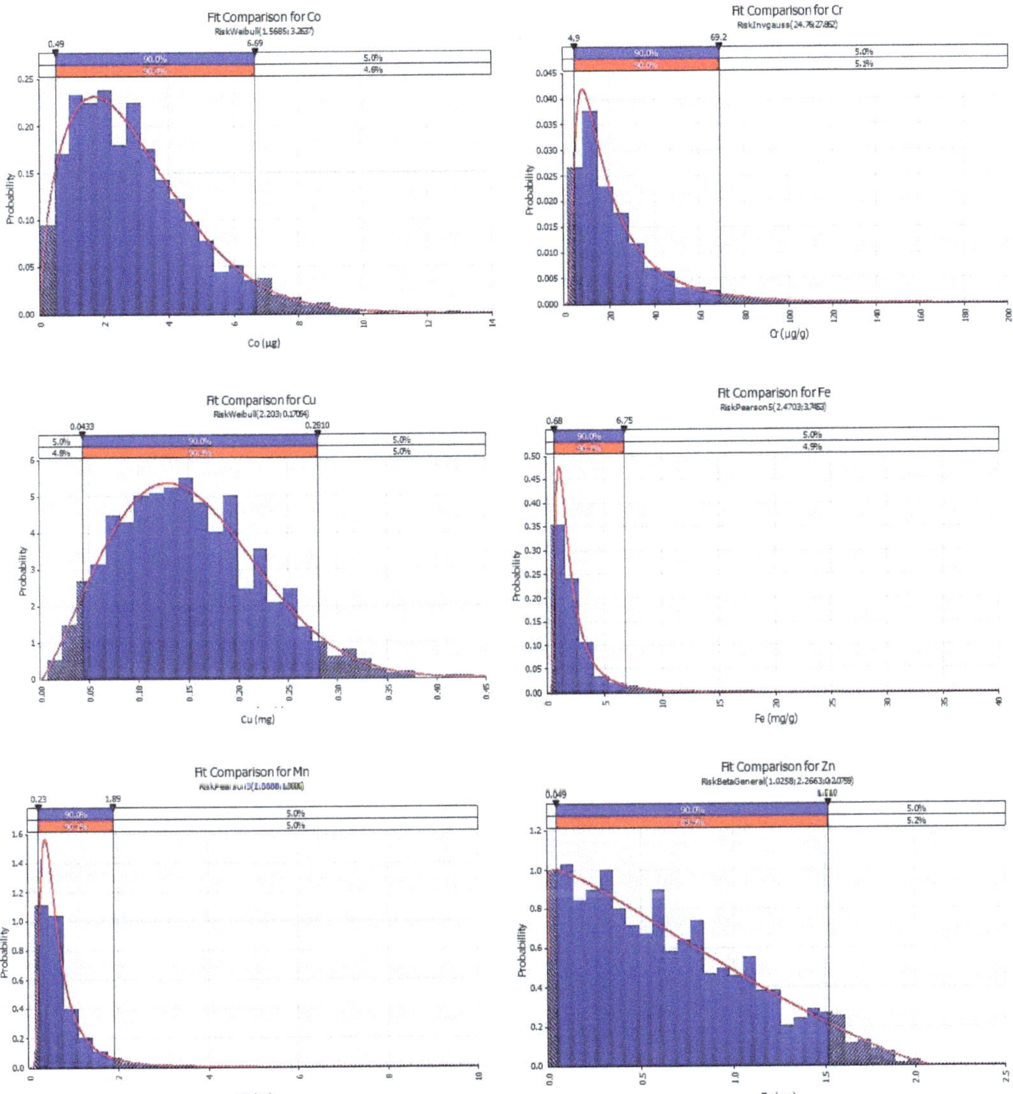

Figure 2. Simulated data and fitted probabilistic distribution for Co, Cr, Cu, Fe, Mn and Zn in conventional *Brassicas*.

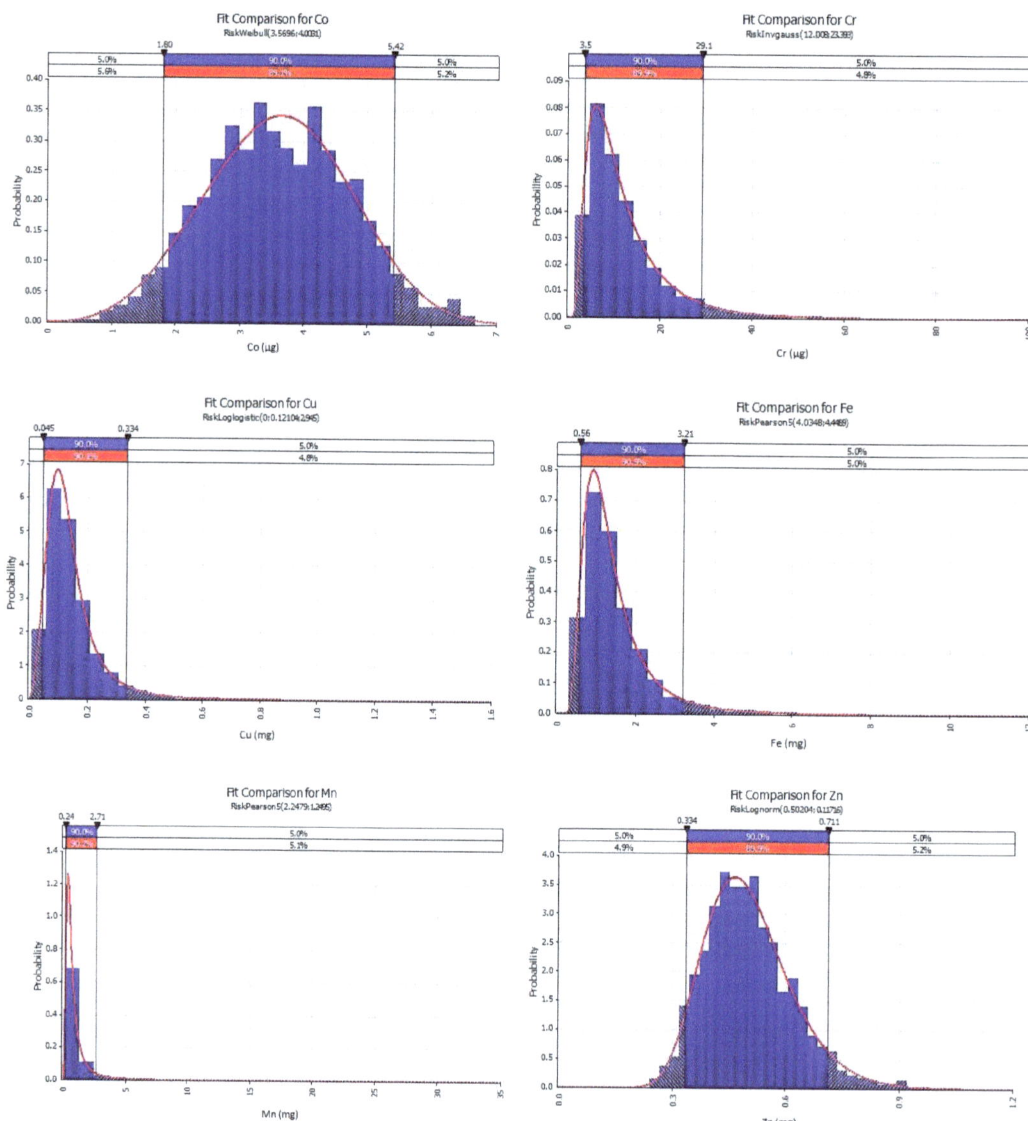

Figure 3. Simulated data and fitted probabilistic distribution for Co, Cr, Cu, Fe, Mn and Zn in organic *Brassicas*.

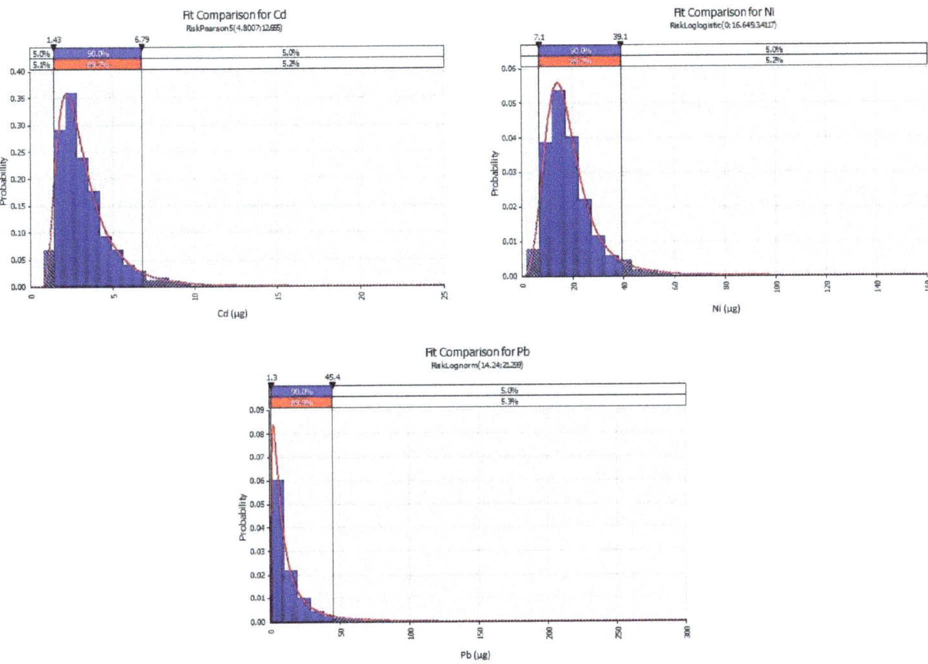

Figure 4. Simulated data and fitted probabilistic distribution for Cd, Ni and Pb in conventional Brassicas.

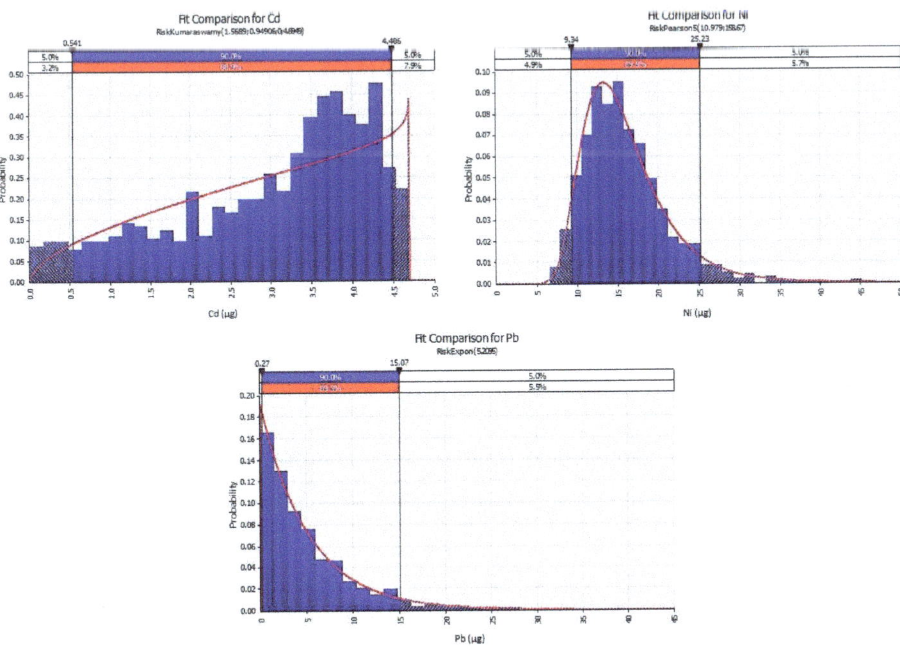

Figure 5. Simulated data and fitted probabilistic distribution for Cd, Ni and Pb in organic Brassicas.

Author Contributions: F.C.-M., conceptualization, methodology, validation, formal analysis, investigation, resources, writing, visualization, supervision, project administration, funding acquisition; J.S.-M., formal analysis; L.R.-P., investigation, resources, supervision; J.B.-H., resources, supervision; A.d.H.-B., conceptualization, methodology, investigation, resources, writing, visualization, supervision, project administration. All authors have read and agreed to the published version of the manuscript.

Funding: This research was funded by the Project "Desarrollo de Alimentos Funcionales y Nutracéuticos Con Efectos Antioxidantes, Antitumorales y Cardiosaludables a partir de una Selección de Crucíferas Mediterráneas" Ref. UCO–1261749-R of the University of Cordoba-Regional Council for the Economic and Knowledge of Junta de Andalucía, which was co-financed by the European Regional Development Fund (ERDF 2014–2020).

Informed Consent Statement: Informed consent was obtained from all subjects involved in the study.

Acknowledgments: The authors thank Diana Badder, English speaker, for linguistic review of the manuscript.

Conflicts of Interest: Declarations of interest from all the authors: none.

References

1. Voorrips, L.E.; Goldbohm, R.A.; van Poppel, G.; Sturmans, F.; Hermus, R.J.J.; van den Brandt, P.A. Vegetable and fruit consumption and risks of colon and rectal cancer in a prospective cohort study—The Netherlands Cohort Study on Diet and Cancer. *Am. J. Epidemiol.* **2000**, *152*, 1081–1092. [CrossRef] [PubMed]
2. Petropoulos, S.A.; Di Gioia, F.; Ntatsi, G. Vegetable organosulfur compounds and their health promoting effects. *Curr. Pharm. Des.* **2017**, *23*, 2850–2875. [CrossRef] [PubMed]
3. Cámara-Martos, F.; Obregón-Cano, S.; Mesa-Plata, O.; Cartea-González, M.E.; de Haro-Bailón, A. Quantification and in vitro bioaccessibility of glucosinolates and trace elements in *Brassicaceae* leafy vegetables. *Food Chem.* **2021**, *339*, 127860. [CrossRef] [PubMed]
4. Schmidt, R.; Bancroft, I. *Genetics and Genomics of the Brassicaceae*, 1st ed.; Springer: Berlin, Germany, 2011; pp. 80–81.
5. FAOSTAT (Food and Agriculture Organization of The United Nation). 2013. Available online: http://faostat3.fao.org/browse/Q/QC/S (accessed on 10 January 2021).
6. Avato, P.; Argentieri, M.P. *Brassicaceae*: A rich source of health improving phytochemicals. *Phytochem. Rev.* **2015**, *14*, 1019–1033. [CrossRef]
7. Rembiałkowska, E. Organic food: Effect on nutrient composition. In *Encyclopedia of Food and Health*, 1st ed.; Caballero, B., Finglas, F., Toldrá, F., Eds.; Academic Press-Elsevier: Oxford, UK, 2016; pp. 178–180.
8. Iglesias-Jiménez, E.; Álvarez, C. Apparent availability of nitrogen in composted municipal refuse. *Biol. Fertil. Soils* **1993**, *16*, 313–318. [CrossRef]
9. Kelly, S.D.; Bateman, A.S. Comparison of mineral concentrations in commercially grown organic and conventional crops—Tomatoes (*Lycopersicon esculentum*) and lettuces (*Lactuca sativa*). *Food Chem.* **2010**, *119*, 738–745. [CrossRef]
10. Bakkali, K.; Martos, N.R.; Souhail, B.; Ballesteros, E. Determination of heavy metal content in vegetables and oils from Spain and Morocco by inductively coupled plasma mass spectrometry. *Anal. Lett.* **2012**, *45*, 907–919. [CrossRef]
11. Hoefkens, C.; Vandekinderen, I.; De Meulenaer, B.; Devlieghere, F.; Baert, K.; Sioen, I.; De Henauw, S.; Verbeke, W.; Van Camp, J. A literature-based comparison of nutrient and contaminat contents between organic and conventional vegetables and potatoes. *Br. Food J.* **2009**, *111*, 1078–1097. [CrossRef]
12. FESNAD. Ingestas dietéticas de referencia (IDR) para la población española. *Act. Diet.* **2010**, *14*, 196–197.
13. European Food Safety Authority (EFSA). Cadmium in food. *EFSA J.* **2009**, *980*, 1–139.
14. European Food Safety Authority (EFSA). Scientific Opinion on the risks to public health related to the presence of nickel in food and drinking water. *EFSA J.* **2015**, *13*, 4002.
15. European Food Safety Authority (EFSA). Panel on Contaminants in the Food Chain (CONTAM); Scientific Opinion on Lead in Food. *EFSA J.* **2010**, *8*, 1570. [CrossRef]
16. Zaccone, C.; Di Caterina, R.; Rotunno, T.; Quinto, M. Soil—Farming system—Food—Health: Effect of conventional and organic fertilizers on heavy metal (Cd, Cr, Cu, Ni, Pb, Zn) content in semolina samples. *Soil Tillage Res.* **2010**, *107*, 97–105. [CrossRef]
17. Douay, F.; Pelfrêne, A.; Planque, J.; Fourrier, H.; Richard, A.; Roussel, H.; Girondelot, B. Assessment of potential health risk for inhabitants living near a former lead smelter. Part 1: Metal concentrations in soils, agricultural crops, and homegrown vegetables. *Environ. Monit. Assess.* **2013**, *185*, 3665–3680. [CrossRef]
18. Hadayat, N.; De Oliveira, L.M.; Da Silva, E.; Han, L.; Hussain, M.; Liu, X.; Ma, L.Q. Assessment of trace metals in five most—consumed vegetables in the US: Conventional vs. organic. *Environ. Pollut.* **2018**, *243*, 292–300. [CrossRef]
19. Krejcová, A.; Návesník, J.; Jicinská, J.; Cernohorský, T. An elemental analysis of conventionally, organically and self-grown carrots. *Food Chem.* **2016**, *192*, 242–249. [CrossRef] [PubMed]

20. Gosling, P.; Hodge, A.; Goodlass, G.; Bending, G.D. Arbuscular mycorrhizal fungi and organic farming. *Agric. Ecosyst. Environ.* **2006**, *113*, 17–35. [CrossRef]
21. Dogan Uluozlu, O.; Tuzen, M.; Soylak, M. Speciation and separation of Cr(VI) and Cr(III) using coprecipitation with Ni2+/2-Nitroso-1-naphthol-4-sulfonic acid and determination by FAAS in water and food samples. *Food Chem. Toxicol.* **2009**, *47*, 2601–2605. [CrossRef]
22. Saraiva, M.; Chekri, R.; Leufroy, A.; Guérin, T.; Sloth, J.J.; Jitary, P. Development and validation of a single run method based on species specific isotope dilution and HPLC-ICP-MS for simultaneous species interconversion correction and speciation analysis of Cr(III)/Cr(VI) in meat and dairy products. *Talanta* **2021**, *222*, 121538. [CrossRef]
23. Mataix-Verdú, J.; Llopis-González, J. Minerales. In *Nutrición y Alimentación Humana. Nutrientes y Alimentos*, 2nd ed.; Mataix-Verdú, J., Ed.; Editorial Ergón: Madrid, Spain, 2015; pp. 265–301.
24. Amari, T.; Lutts, S.; Taamali, M.; Lucchini, G.; Sacchi, G.A.; Abdelly, C.; Ghnaya, T. Implication of citrate, malate and histidine in the accumulation and transport of nickel in *Mesembryanthemum crystallinum* and *Brassica juncea*. *Ecotoxicol. Environ. Saf.* **2016**, *126*, 122–128. [CrossRef]
25. Karavoltsos, S.; Sakellari, A.; Dassenakis, M.; Scoullos, M. Cadmium and lead in organically produced foodstuffs from the Greek Market. *Food Chem.* **2008**, *106*, 843–851. [CrossRef]
26. Cámara-Martos, F.; Ramírez-Ojeda, A.; Jiménez-Mangas, M.; Sevillano-Morales, J.; Moreno-Rojas, R. Selenium and cadmium in bioaccessible fraction of organic weaning food: Risk assessment and influence of dietary components. *J. Trace Elem. Med. Biol.* **2019**, *56*, 116–123. [CrossRef] [PubMed]
27. Cámara-Martos, F.; Moreno-Rojas, R. Cobalt: Toxicology. In *Encyclopedia of Food and Health*, 1st ed.; Caballero, B., Finglas, F., Toldrá, F., Eds.; Academic Press-Elsevier: Oxford, UK, 2016; Volume 2, pp. 172–178.

Review

The Beneficial Health Effects of Vegetables and Wild Edible Greens: The Case of the Mediterranean Diet and Its Sustainability

Elena Chatzopoulou [1], Márcio Carocho [2], Francesco Di Gioia [3] and Spyridon A. Petropoulos [4,*]

1 Kent Business School, University of Kent, Canterbury CT2 7NZ, Kent, UK; e.chatzopoulou@kent.ac.uk
2 Centro de Investigação de Montanha (CIMO), Instituto Politécnico de Bragança, Campus de Santa Apolónia, 5300-253 Bragança, Portugal; mcarocho@ipb.pt
3 Department of Plant Science, Pennsylvania State University, 207 Tyson Building, University Park, PA 16802, USA; fxd92@psu.edu
4 Department of Agriculture Crop Production and Rural Environment, University of Thessaly, Fytokou Street, 38446 Volos, Greece
* Correspondence: spetropoulos@uth.gr; Tel.: +30-2421-09-3196

Received: 20 November 2020; Accepted: 17 December 2020; Published: 21 December 2020

Abstract: The Mediterranean diet (MD) concept as currently known describes the dietary patterns that were followed in specific regions of the area in the 1950s and 1960s. The broad recognition of its positive effects on the longevity of Mediterranean populations also led to the adoption of this diet in other regions of the world, and scientific interest focused on revealing its health effects. MD is not only linked with eating specific nutritional food products but also with social, religious, environmental, and cultural aspects, thus representing a healthy lifestyle in general. However, modern lifestyles adhere to less healthy diets, alienating people from their heritage. Therefore, considering the increasing evidence of the beneficial health effects of adherence to the MD and the ongoing transitions in consumers' behavior, the present review focuses on updating the scientific knowledge regarding this diet and its relevance to agrobiodiversity. In addition, it also considers a sustainable approach for new marketing opportunities and consumer trends of the MD.

Keywords: agrobiodiversity; antioxidant activity; bioactive compounds; health benefits; medicinal properties; Mediterranean diet; market food trends; sustainability; wild edible greens

---

1. Introduction

Our diet and the proper selection of dietary sources of nutrients are pivotal for our body condition. According to numerous medical studies, these choices can highly influence health conditions [1–4]. The significance of food consumption for human health was first recognized by Hippocrates who quoted "Let food be thy medicine and medicine be thy food". Now, several traditional and/or regional dietary patterns exist throughout the world which are interrelated with cultural, religious, and social beliefs, and are handed from generation to generation. The recent food-based dietary guidelines (FBDGs) for European countries produced by the Food and Agricultural Organization (FAO) and the World Health Organization (WHO) highlight the importance of planning a strategy to promote healthy diets and prevent nutritional deficiencies and/or excesses in the global population [5]. Although the Mediterranean diet (MD) is one of the most well-known dietary patterns that is highly appreciated for its beneficial health effects, a contemporary overview that links the shifts of consumers' behavior towards modern diets and the effects of the MD on diseases is lacking. The health benefits due to adherence to the MD were first introduced in the late 1950's in the Seven Countries Study. The MD dietary pattern was probably first referred to using this term by Keys, and it was suggested that the residents of Crete in Greece had the highest life expectancy rates [6–9]. Subsequently, numerous studies have reached the conclusion that the relatively low mortality rates are due to the adherence to the MD, and a large number of scientists have suggested this diet as a prevention factor for cancer, diabetes, and other chronic diseases [10–13]. However, a recent follow-up of the Seven Countries Study revealed that populations now follow dietary patterns that deviate from those followed 50 years ago, and that these changes have resulted in high incidences of obesity, high systolic blood pressure, and hypercholesterolemia [9].

Since 2013, this diet has been considered an Intangible Cultural Heritage by UNESCO based on its values of hospitality, neighborliness, intercultural dialogue, and creativity, and bases its successes on a set of skills, rituals, knowledge, and traditions concerning crops, harvesting, fishing, animal husbandry, conservation, processing, cooking, and the sharing and consumption of food [14].

The MD, as it is now known, is based on a longitudinal study conducted in the early 1960s, and is structured on the food habits of the island of Crete and southern Italy [14]. According to the research of Keys et al. [15], the life expectancy of adult Greek men was the highest in the world, and breast cancer rates in women were less than one-half of those in the United States; however, it should be noted that the existing medical services were limited at that time [16]. The definition of the MD has been subject to continuous update, starting from the early 1960s, when it was suggested to be a diet comprising low consumption of saturated fatty acids and high intake of vegetable oils, followed mostly in Greece and southern Italy [17]. Subsequently, several dietary patterns described as intervention diets and based on the traditional MD, known also as the "modern MedDiets", have been observed. These diets integrate Westernized eating patterns into traditional eating habits [6,18]. Moreover, the local populations of the broader Mediterranean region have continuously interacted with each other throughout a period of centuries, resulting in several common eating habits and food ingredients among the countries around the Mediterranean Sea [19,20]. This continuous evolution and adaptation of dietary patterns to new conditions requires a constant reconsideration of the definition of the MD, particularly when considering the recent advances in analytical techniques that allow the effects and the mechanisms responsible for the reported and confirmed health effects to be revealed.

The MD cannot be simply described as a list of foods, because it integrates and constitutes a social, cultural, and rural way of living, e.g., sharing and exchanging food for festivities, celebrations, social, and religious purposes [21]. The MD can be described as a dietary pattern that includes all of the food groups with daily consumption of large portions of cereals, fruits, vegetables, dairy products, and olive oil (as the primary source of fat) [22,23]. Consumption of fish, poultry, and eggs is limited to a few portions per week, and the eating of red meat is restricted to once or twice per month [24]. Vegetables and fruits, in particular, provide one-tenth or more of total calories on a daily basis. However, regarding the consumption of vegetables and fruits in the ancient MD, our knowledge is limited and relies on historical records [25]. The key ingredient of the MD is olive oil, which is the main source of fat. Additionally, olive oil and, in particular, extra-virgin olive oil, is a source of the potent antioxidant vitamin E and polyphenols, it has antioxidant and antithrombotic properties, and according to numerous clinical, meta-analysis, and cohort studies it reduces the risk of coronary disease, diabetes, hypertension, cardiovascular disease-related mortality, and several types of cancer [26–32]. Olive oil has been systematically consumed by the Mediterranean inhabitants for thousands of years with no evidence of harm, although it comprises 28% and 40% of energy intake from total fat in southern Italy and Crete, respectively [33]. In addition to the consumption of olive oil, another key element of the traditional MD common to many populations around the Mediterranean basin is the daily consumption of a wide variety of seasonal fresh fruit and vegetables, including underutilized local genetic resources and wild edible plants, which together constitute a rich source of functional food [34,35]. Although the consumption of wild edible plants and local varieties or landraces is strictly linked to the traditional knowledge and rural lifestyle of the Mediterranean populations, the culinary use of food products from such genetic resources is persistent today and offers new opportunities to rediscover the value of the MD. From this perspective, it is important to examine the role that agrobiodiversity played in the traditional MD and to consider how it may continue to play a key role in the modern MD.

The relatively recent rapid urbanization in the Mediterranean countries has led to the abandonment of rural life, and the adoption of modern lifestyles and Westernized diets has been followed by increased consumption of meat, animal fats, and dairy products. However, despite these changes, the overall consumption of sugar, meat, and other specific "harmful" food products is less frequent among the citizens of Mediterranean countries compared to North Europeans, although significant differences in the degree of adherence to MD are reported among populations of different Mediterranean countries [36]. Considering the important health effects of the MD and the ongoing transitions in consumers' behavior, the present review focuses on updating the knowledge regarding the health effects related to the adherence to this particular diet, its relevance to agrobiodiversity and the sustainable approach of a healthy diet, and, finally, the new marketing opportunities that have emerged. The compiled information was collected from various databases, such as Scopus, ScienceDirect, PubMed, Google Scholar, and ResearchGate, after searching for specific terms and keywords, e.g., Mediterranean diet and health effects, and agrobiodivesity.

2. Mediterranean Diet and Agrobiodiversity

Agrobiodiversity as part of overall biodiversity can be defined as the diversity of living forms within agricultural ecosystems and is strongly linked with diversity in food and agricultural production and, thus, with nutrition and human health. In addition to the diversity of common crop species, Mediterranean agrobiodiversity resources also include wild edible plants [37–42] and the intra-specific diversity expressed by the cultivation of different varieties of the same species, which are often selected locally and called "local landraces" or "local varieties" [43].

In rural areas across the Mediterranean region, it was a usual practice to gather wild plants for food purposes, and many native species are essential ingredients for several traditional dishes. Furthermore, several wild edible plants have been domesticated, and generation after generation, farmers have wisely selected and retained vegetable genotypes with distinct traits, considered to have particular value or utility [43–45].

As we rediscover the value of such agrobiodiversity heritage and its potential contribution to our nutrition security and health, young consumers are reconsidering the virtues of the MD primarily based on the regular consumption of fruit and vegetables, and local landraces of fruit and vegetables and wild edible plants are increasingly appreciated for their nutritional properties and content of bioactive compounds [34,45–52].

During recent decades, the scientific community has greatly contributed to highlighting the value of underutilized genetic resources and to the characterization of their agronomic, nutritional, and functional properties [53]. Furthermore, the local food service industry is sourcing such genetic resources from rural communities and small farms and, with the support of renowned chefs, is contributing to the rediscovery of the use of local landraces and wild edible plants. This is helping to revive and repropose modern versions of traditional dishes that are cornerstone of the MD, thereby contributing to the generation of value and the establishment of a link between agrobiodiversity, local tradition, and the MD. The knowledge generated by the scientific community in synergy with the initiative of the local food service industry is raising the interest of new generations of consumers that are increasingly health conscious and attentive to the quality, origin, and sustainability of their food and diet. These trends are generating new market opportunities, particularly for small-scale farms that often are the custodians of agrobiodiversity resources and local rural traditions in the Mediterranean basin [34,54,55]. Linking together local rural traditions, the traditional MD, and smallholder farms that are investing in the cultivation of local landraces or wild edible plants with the modern MD, it is possible to generate community driven "Mediterranean food and cultural hubs", which are capable of educating and sensitizing modern consumers about the role that agrobiodiversity plays or may play in our diet and nutrition. The increasing knowledge and education of new generations about how to apply today the principles of the MD, with the aim of improving their health, will also contribute to preserving and valorizing local rural communities and economies, in addition to protecting the patrimony of agrobiodiversity and agricultural ecosystems.

The inestimable value of the Mediterranean agrobiodiversity, and its strong link with the MD and its beneficial effects on human health, are increasingly emerging due to the collaborative work of ethnobotanists, horticulturists, plant breeders, food chemists, and nutritionists, who together are recovering and generating new knowledge and genetic material, and are characterizing the nutritional and bioactive profile of many fruit and vegetable local landraces and wild edible species. Agrobiodiversity resources typical of the Mediterranean region, rich in bioactive compounds and commonly included as ingredients of the MD, are, for example, the multiple local landraces and selections of vegetables characterized by different shapes, harvest times, pigmentation, and associated nutritional and bioactive profiles [56,57], such as artichoke and cardoon [58–63], carrots [64–66], chicory and lettuce [67,68] among the Asteraceae family; broccoli and cauliflower [69] and kale [70] among the Brassicaceae family; cultivated and wild garlic [47,48,71–73], and onion [50,74–76] among the Liliaceae family; common beans, cowpea, fabae, and lentils among the Leguminosae family [77–79], melons and watermelons [80–82], or zucchini and squashes [83,84] among the Cucurbitaceae family; and tomato [85–90], pepper [91], and eggplants [92] among the Solanaceous family. Other valuable genetic resources commonly used in the MD and considered rich sources of bioactive compounds include minor vegetable species such as arugula [93–95] and broccoli raab or rapini [54,96,97], and numerous wild edible plants [39,42,44,98–103]. Moreover, the link between agrobiodiversity and MD is also expressed by the variety of plant parts of different species used in the kitchen, which range from roots and other below-ground organs, to stems, leaves or leaf petioles, fruit, and sometimes edible flowers or inflorescences. Exemplary are the cases of the zucchini squash for which, in Southern Italy, stems, young leaves, and flowers are traditionally used, in addition to the courgette [81]; artichoke, for which plant off-shoots are used in addition to the heads; or, the case of rapini, for which leaves and/or inflorescence are commonly used (Figure 1). Similarly, among wild species, the edible parts include leaves and flowers in *Borago officinalis*; leaves, inflorescence, and roots in *Cichorium intybus* L.; and in the case of *Scolymus hispanicus* L., the midribs are used after discarding the spiny leaf-lamina (Figure 2).

Figure 1. (**a**) Whole zucchini squash plants harvested to prepare a traditional Italian rural first dish "spaghetti e cime di zucchine" using every portion of the plant, including small fruits, leaves, stems, and flowers. (**b**) A field of rapini plants and harvested product characterized by the different size of inflorescence (associated with early harvest) in which leaves and inflorescence or only the large inflorescence is used. (**c**) An artichoke field of the selection "Violet de Provence" with heads ready to harvest, followed by an image of plant off-shoots called "carducci", generated and harvested at the base of the artichoke plant, being used to prepare fried stems. Photo credits: Francesco Di Gioia.

The increasing demands for healthy and natural food ingredients, in addition to the rediscovery of MD from new generations and the adoption of this particular diet in other regions of the world, necessitates the sustainable management of plant genetic resources [42,98,101,104]. This approach involves various practices, including the cultivation and domestication of wild species and their integration in production systems to promote the sustainability of agro-ecosystems and allow the conservation of valuable genetic resources [37]. The compilation and constant update of FBDGs is also a useful means to promote agrobiodiversity and sustainability through policies and dietary recommendations [5]. Moreover, according to Tarsitano et al. [105] and Burlingame and Dernini [106], the MD or the Mediterranean way encompasses dietary habits and lifestyles, and considers socioeconomic drivers that could be used as a means to achieve the goals set for the 2030 Agenda for Sustainable Development and thus promote biocultural diversity.

Figure 2. (**a**) Wild chicory (*Cichorium intybus* L.) leaves sold in a local farmer's market in Puglia (Italy), whole plant appearance, and a traditional Mediterranean dish prepared combining wild chicory leaves characterized by bitter flavor and dry fava beans characterized by mild taste. (**b**) Example of wild plants of *Scolymus hispanicus* L, and midribs cleaned and separated from the spiny lamina and used to prepare a traditional Easter dish with other wild plants and herbs typical of the Murgia (Puglia, Italy), combined with eggs and occasionally lamb. (**c**) Plants of wild *Sinapis alba* L. harvested, cleaned, and used to prepare "spaghetti and cime amarelle" a traditional rural dish in Puglia (Italy), revisited according to today's habits using processed pasta instead of traditional whole-grain pasta. Photo credits: Francesco Di Gioia.

Plant-based diets, including MD, are not only beneficial to human health due to reduced risk factors for several chronic diseases, but they also contribute to environmental sustainability through the adoption of healthy lifestyles [107]. Although MD is sustainable in its conception, the recently identified social and economic trends make it necessary to reevaluate the whole food chain, including crop production, food production, and consumption, also considering environmental, nutritional, economic, and social aspects [106]. Recently, several reports highlighted the importance of commercial cultivation of wild edible greens within the framework of sustainable management of native Mediterranean species and the exigent need for an increase in agrobiodiversity in farming systems, which is threatened by ongoing climate change and monocropping [108–111]. According to a recent report published by the Food and Agriculture Organization of the United Nations, only nine crops contribute 66% of the global total production of food [37]. The existing cropping systems in most of the Mediterranean countries consist of small-scale farms, which are ideal for the commercial cultivation of high added value crops, such as underutilized wild edible greens [41,55,99,100,102,112–115]. Moreover, the socioeconomic conditions faced by southern European countries during the last decade resulted in limited access to healthy and natural food ingredients that were traditionally used by poor people on a daily basis, because the prices of these food products increased making them affordable only to citizens with high incomes [116]. Several studies revealed that the adherence to MD was higher in social strata that belonged to high socioeconomic status, a finding that was also reflected in higher mortality rates and higher frequencies of chronic diseases among lower social strata [117–119]. Therefore, commercial cultivation of wild species could be a useful means to reduce their production cost, to increase their availability throughout the year, and, finally, to make them more affordable to the general public [110]. Moreover, the domestication and commercial cultivation of wild edible plants allow small-scale farms to obtain more stable yields and guarantee the origin of the product and its hygienic quality, in addition to the conservation of the agro-ecosystem, which may be an issue when collecting spontaneous plants. Although the commercial cultivation of wild edible species may be challenging in many aspects, farmers have successfully repeated this process for a large number of crops. An example of a Mediterranean wild edible species that has been successfully domesticated in recent decades is wild arugula (*Diplotaxis tenuifolia* (L.) DC.). Today, wild rocket is cultivated globally in open fields and protected environments using both soil and soilless systems. It is available to buy year-round, mostly as a fresh-cut leafy vegetable, and is increasingly appreciated by consumers for its pungent taste and interesting nutritional and bioactive profile [94,95,120]. The same path toward domestication and commercial cultivation of similar species could take place and is currently being investigated for many other Mediterranean wild edible species [121–126]. Using advanced sustainable cultivation methods, including soilless cultivation systems and agronomic biofortification techniques, it is possible to use underutilized genetic resources, such as domesticated wild edible plants and vegetable local landraces, to produce functional vegetable products [42,57,94,98,127,128] (Figure 3). In response to new food market trends, some specialized small-scale farms are exploiting Mediterranean wild edible plants and local landraces to produce high-value novel fresh functional products, such as fresh-cut baby leaf, microgreens, and edible flowers, which are increasingly used by chefs to add value, color, and taste to their plates, in addition to being recognized by consumers for their nutritional value and functional properties [129–134] (Figure 3). In addition to creating new market opportunities for small-scale farms, the cultivation of wild edible species and local landraces highly adapted to local environmental conditions may allow the recovery and exploitation of marginal land areas affected by low fertility or high salinity that otherwise could not be used for the production of conventional crops, thereby contributing to preserve marginal Mediterranean agroecosystems and genetic resources at risk of being lost [41,54,55].

Figure 3. Example of typical Mediterranean wild edible plants and vegetable local landraces suitable to produce edible flowers of (**a**) *Cichorium intybus* L., (**b**) *Malva sylvestris* L., (**c**) *Borago officinalis* L.; microgreens of (**d**) *Borago officinalis* L., (**e**) *Salicornia patula* Duval-Jouve, (**f**) *Brassica rapa* L. Broccoletto group; (**g**) baby-leaf of *Brassica rapa* L. selection "Sessantina"; (**h**) fresh-cut zucchini squash flowers, and (**i**) fresh-cut stems of *Cichorium intybus* L. selection 'Molfettese'. Photo credits: Francesco Di Gioia.

3. The Modern Mediterranean Diet and Its Health Benefits

Numerous reports highlighted the importance of adopting healthy dietary habits for benefiting the condition and well-being of the human body, focusing on various chronic diseases that impact developed countries, e.g., cancer, cardiovascular diseases, immunomodulatory diseases, diabetes, neurogenerative diseases, myocardial infarctions, and cognitive disorders [135]. The MD has been at the epicenter of research in recent decades, with several beneficial health effects being attributed to the adherence to this particular diet [11,136–139]. Moreover, according to the study of van den Brandt [140], the MD has been highly associated with significant reduction in mortality rates when combined with other lifestyle factors, such as body mass index, smoking, and physical activity.

The following sections provide information regarding the most important health effects of the MD based on clinical and cohort study results.

3.1. Antioxidant Activity

The ingredients of the MD contain several antioxidant compounds that have the capacity to protect against cancer, metabolic disorders, aging, and cardiovascular disease, as mentioned in several reports [12,111,141,142]. Total antioxidant capacity (TAC) consists of all of the antioxidants that are present in plasma and body fluids, and provides an integrated measurement rather than the simple sum of measurable antioxidants. For example, in the study of Pitsavos et al. [141] TAC was also positively correlated with the consumption of olive oil and fruit and vegetables, whereas it was inversely associated with the consumption of red meat and low oxidized LDL-cholesterol concentrations, which may explain the beneficial role of MD on the cardiovascular system. Moreover, according to observational studies, fruit and vegetable consumption showed protective effects against cardiovascular diseases, and possible evidence exists for decreased risk of colon cancer, depression, and pancreatic diseases from fruit intake; and colon and rectal cancer, hip fracture, stroke, depression, and pancreatic diseases from vegetable intake [135]. Mounting reports of the high quantity of antioxidants in the MD have been the subject of cohort studies that attempt to prove the strong correlation between the antioxidants in the diet and the lower incidence of diseases in Mediterranean countries [7,136,143,144]. One example is the study of Billingsley and Carbone [145], which attempted to reveal how the antioxidants may reduce the incidence of cardiovascular diseases in high risk patients. The main conclusion of the study is that the high quantity of polyphenols may in part explain the lower risk for patients following the MD, however, there are limitations to their efficacy. Polyphenols are secondary metabolites which act as free radical scavengers and metal chelators, while they also affect the antioxidant activity of various enzymes [146]. Therefore, a high intake of polyphenols through the diet may improve the defense mechanisms of the human body against several diseases related to oxidative stress [10,29,147]. The adherence to Mediterranean-style dietary habits has been associated with reduced risk of diabetes in cohort studies conducted in U.S.A., which highlights the health benefits of the MD [144]. However, further studies are needed to understand if the antioxidants play a major role in the prevention of these cardiovascular diseases, or if it is the result of the combination of various factors (climate, lifestyle, daylight, overall country wealth, socioeconomic status, among others).

3.2. Life Expectancy Rates

According to the existing literature, there is an inverse association between mortality and intake of vegetable fats and plant proteins, whereas there is evident positive correlation between mortality and high intake of animal fats, monounsaturated fats of animal origins, and sugar [148]. These coincide with food-based dietary guidelines of the traditional MD and are associated with lower mortality rates, not only for deaths related to chronic diseases, but also overall mortality rates [148]. The data meta-analysis also showed that legumes and olive oil are major food items discriminating subjects between lower or higher mortality rates, and the lipid-lowering effects and/or reduced thrombosis risk, in addition to dietary fiber or vitamin E intake, could also account for the reduction in mortality rates [136,149]. Other cohort studies and clinical trials have attributed the low mortality rates associated with the MD to the preventive effects of dietary sources against the various types of cardiovascular diseases [136]. The analysis of two epidemiological studies conducted in 2020 involving over 2000 Greek citizens over 50 years old showed that a higher adherence to the MD was associated with a higher level of successful aging [13].

3.3. Type 2 Diabetes

According to a study conducted in United States and Spain of 7018 participants, it was suggested that meaningful changes to eating habits, such as low-fat diets and the adoption of variants of the MD including vegetables, fish, legumes, and extra-virgin oil or nuts, may reduce the risks of chronic diseases [150]. In the same study, it was also reported that even individuals who have a genetic predisposition to diabetes due to carrying two copies of a certain gene variant associated with a high risk of developing type 2 diabetes may reduce the risk of developing this disease by following MD eating habits. Furthermore, the MD is considered a healthy approach to control obesity and complications of the cardiovascular system, especially in women in menopause [147]. The mechanisms behind the benefits of the MD towards patients of type 2 diabetes were tentatively explained by Esposito et al. [142], who analyzed the data from eight meta-analyses and five randomized control trials. The authors stated that the benefits of this particular diet are associated with the anti-inflammatory and antioxidative effects from high-quality foods that down-regulate the activation of the innate immune system by reducing the production of proinflammatory cytokines, inducing the production of anti-inflammatory cytokines, thus improving the sensitivity to insulin in peripheral tissues. Moreover, the adherence to MD is associated with lower glycosylated hemoglobin levels and improved cardiovascular risk factors compared to diets that have low fat content. Vitale et al. [151] performed a cohort study and investigated the effect of polyphenols intake on cardiovascular risk factors of 2573 type 2 diabetes patients. The results of this study showed that diets rich in polyphenols may improve the cardiovascular risk factor, especially in patients with type 2 diabetes, and they suggested that beneficial effects of such diets could also be extrapolated to the general population [151].

3.4. Parkinson's Disease (PD)

Parkinson's disease is a common neurodegenerative disorder which affects approximately 1% of individuals over the age of 60 years old in North America and Europe. According to literature reports, the association of Parkinson's disease status with MD adherence was significant. Alcalay et al. [152] analyzed a Willett semi-quantitative questionnaire completed by 257 PD participants to score their diet during the preceding year. The findings of this study revealed a positive correlation of the adherence to a Mediterranean-type diet and the reduced rates of PD, and the scores of the Mediterranean-type diet were associated with the earlier onset of PD. Although the mechanisms behind these activities are unknown, it is hypothesized that MD ingredients contain various bioactive phytochemicals that lower inflammation and oxidative stress, which both induce PD pathogenesis [153].

According to the study of Maraki et al. [154], there is a lower probability of prodromal PD in higher MD adherence in older people based on the evaluation of the MD adherence scores. A recent randomized controlled trial involving 80 patients with PD analyzed the effect of the adherence to the MD for 10 weeks compared to a control group [155]. The findings of this trial suggested a significant improvement in cognitive functions in patients that followed the MD, although the sample size and the short duration of the study do not allow for explicit conclusions.

3.5. Alzheimer's Disease

Dietary habits may affect the onset and development of neurodegenerative diseases, such as Alzheimer's disease through the increased resistant of neurons to degeneration [156]. The analysis of dietary patterns and their contribution to cognitive functions may provide useful information regarding the degenerative mechanisms and the prevention of mental dysfunctions, particularly because drugs available to treat Alzheimer's disease have only symptomatic effects and do not delay its development [157]. In the study of Scarmeas et al. [158], the evaluation of a total of 192 individuals diagnosed with Alzheimer's disease showed that adherence to the MD may reduce the mortality rates of patients, and the same dietary pattern may also reduce the risk of this particular disease [156]. The systematic review of Masan et al. [159] evaluated the results of twenty-four studies and suggested that the high intake of n-3 fatty acids and the adherence to MD may have protective effects against age-related cognitive decline. A more recent review identified through numerous studies the positive correlation of the MD and the improvement in the overall cognitive health of middle-aged and older people, although more studies are needed to overcome the problems related with the heterogeneity of scoring systems, in addition to the variability in the evaluation of cognitive outcomes among the various studies [160].

3.6. Heart Diseases

The first strong evidence of the beneficial health effects of MD was associated with the lower mortality rates from cardiovascular diseases observed in Greece [6]. Plant-based diets such as MD may contribute to better control of dyslipidemia management in addition to the prevention of cardiovascular diseases (CVD) [4,107]. The effects of MD on risk factors for heart diseases are outlined in several research reports, which conclude that this particular style of diet may decrease the death rates from heart diseases that constitute the main cause of death in developed countries, accounting for about 17 million deaths per year [161]. Carbone et al. [162], who conducted a cross-section animal model study, suggested that high-sugar and high-fat diets such as the modern Western diet resulted in cardiac systolic and diastolic dysfunction, which was reversed when subjects were fed with a standard diet. Despite issues with the methodological approaches followed in intervention trials, and meta-analyses reports that raised concerns regarding the postulated beneficial effects of MD against CVD, a recent critical review report by Martínez-González et al. [163] concluded that these potential controversies are not sufficient to compromise the positive impact of MD on the functions of the cardiovascular system. They also suggested that the existing evidence is strong and consistent, and justifies that the adherence to the MD improves cardiovascular health outcomes through the reductions in rates of coronary heart disease, ischemic strokes, and total cardiovascular diseases. The first evidence of the beneficial effects of the MD was reported in 1950s. Subsequently, numerous clinical trials and epidemiological studies have confirmed the association of this diet with the low incidence of CVD through the improvement of the blood lipid profile [6,164]. A recent clinical study suggested that the consumption of olive oil, which is a key ingredient of the MD, may improve several risk factors of CVD via the diet's antithrombotic and anti-inflammatory activities [27]. Moreover, Estruch et al. [165] compared the short-term impact of two MD ingredients (olive-oil or nuts groups) with a low-fat diet on 772 asymptomatic individuals with high risk of cardiovascular diseases and concluded that, although no clinical outcomes were evaluated, both MDs had beneficial effects against cardiovascular diseases. The short-term effects were associated with the improvement in systolic blood pressure, plasma glucose levels, and the cholesterol–high-density lipoprotein cholesterol ratio [165]. Similar findings were observed by Lee et al. [166], who carried out a controlled crossover study and suggested that following an MD for a 10-day period may improve mood and functions of the cardiovascular system, such as blood systolic pressure. Anagnostis et al. [167] also attributed the beneficial effects of the MD to better blood pressure control and reduction of metabolic syndrome. Long-term cohort studies have also been undertaken, such as that conducted by Buckland et al. [168], who carried out a cohort study with 609 participants who experienced a fatal or nonfatal confirmed acute myocardial infarction or unstable angina requiring revascularization. The authors suggested that high adherence to the MD reduces the risk factors of CVD, particularly the incidents of fatal or nonfatal coronary heart disease events, and the diet should play a significant role in the prevention of coronary heart disease in healthy populations. In the same context, Panagiotakos et al. [169] mentioned the high risk of elevated blood pressure for acute coronary syndromes and the positive effects of the MD in controlling hypertension. Dontas et al. [170] suggested that MD may significantly reduce the development of these syndromes, particularly in the elder population, via the reduction in blood pressure levels and in markers of vascular inflammation. In addition to clinical and epidemiological studies, several other reports highlighted the beneficial effects of various wild edible plants that are commonly consumed in healthy dietary patterns against specific risk factors that may induce cardiovascular diseases, such as *Portulaca oleracea* [171], *Borago officinalis* [172], various Asteraceae species [98], and *Urtica dioica* [173].

3.7. Cancer

The risk percentages of numerous types of cancer can be reduced by following a Mediterranean healthy diet [174–177]. Breast cancer is the most common cancer and cause of cancer death in women, accounting for 23% of all cancers and 13.7% of cancer deaths [178], and prostate cancer is the second most frequent cancer among men [176]. For both cancer types, the adherence to the MD has been associated with reduced risk and mortality rates [174–176]. The epidemiological data from a case-control study conducted by Turati et al. [174] revealed that high adherence to the MD significantly reduced breast cancer risk, although no adequate information on the breast cancer subtype was available. Moreover, Laudisio et al. [175] suggested that beneficial effects of the MD are associated with the regulation of cell proliferation, induction of apoptosis, and antioxidant activities of single dietary components or the synergistic effects of multiple components. However, although these results are confirmed in postmenopausal women, for premenopausal women, for whom a significantly increased risk of cancer recurrence and higher mortality rate exist, more long-term studies are needed to confirm these results. In the same context, Capurso and Vendemiale [176] analyzed the data from observational and case-control studies related to prostate cancer and suggested that MD benefits derive from the high consumption of olive oil, which exerts antioxidant effects and activities on cancer cell signaling and cell cycle progression and proliferation, in addition to tomato sauce, which is rich in lycopene and also exhibits antioxidant properties through the modulation of down-regulation mechanisms of the inflammatory response. According to Coughlin et al. [177] and Di Gioia and Petropoulos [179], plant-based diets are rich in phytoestrogens, which compete with estrogens by binding in estrogen receptors. Moreover, as mentioned by Buckland et al. [143], the MD reduced the risk of breast cancer by 6% overall, and by 7% in postmenopausal women, and Bosetti et al. [180] highlighted the role of the MD in reducing the risk factors of pancreatic cancer. According to the review report carried out by Di Daniele et al. [139], the healthy lifestyle associated with MD may reduce the incidences of obesity-related cancer types.

These beneficial activities have been associated with the presence of specific compounds in plants that are commonly used in this dietary pattern, including phenolic compounds, such as hydroxycinnamic and chlorogenic acid [181], saponins [182], flavones and flavonoids [99,101,183,184], tannins [126], polysaccharides [185], glucosinolates [186,187], and other organosulfur compounds [71,188]. The mechanisms involved in the bioactive properties of MD ingredients against cancer have not yet been thoroughly revealed, however, it is hypothesized that they modify the expression of hormones and growth factors that induce cancer pathogenesis [12].

3.8. Behavioral Effects

Olive oil, fruits, and vegetables, which are key ingredients of the MD, have shown significant protective effects against cognitive decline and dementia through the attenuation of oxidative stress [153,189]. An additional advantage of the MD is its high content in polyunsaturated fatty acids, which are essential for the formation and development of brain tissue [190–192]. According to McMillan et al. [193], who studied the short-term effects of MD, even the adoption of a healthy diet for a 10-day interval may improve the mood and cognitive performance in healthy individuals. This evidence was also supported by the study of Wade et al. [189], who evaluated the MD scores of populations outside the Mediterranean basin over a 5 year period, and concluded that the adherence to the MD may improve the global cognitive health and specific cognitive functions. Moreover, Ferrer-Cascales et al. [194] associated the adherence to the MD with the attenuation of loneliness perception and isolation through the mediation of stress. The study of Ferrer-Cascales et al. [194] also revealed an interrelationship between the degree of adherence to the MD, alcohol intake, emotional eating, and anxiety in student subjects, highlighting the importance of the implementation of educational programs to promote healthy habits among young populations.

Table 1 summarizes different studies, namely, observational, cohort, meta-analysis, randomized control analysis (RCA), systematic reviews, and cross-sectional analysis, in which the benefits of adherence to the MD in human health are identified. Overall, these references prove that this diet is beneficial against many diseases, namely, cardiovascular, neurological, cancer, renal, and others. Nonetheless, a large number of these studies also state that, although evidence of the benefits of the diet has mounted, other tools should be used to prove beyond a doubt its influence, namely, studies in countries that do not adhere to the MD, or regions where the genetic background is different from that of the population in the Mediterranean area.

Table 1. Representation of different types of studies involving humans, in which positive effects of the consumption of the Mediterranean diet were found. Some references focus on various published studies and thus the number of people and countries where the study was carried out are not stated.

Type of Study	Country	No. of People	Health Effects	Reference
Observational study	-	-	Menopausal diseases	[195]
Observational study	Chile	4348	Chronic disease risk	[196]
Meta-analysis	-	-	Metabolic syndrome	[197]
Cohort study	Spain	9408 men	Hypertension	[198]
Cohort study	Italy	13,597	Hypertension	[199]
Meta-analysis	-	7000	Hypertension	[200]
Randomized control analysis (RCA)	-	296	Atherosclerosis	[201]
RCA	Spain	187	Atherosclerosis	[202]
Cohort study	10 European countries	373,803	Obesity	[203]
Cohort study	8 European countries	15,798	Type 2 diabetes	[204]
Cohort studies	-	-	Breast cancer	[205]
Cohort study	United States	51,529	Prostate cancer	[206]
Population study	Greece	22,043	Mortality	[207]
Cohort study	-	51,529	Cognitive decline	[208]
Cohort study	United Kingdom	74,886	Stroke	[209]
RCA	5 European countries	1294	Osteoporosis	[210]
Observational studies	-	-	Depression	[211]
Cohort study	Spain	22,786	Obesity, cardiovascular disease, type 2 diabetes	[212]
Observational study	Greece	1,865	Dementia	[213]
Observational study	Italy	480	Osteoporosis	[214]
RCA	Spain	288	Cardiovascular events	[215]
Meta-analysis	-	72 studies	Degenerative diseases	[216]
Longitudal study	-	3,316,633	Kidney stones	[217]
Randomized control trial	Iran	40	Parkinson's disease	[155]
Cross-sectional analysis	Germany	340	Cognitive performance	[218]
Cross-sectional analysis	United Kingdom	511	Cognitive function	[219]
Systematic review and meta-analysis	-	41,963	Cognitive function	[220]
Systematic review and meta-analysis	-	34,168	Cognitive function	[221]
Cohort study	Italy	1936	Sleep behavior	[222]
Cohort study	Italy	185	Sleep behavior	[223]
Cohort study	Italy	1596	Sleep behavior	[224]

4. Consumers' Behavior, Market Opportunities and Trends

Two of the major 21st century food policy challenges are the reduction of diet-related diseases and the improvement of environmental sustainability. Mediterranean dietary ingredients contain more energy, fiber, and nutrients per liter of water used; therefore, a shift towards the MD could reduce the consumptive water footprint by about 750 L/capita/day [225]. Consumers' perceptions about sustainability and food consumption has a highly significant effect on attitudes and personal and social norms related to waste reduction [226]. The most important drivers of food consumption are price, taste, convenience, and social and cultural factors and habits, in addition to the ethical aspects related to food [227]. The MD, as a plant-oriented dietary approach, appears to be an appropriate choice for humans and environmental sustainability because it reduces the environmental impact associated with the food value chain [228]. Moreover, the trend of decreasing adherence to the MD necessitates cross-disciplinary studies regarding its environmental, socioeconomic, cultural, and sustainability aspects, to preserve and transfer this sustainable dietary pattern to the next generations [229].

The MD may represent an ideal example of a sustainable diet and a part of a sustainability-oriented lifestyle; however, its adherence is significantly affected by consumers' income and education, and other socio-economic drivers [225]. Households with children and those with a higher degree of adherence to the MD have higher probabilities of buying complementary and healthy products, and increasing household size reduces the likelihood of buying niche products [230]. Similar results are presented in a study conducted in Italy by Carlos et al. [230], who identified three different consumer groups: (i) families with MD eating habits (26%); (ii) families in transition who do not adhere to the MD (37%); (iii) families with a less healthy diet. According to this definition, almost 26% of the Italian families adhered to the MD. Concerning Italian households' spending behavior, families with an older and more educated background are more likely to spend more on MD products compared to young and less educated consumers [231,232].

Because of its palatability and long-term sustainability, the MD combined with physical activity has shown promising results in terms of weight loss in individuals with obesity, in addition to similar beneficial effects in menopause-related obesity; the adherence to the MD in menopause reduced the risk of becoming obese, and improved the cardiometabolic profile and the menopausal symptoms [147]. Malnutrition (e.g., both obesity and undernutrition) is considered to be, in addition to climate change, a global threat, indicating an urgent need for a healthier and more sustainable food system [8]. Because MD adherence is affected by socio-economic factors, significant concerns have been raised about the abandonment of healthy diets and the shift towards Westernized and less sustainable eating patterns [233]. Moreover, recent research shows that the current Spanish diet is considerably different from the recommended Mediterranean diet, with three times more meat, dairy, and sugar product consumption, and one-third fewer fruits, vegetables, and cereals [234]. Another study of Spanish adults proved the direct association between the adherence to the MD and all of the physical, and most of the mental, health domains, including vitality and social and emotional functioning [235].

Many of the major post-modern chronic diseases, such as heart diseases, hypertension, obesity, and various types of cancer, have a dietary basis and are highly influenced by changes in dietary patterns due to changes in lifestyles. For example, from 1985 to 2005, adherence of young people to the MD decreased in southern Italy mainly due to the reduced consumption of olive oil [236]. To confront the recent decreased adherence to the MD and broaden the implementation of the MD outside the Mediterranean region requires focused communication activities and interventions in public health by governmental entities and stakeholders involved in policy making, the food industry, education sector, etc. [21].

The recent consumer trend for diversified diets suggests the consumption of wild greens may be fulfilled with the consumption of novel Mediterranean food products, such as wild halophytes or other wild edible species of the Mediterranean basin [34,41,54,98,237,238]. These wild species are considered a valuable genetic resource with the ability to adapt to severe conditions, such as soil and irrigation water salinity, thus facilitating the design and production process. Additionally, they provide significant functional and health beneficial properties that could be implemented in various food products, such as beverages, vegetable extracts, leafy greens, micro-encapsulated oils, food additives, and antimicrobial agents [41]. Furthermore, to combat the unfortunate trend of reduced adherence to the MD by the young Mediterranean population, a fruitful solution might be to improve dietary habits in the workplace, by altering eating patterns and applying diverse group and individual strategies, such as the installation of vending machines and the supplementation of workplace canteens with healthy food products [239]. In contrast, there is also an increasing trend of alternative dietary patterns, such as vegetarianism and veganism, in which higher MD adherence scores are observed compared to those of omnivores [240].

Another aspect to be considered about marketing trends and consumers' behavior is organic farming and its relevance to the MD. Regarding organic farming and the consumption of healthy food products, a recent study of Cypriot consumers revealed that 99% of respondents were aware of organic vegetables, but only 69% and 49% consumed or had knowledge of organically cultivated vegetables, respectively [241]. Moreover, consumers show great willingness to not only be properly informed via different means, but also to pay higher prices for organic vegetables, thus indicating a sustainable and profitable market opportunity. An emerging trend of healthy food consumption can also be seen in the recent opening of organic food restaurants in which consumers perceived the dishes to have good food quality, thus positively influencing price fairness and perceived value, leading to customer satisfaction and intention to return [242]. The trend for consumption of healthy food and sustainable cultivation has led niche market farmers of the World Wide Opportunities on Organic Farms movement (WWOOF) to open the farms to share their organic farming knowledge with guests, as a measure to attain the social, environmental, cultural and economic values that motivated their participation in the movement [243]. The positive exchange outcomes for the farmers were reduced uncertainty for their occupation, and enhanced equality and fairness, in addition to joint responsibility between the farmers and the citizens. In addition to the intangible benefits, organic food supply chains, when designed and implemented appropriately, are capable of delivering significant financial gains with higher sustainable development benefits, particularly when a sharing economy mechanism is applied [244].

5. Conclusions and Future Directions

The MD is highly recognized as a healthy diet with several extensions in aspects of daily life among inhabitants of the broader Mediterranean basin and other parts of the world. Recent evidence has shown a rapid transition to Western dietary habits with direct consequences for the health status of the general public, as reflected in the high incidence rates of chronic diseases. However, there is increasing interest among consumers for food safety, and healthy or functional food products, which has led to the rediscovery of traditional eating habits and the adherence to traditional diets such as the MD. Moreover, the interrelation of the MD with cultural, religious, and other life aspects is the driving force for the evolution of the traditional MD to a more modern approach to eating habits, thus allowing consumers to adapt to a healthy lifestyle. Considering these trends, the scientific community should further explore how these new alternative and modern "MedDiets" affect human health, to evidence the health effects with clinical studies and to reveal the protective mechanisms against various chronic diseases. Moreover, the agrifood industry must adapt to this new era by designing new and attractive food products that fulfill current consumer needs, while preserving the socioeconomic aspects of the MD and sustaining agrobiodiversity.

Author Contributions: E.C., M.C., F.D.G., S.A.P. contributed equally to the preparation of the draft manuscript and the review and editing of the final manuscript. All authors have read and agreed to the published version of the manuscript.

Funding: M. Carocho would like to thank the Portuguese Foundation for Science and Technology for funding the Centro de Investigação de Montanha (UIDB/00690/2020) and for his individual junior research contract (CEECIND/00831/2018). F. Di Gioia contribution was supported by the USDA National Institute of Food and Agriculture and Hatch Appropriations under Project #PEN04723 and Accession #1020664. S. A. Petropoulos contribution was supported by the General Secretariat for Research and Technology of Greece and PRIMA foundation under the project Valuefarm (Prima2019-11).

Acknowledgments: The authors wish to thank Di Gioia Michele, Larosa Anna, Rossella Carone, Losito Mario and Giuseppe Lops for their assistance in the collection of local vegetables and wild edible greens and/or the preparation of rural dishes typical of the traditional Mediterranean diet used for the pictures presented in this manuscript.

Conflicts of Interest: The authors declare no conflict of interest.

References

1. Cena, H.; Calder, P.C. Defining a healthy diet: Evidence for the role of contemporary dietary patterns in health and disease. *Nutrients* **2020**, *12*, 334. [CrossRef] [PubMed]
2. Murakami, K.; Livingstone, M.B.E.; Fujiwara, A.; Sasaki, S. Breakfast in Japan: Findings from the 2012 national health and nutrition survey. *Nutrients* **2018**, *10*, 1551. [CrossRef] [PubMed]
3. Fernstrom, J.D. Effects of the diet on brain function. *Acta Astronaut.* **1981**, *8*, 1035–1042. [CrossRef]

4. Rodríguez-Martín, C.; Garcia-Ortiz, L.; Rodriguez-Sanchez, E.; Martin-Cantera, C.; Soriano-Cano, A.; Arietaleanizbeaskoa, M.S.; Magdalena-Belio, J.F.; Menendez-Suarez, M.; Maderuelo-Fernandez, J.A.; Lugones-Sanchez, C.; et al. The relationship of the atlantic diet with cardiovascular risk factors and markers of arterial stiffness in adults without cardiovascular disease. *Nutrients* **2019**, *11*, 742. [CrossRef]
5. Montagnese, C.; Santarpia, L.; Buonifacio, M.; Nardelli, A.; Caldara, A.R.; Silvestri, E.; Contaldo, F.; Pasanisi, F. European food-based dietary guidelines: A comparison and update. *Nutrition* **2015**, *31*, 908–915. [CrossRef]
6. Lorgeril, M. De Mediterranean Diet and Cardiovascular Disease: Historical Perspective and Latest Evidence. *Curr. Atheroscler. Rep.* **2013**, *15*, 4–8. [CrossRef]
7. Trichopoulou, A.; Bamia, C.; Trichopoulos, D. Anatomy of health effects of Mediterranean diet: Greek EPIC prospective cohort study. *BMJ* **2009**, *339*, 26–28. [CrossRef]
8. Hidalgo-Mora, J.J.; García-Vigara, A.; Sánchez-Sánchez, M.L.; García-Pérez, M.Á.; Tarín, J.; Cano, A. The Mediterranean diet: A historical perspective on food for health. *Maturitas* **2020**, *132*, 65–69. [CrossRef]
9. Hatzis, C.M.; Papandreou, C.; Patelarou, E.; Vardavas, C.I.; Kimioni, E.; Sifaki-Pistolla, D.; Vergetaki, A.; Kafatos, A.G. A 50-year follow-up of the Seven Countries Study: Prevalence of cardiovascular risk factors, food and nutrient intakes among Cretans. *Hormones* **2013**, *12*, 379–385. [CrossRef]
10. Mazzocchi, A.; Leone, L.; Agostoni, C.; Pali-Schöll, I. The secrets of the mediterranean diet. Does [only] olive oil matter? *Nutrients* **2019**, *11*, 2941. [CrossRef]
11. DellaGreca, M. Nutraceuticals and Mediterranean Diet. *Med. Aromat. Plants* **2012**, *1*, 1–3. [CrossRef]
12. Tosti, V.; Bertozzi, B.; Fontana, L. Health Benefits of the Mediterranean Diet: Metabolic and Molecular Mechanisms. *J. Gerontol. Ser. A Biol. Sci. Med. Sci.* **2018**, *73*, 318–326. [CrossRef] [PubMed]
13. Foscolou, A.; D'Cunha, N.M.; Naumovski, N.; Tyrovolas, S.; Chrysohoou, C.; Rallidis, L.; Polychronopoulos, E.; Matalas, A.L.; Sidossis, L.S.; Panagiotakos, D. The association between the level of adherence to the Mediterranean diet and successful aging: An analysis of the ATTICA and MEDIS (MEDiterranean Islands Study) epidemiological studies. *Arch. Gerontol. Geriatr.* **2020**, *89*, 104044. [CrossRef] [PubMed]
14. Saulle, R.; La Torre, G. The Mediterranean Diet, recognized by UNESCO as a cultural heritage of humanity. *Ital. J. Public Health* **2010**, *7*, 414–415.
15. Keys, A.; Mienotti, A.; Karvonen, M.J.; Aravanis, C.; Blackburn, H.; Buzina, R.; Djordjevic, B.S.; Dontas, A.S.; Fidanza, F.; Keys, M.H.; et al. The diet and 15-year death rate in the seven countries study. *Am. J. Epidemiol.* **1986**, *124*, 903–915. [CrossRef]
16. Nestle, M. Mediterranean diets: Historical and research overview. *Am. J. Clin. Nutr.* **1995**, *61*, 1313S–1320S. [CrossRef]
17. Davis, C.; Bryan, J.; Hodgson, J.; Murphy, K. Definition of the mediterranean diet: A literature review. *Nutrients* **2015**, *7*, 9139–9153. [CrossRef]
18. Varela-Moreiras, G.; Ávila, J.M.; Cuadrado, C.; del Pozo, S.; Ruiz, E.; Moreiras, O. Evaluation of food consumption and dietary patterns in Spain by the Food Consumption Survey: Updated information. *Eur. J. Clin. Nutr.* **2010**, *64*, S37–S43. [CrossRef]
19. Karamanos, B.; Thanopoulou, A.; Angelico, F.; Assaad-Khalil, S.; Barbato, A.; Del Ben, M.; Dimitrijevic-Sreckovic, V.; Djordjevic, P.; Gallotti, C.; Katsilambros, N.; et al. Nutritional habits in the Mediterranean Basin. The macronutrient composition of diet and its relation with the tradiational Mediterranean diet. Multi-centre study of the Mediterranean Group for the study of diabetes (MGSD). *Eur. J. Clin. Nutr.* **2002**, *56*, 983–991. [CrossRef]
20. Pfeilstetter, R. Heritage entrepreneurship. Agency-driven promotion of the Mediterranean diet in Spain. *Int. J. Herit. Stud.* **2015**, *21*, 215–231. [CrossRef]
21. Murphy, K.J.; Parletta, N. Implementing a Mediterranean-Style Diet outside the Mediterranean Region. *Curr. Atheroscler. Rep.* **2018**, *20*, 1–10. [CrossRef] [PubMed]
22. Bach-Faig, A.; Berry, E.M.; Lairon, D.; Reguant, J.; Trichopoulou, A.; Dernini, S.; Medina, F.X.; Battino, M.; Belahsen, R.; Miranda, G.; et al. Mediterranean diet pyramid today. Science and cultural updates. *Public Health Nutr.* **2011**, *14*, 2274–2284. [CrossRef] [PubMed]
23. D'Alessandro, A.; Lampignano, L.; De Pergola, G. Mediterranean diet pyramid: A proposal for Italian people. a systematic review of prospective studies to derive serving sizes. *Nutrients* **2019**, *11*, 1296. [CrossRef]
24. Willett, W.C.; Sacks, F.; Trichopoulou, A.; Drescher, G.; Ferro-Luzzi, A.; Helsing, E.; Trichopoulos, D. Mediterranean diet pyramid: A cultural model for healthy eating. *Am. J. Clin. Nutr.* **1995**, *61*, 1402S–1406S. [CrossRef] [PubMed]

25. Lăcătușu, C.M.; Grigorescu, E.D.; Floria, M.; Onofriescu, A.; Mihai, B.M. The mediterranean diet: From an environment-driven food culture to an emerging medical prescription. *Int. J. Environ. Res. Public Health* **2019**, *16*, 942. [CrossRef] [PubMed]
26. Mattson, F.H.; Grundy, S.M. Comparison of effects of dietary saturated, mono- unsaturated, and polyunsaturated fatty acids on plasma lipids and lipoproteins in man. *J. Lipid Res.* **1985**, *26*, 194–202. [PubMed]
27. Rus, A.; Molina, F.; Martínez-Ramírez, M.J.; Aguilar-Ferrándiz, M.E.; Carmona, R.; Moral, M.L. Del Effects of olive oil consumption on cardiovascular risk factors in patients with fibromyalgia. *Nutrients* **2020**, *12*, 918. [CrossRef]
28. Zamora-Zamora, F.; Martínez-Galiano, J.M.; Gaforio, J.J.; Delgado-Rodríguez, M. Effects of olive oil on blood pressure: A systematic review and meta-analysis. *Grasas Aceites* **2018**, *69*. [CrossRef]
29. Godos, J.; Vitale, M.; Micek, A.; Ray, S.; Martini, D.; Del Rio, D.; Riccardi, G.; Galvano, F.; Grosso, G. Dietary polyphenol intake, blood pressure, and hypertension: A systematic review and meta-analysis of observational studies. *Antioxidants* **2019**, *8*, 152. [CrossRef]
30. Grosso, G.; Micek, A.; Godos, J.; Pajak, A.; Sciacca, S.; Galvano, F.; Giovannucci, E.L. Dietary Flavonoid and Lignan Intake and Mortality in Prospective Cohort Studies: Systematic Review and Dose-Response Meta-Analysis. *Am. J. Epidemiol.* **2017**, *185*, 1304–1316. [CrossRef]
31. Xu, H.; Luo, J.; Huang, J.; Wen, Q. Flavonoids intake and risk of type 2 diabetes mellitus: A meta-analysis of prospective cohort studies. *Medicine* **2018**, *97*, 1–7. [CrossRef]
32. Grosso, G.; Godos, J.; Lamuela-Raventos, R.; Ray, S.; Micek, A.; Pajak, A.; Sciacca, S.; D'Orazio, N.; Del Rio, D.; Galvano, F. A comprehensive meta-analysis on dietary flavonoid and lignan intake and cancer risk: Level of evidence and limitations. *Mol. Nutr. Food Res.* **2017**, *61*, 1600930. [CrossRef] [PubMed]
33. Ferro-Luzzi, A.; James, W.P.T.; Kafatos, A. The high-fat Greek diet: A recipe for all? *Eur. J. Clin. Nutr.* **2002**, *56*, 796–809. [CrossRef] [PubMed]
34. Corrêa, R.C.G.; Di Gioia, F.; Ferreira, I.C.F.R.; Petropoulos, S.A. Wild greens used in the Mediterranean diet. In *The Mediterranean Diet: An Evidence-Based Approach*; Preedy, V., Watson, R., Eds.; Academic Press: London, UK, 2020; pp. 209–228. ISBN 9788578110796.
35. Motti, R.; Bonanomi, G.; Lanzotti, V.; Sacchi, R. The Contribution of Wild Edible Plants to the Mediterranean Diet: An Ethnobotanical Case Study along the Coast of Campania (Southern Italy). *Econ. Bot.* **2020**, *74*, 249–272. [CrossRef]
36. Benhammou, S.; Heras-González, L.; Ibáñez-Peinado, D.; Barceló, C.; Hamdan, M.; Rivas, A.; Mariscal-Arcas, M.; Olea-Serrano, F.; Monteagudo, C. Comparison of Mediterranean diet compliance between European and non-European populations in the Mediterranean basin. *Appetite* **2016**, *107*, 521–526. [CrossRef] [PubMed]
37. FAO. The State of the World's Biodiversity for Food and Agriculture. In *The State of the World's Biodiversity for Food and Agriculture*; Pilling, D., Bélanger, J., Eds.; FAO: Rome, Italy, 2019; 572p, ISBN 9789251312704.
38. Hadjichambis, A.C.; Paraskeva-Hadjichambi, D.; Della, A.; Elena Giusti, M.; De Pasquale, C.; Lenzarini, C.; Censorii, E.; Reyes Gonzales-Tejero, M.; Patricia Sanchez-Rojas, C.; Ramiro-Gutierrez, J.M.; et al. Wild and semi-domesticated food plant consumption in seven circum-Mediterranean areas. *Int. J. Food Sci. Nutr.* **2008**, *59*, 383–414. [CrossRef]
39. De Cortes Sánchez-Mata, M.; Tardío, J. *Mediterranean Wild Edible Plants*; Springer: New York, NY, USA, 2016; ISBN 978-1-4939-3327-3.
40. Petropoulos, S.; Fernandes, A.; Barros, L.; Ferreira, I. A comparison of the phenolic profile and antioxidant activity of different *Cichorium spinosum* L. ecotypes. *J. Sci. Food Agric.* **2017**, *98*, 183–189. [CrossRef]
41. Petropoulos, S.A.A.; Karkanis, A.; Martins, N.; Ferreira, I.C.F.R. Edible halophytes of the Mediterranean basin: Potential candidates for novel food products. *Trends Food Sci. Technol.* **2018**, *74*, 69–84. [CrossRef]
42. Petropoulos, S.; Ntatsi, G.; Levizou, E.; Barros, L.; Ferreira, I. Nutritional profile and chemical composition of *Cichorium spinosum* ecotypes. *LWT Food Sci. Technol.* **2016**, *73*, 95–101. [CrossRef]
43. Elia, A.; Santamaria, P. Biodiversity in vegetable crops, a heritage to save: The case of Puglia region. *Ital. J. Agron.* **2013**, *8*, 21–34. [CrossRef]
44. Heinrich, M.; Müller, W.E.; Galli, C. *Local Mediterranean Food Plants and Nutraceuticals*; Karger: Basel, Switzerland, 2006; ISBN 978-3-8055-8124-0.

45. Sánchez-Mata, M.C.; Loera, R.D.C.; Morales, P.; Fernández-Ruiz, V.; Cámara, M.; Marqués, C.D.; Pardo-de-Santayana, M.; Tardío, J.; Cabrera Loera, R.D.; Morales, P.; et al. Wild vegetables of the Mediterranean area as valuable sources of bioactive compounds. *Genet. Resour. Crop Evol.* **2012**, *59*, 431–443. [CrossRef]
46. Boari, F.; Cefola, M.; Di Gioia, F.; Pace, B.; Serio, F.; Cantore, V. Effect of cooking methods on antioxidant activity and nitrate content of selected wild Mediterranean plants. *Int. J. Food Sci. Nutr.* **2013**, *64*, 870–876. [CrossRef] [PubMed]
47. Bonasia, A.; Conversa, G.; Lazzizera, C.; Loizzo, P.; Gambacorta, G.; Elia, A. Evaluation of Garlic Landraces from Foggia Province (Puglia Region; Italy). *Foods* **2020**, *9*, 850. [CrossRef] [PubMed]
48. Petropoulos, S.A.; Fernandes, Â.; Ntatsi, G.; Petrotos, K.; Barros, L.; Ferreira, I.C.F.R. Nutritional value, chemical characterization and bulb morphology of Greek Garlic landraces. *Molecules* **2018**, *23*, 319. [CrossRef] [PubMed]
49. Petropoulos, S.A.S.; Fernandes, Â.; Barros, L.; Ferreira, I.C.F.R.; Ntatsi, G. Morphological, nutritional and chemical description of "Vatikiotiko", an onion local landrace from Greece. *Food Chem.* **2015**, *182*, 156–163. [CrossRef]
50. Siracusa, L.; Avola, G.; Patanè, C.; Riggi, E.; Ruberto, G. Re-evaluation of traditional mediterranean foods. the local landraces of "cipolla di giarratana" (*Allium cepa* L.) and long-storage tomato (*Lycopersicon esculentum* L.): Quality traits and polyphenol content. *J. Sci. Food Agric.* **2013**, *93*, 3512–3519. [CrossRef]
51. Riggi, E.; Avola, G.; Siracusa, L.; Ruberto, G. Flavonol content and biometrical traits as a tool for the characterization of "cipolla di Giarratana": A traditional Sicilian onion landrace. *Food Chem.* **2013**, *140*, 810–816. [CrossRef]
52. Marrelli, M.; Statti, G.; Conforti, F. A review of biologically active natural products from Mediterranean wild edible plants: Benefits in the treatment of obesity and its related disorders. *Molecules* **2020**, *25*, 649. [CrossRef]
53. Petropoulos, S.A.; Barros, L.; Ferreira, I.C.F.R. Editorial: Rediscovering Local Landraces: Shaping Horticulture for the Future. *Front. Plant Sci.* **2019**, *10*, 1–2. [CrossRef]
54. Conversa, G.; Lazzizera, C.; Bonasia, A.; Cifarelli, S.; Losavio, F.; Sonnante, G.; Elia, A. Exploring on-farm agro-biodiversity: A study case of vegetable landraces from Puglia region (Italy). *Biodivers. Conserv.* **2020**, *29*, 747–770. [CrossRef]
55. Correa, R.C.G.; Di Gioia, F.; Ferreira, I.; SA, P. Halophytes for Future Horticulture: The Case of Small-Scale Farming in the Mediterranean Basin. In *Halophytes for Future Horticulture: From Molecules to Ecosystems towards Biosaline Agriculture*; Grigore, M.N., Ed.; Springer Nature Switzerland AG: Cham, Switzerland, 2020; pp. 1–28, ISBN 9783030178543.
56. Petropoulos, S.A.; Sampaio, S.L.; Di Gioia, F.; Tzortzakis, N.; Rouphael, Y.; Kyriacou, M.C.; Ferreira, I. Grown to be Blue—Antioxidant Properties and Health Effects of Colored Vegetables. Part I: Root Vegetables. *Antioxidants* **2019**, *8*, 617. [CrossRef]
57. Di Gioia, F.; Tzortzakis, N.; Rouphael, Y.; Kyriacou, M.C.; Sampaio, S.L.; Ferreira, I.C.F.R.; Petropoulos, S.A. Grown to be blue—Antioxidant properties and health effects of colored vegetables. Part II: Leafy, fruit, and other vegetables. *Antioxidants* **2020**, *9*, 97. [CrossRef] [PubMed]
58. D'Antuono, I.; Di Gioia, F.; Linsalata, V.; Rosskopf, E.N.; Cardinali, A. Impact on health of artichoke and cardoon bioactive compounds: Content, bioaccessibility, bioavailability, and bioactivity. In *Phytochemicals in Vegetables: A Valuable Source of Bioactive Compounds*; Petropoulos, S.A., Ferreira, I.C.F.R., Barros, L., Eds.; Bentham Science Publishers Ltd.: Sharjah, UAE, 2018; 373p.
59. Pavan, S.; Curci, P.L.; Zuluaga, D.L.; Blanco, E.; Sonnante, G. Genotyping-by-sequencing highlights patterns of genetic structure and domestication in artichoke and cardoon. *PLoS ONE* **2018**, *13*. [CrossRef] [PubMed]
60. Mauro, R.; Portis, E.; Acquadro, A.; Lombardo, S.; Mauromicale, G.; Lanteri, S. Genetic diversity of globe artichoke landraces from Sicilian small-holdings: Implications for evolution and domestication of the species. *Conserv. Genet.* **2009**, *10*, 431–440. [CrossRef]
61. Petropoulos, S.A.; Pereira, C.; Ntatsi, G.; Danalatos, N.; Barros, L.; Ferreira, I.C.F.R. Nutritional value and chemical composition of Greek artichoke genotypes. *Food Chem.* **2017**. [CrossRef]
62. Petropoulos, S.A.; Pereira, C.; Tzortzakis, N.; Barros, L.; Ferreira, I.C.F.R. Nutritional value and bioactive compounds characterization of plant parts from *Cynara cardunculus* L. (Asteraceae) cultivated in central Greece. *Front. Plant Sci.* **2018**, *9*, 1–12. [CrossRef]

63. Petropoulos, S.; Fernandes, Â.; Pereira, C.; Tzortzakis, N.; Vaz, J.; Soković, M.; Barros, L.; Ferreira, I.C.F.R. Bioactivities, chemical composition and nutritional value of *Cynara cardunculus* L. seeds. *Food Chem.* **2019**, *289*, 404–412. [CrossRef]
64. Scarano, A.; Gerardi, C.; D'Amico, L.; Accogli, R.; Santino, A. Phytochemical analysis and antioxidant properties in colored tiggiano carrots. *Agriculture* **2018**, *8*, 102. [CrossRef]
65. Gracia, A.; Sánchez, A.M.; Jurado, F.; Mallor, C. Making use of sustainable local plant genetic resources: Would consumers support the recovery of a traditional purple carrot? *Sustainability* **2020**, *12*, 6549. [CrossRef]
66. Cefola, M.; Pace, B.; Renna, M.; Santamaria, P.; Signore, A.; Serio, F. Compositional analysis and antioxidant profile of yellow, orange and purple polignano carrots. *Ital. J. Food Sci.* **2012**, *24*, 284–291.
67. Innocenti, M.; Gallori, S.; Giaccherini, C.; Ieri, F.; Vincieri, F.F.; Mulinacci, N. Evaluation of the phenolic content in the aerial parts of different varieties of *Cichorium intybus* L. *J. Agric. Food Chem.* **2005**, *53*, 6497–6502. [CrossRef] [PubMed]
68. Missio, C.J.; Rivera, A.; Figàs, M.R.; Casanova, C.; Camí, B.; Soler, S.; Simó, J. A comparison of landraces vs. modern varieties of lettuce in organic farming during the winter in the Mediterranean area: An approach considering the viewpoints of breeders, consumers and farmers. *Front. Plant Sci.* **2018**, *9*, 1–15. [CrossRef] [PubMed]
69. Branca, F.; Chiarenza, G.L.; Cavallaro, C.; Gu, H.; Zhao, Z.; Tribulato, A. Diversity of Sicilian broccoli (*Brassica oleracea* var. *italica*) and cauliflower (*Brassica oleracea* var. *botrytis*) landraces and their distinctive bio-morphological, antioxidant, and genetic traits. *Genet. Resour. Crop Evol.* **2018**, *65*, 485–502. [CrossRef]
70. Lotti, C.; Iovieno, P.; Centomani, I.; Marcotrigiano, A.R.; Fanelli, V.; Mimiola, G.; Summo, C.; Pavan, S.; Ricciardi, L. Genetic, bio-agronomic, and nutritional characterization of kale (*Brassica oleracea* L. var. *acephala*) diversity in Apulia, Southern Italy. *Diversity* **2018**, *10*, 25. [CrossRef]
71. Petropoulos, S.A.; Di Gioia, F.; Polyzos, N.; Tzortzakis, N. Natural antioxidants, health effects and bioactive properties of wild *Allium* species. *Curr. Pharm. Des.* **2020**, *26*, 1816–1837. [CrossRef]
72. Petropoulos, S.; Fernandes, Â.; Barros, L.; Ciric, A.; Sokovic, M.; Ferreira, I.C.F.R. Antimicrobial and antioxidant properties of various Greek garlic genotypes. *Food Chem.* **2018**, *245*, 7–12. [CrossRef]
73. Avgeri, I.; Zeliou, K.; Petropoulos, S.A.; Bebeli, P.J.; Papasotiropoulos, V.; Lamari, F.N. Variability in bulb organosulfur compounds, sugars, phenolics, and pyruvate among greek garlic genotypes: Association with antioxidant properties. *Antioxidants* **2020**, *9*, 967. [CrossRef]
74. Mallor, C.; Balcells, M.; Mallor, F.; Sales, E. Genetic variation for bulb size, soluble solids content and pungency in the Spanish sweet onion variety Fuentes de Ebro. Response to selection for low pungency. *Plant Breed.* **2011**, *130*, 55–59. [CrossRef]
75. Petropoulos, S.A.; Ntatsi, G.; Fernandes, Â.; Barros, L.; Barreira, J.C.M.; Ferreira, I.C.F.R.; Antoniadis, V. Long-term storage effect on chemical composition, nutritional value and quality of Greek onion landrace "Vatikiotiko". *Food Chem.* **2016**, *201*, 168–176. [CrossRef]
76. Ricciardi, L.; Mazzeo, R.; Marcotrigiano, A.R.; Rainaldi, G.; Iovieno, P.; Zonno, V.; Pavan, S.; Lotti, C. Assessment of genetic diversity of the "acquaviva red onion" (*Allium cepa* L.) apulian landrace. *Plants* **2020**, *9*, 260. [CrossRef]
77. Oliveira, H.R.; Tomás, D.; Silva, M.; Lopes, S.; Viegas, W.; Veloso, M.M. Genetic diversity and population structure in Vicia faba L. landraces and wild related species assessed by nuclear SSRs. *PLoS ONE* **2016**, *11*, 1–18. [CrossRef] [PubMed]
78. Lioi, L.; Zuluaga, D.L.; Pavan, S.; Sonnante, G. Genotyping-by-sequencing reveals molecular genetic diversity in Italian common bean landraces. *Diversity* **2019**, *11*, 154. [CrossRef]
79. Lioi, L.; Morgese, A.; Cifarelli, S.; Sonnante, G. Germplasm collection, genetic diversity and on-farm conservation of cowpea [*Vigna unguiculata* (L.) Walp.] landraces from Apulia region (southern Italy). *Genet. Resour. Crop Evol.* **2019**, *66*, 165–175. [CrossRef]
80. Hammer, K.; Hanelt, P.; Perrino, P. Carosello and the taxonomy of *Cucumis melo* L. especially of its vegetable races. *Die Kult.* **1986**, *34*, 249–259. [CrossRef]
81. Paris, H.; Janick, J. Early evidence for the culinary use of squash flowers in Italy. *Chron. Horticult.* **2005**, *45*, 20–21.
82. Staub, J.E.; López-Sesé, A.I.; Fanourakis, N. Diversity among melon landraces (*Cucumis melo* L.) from Greece and their genetic relationships with other melon germplasm of diverse origins. *Euphytica* **2004**, *136*, 151–166. [CrossRef]

83. Formisano, G.; Roig, C.; Esteras, C.; Ercolano, M.R.; Nuez, F.; Monforte, A.J.; Picó, M.B. Genetic diversity of Spanish *Cucurbita pepo* landraces: An unexploited resource for summer squash breeding. *Genet. Resour. Crop Evol.* **2012**, *59*, 1169–1184. [CrossRef]
84. Paris, H.S.; Lebeda, A.; Křistkova, E.; Andres, T.C.; Nee, M.H. Parallel Evolution Under Domestication and Phenotypic Differentiation of the Cultivated Subspecies of *Cucurbita pepo* (Cucurbitaceae). *Econ. Bot.* **2012**, *66*, 71–90. [CrossRef]
85. Carillo, P.; Kyriacou, M.C.; El-Nakhel, C.; Pannico, A.; dell'Aversana, E.; D'Amelia, L.; Colla, G.; Caruso, G.; De Pascale, S.; Rouphael, Y. Sensory and functional quality characterization of protected designation of origin 'Piennolo del Vesuvio' cherry tomato landraces from Campania-Italy. *Food Chem.* **2019**, *292*, 166–175. [CrossRef]
86. Conesa, M.; Fullana-Pericàs, M.; Granell, A.; Galmés, J. Mediterranean Long Shelf-Life Landraces: An Untapped Genetic Resource for Tomato Improvement. *Front. Plant Sci.* **2020**, *10*, 1–21. [CrossRef]
87. Figàs, M.R.; Prohens, J.; Raigón, M.D.; Pereira-Dias, L.; Casanova, C.; García-Martínez, M.D.; Rosa, E.; Soler, E.; Plazas, M.; Soler, S. Insights into the adaptation to greenhouse cultivation of the traditional mediterranean long shelf-life tomato carrying the alc mutation: A multi-trait comparison of landraces, selections, and hybrids in open field and greenhouse. *Front. Plant Sci.* **2018**, *871*, 1–16. [CrossRef] [PubMed]
88. Renna, M.; Durante, M.; Gonnella, M.; Buttaro, D.; D'Imperio, M.; Mita, G.; Serio, F. Quality and nutritional evaluation of regina tomato, a traditional long-storage landrace of puglia (Southern Italy). *Agriculture* **2018**, *8*, 83. [CrossRef]
89. Di Gioia, F.; Serio, F.; Buttaro, D.; Ayala, O.; Santamaria, P. Influence of rootstock on vegetative growth, fruit yield and quality in "Cuore di Bue", an heirloom tomato. *J. Hortic. Sci. Biotechnol.* **2010**, *85*, 477–482. [CrossRef]
90. Figás, M.R.; Prohens, J.; Díez, M.J.; Soler, S. Recovering and enhancing the local tomatoes of the Vall d'Albaida, an inland district in the region of València (Spain). *Landraces* **2019**, *4*, 24–27.
91. Pereira-Dias, L.; Vilanova, S.; Fita, A.; Prohens, J.; Rodríguez-Burruezo, A. Genetic diversity, population structure, and relationships in a collection of pepper (*Capsicum* spp.) landraces from the Spanish centre of diversity revealed by genotyping-by-sequencing (GBS). *Hortic. Res.* **2019**, *6*. [CrossRef]
92. Gramazio, P.; Chatziefstratiou, E.; Petropoulos, C.; Chioti, V.; Mylona, P.; Kapotis, G.; Vilanova, S.; Prohens, J.; Papasotiropoulos, V. Multi-level characterization of eggplant accessions from Greek islands and the mainland contributes to the enhancement and conservation of this germplasm and reveals a large diversity and signatures of differentiation between both origins. *Agronomy* **2019**, *9*, 887. [CrossRef]
93. Bonasia, A.; Conversa, G.; Lazzizera, C.; Elia, A. Post-harvest performance of ready-to-eat wild rocket salad as affected by growing period, soilless cultivation system and genotype. *Postharvest Biol. Technol.* **2019**, *156*, 110909. [CrossRef]
94. Di Gioia, F.; Avato, P.; Serio, F.; Argentieri, M.P. Glucosinolate profile of *Eruca sativa*, *Diplotaxis tenuifolia* and *Diplotaxis erucoides* grown in soil and soilless systems. *J. Food Compos. Anal.* **2018**, *69*, 197–204. [CrossRef]
95. Guijarro-Real, C.; Rodríguez-Burruezo, A.; Fita, A. Volatile profile of wall rocket baby-leaves (*Diplotaxis erucoides*) grown under greenhouse: Main compounds and genotype diversity. *Agronomy* **2020**, *10*, 802. [CrossRef]
96. Di Gioia, F.; Santamaria, P. Ai mercati piace la cima di rapa pugliese ortagio "antico". *Ortofrutta Ital.* **2009**, *2*, 102–106.
97. Mazzeo, R.; Morgese, A.; Sonnante, G.; Zuluaga, D.L.; Pavan, S.; Ricciardi, L.; Lotti, C. Genetic Diversity in broccoli rabe (*Brassica rapa* L. subsp. *sylvestris* (L.) Janch.) from Southern Italy. *Sci. Hortic.* **2019**, *253*, 140–146.
98. Petropoulos, S.A.; Fernandes, Â.; Tzortzakis, N.; Sokovic, M.; Ciric, A.; Barros, L.; Ferreira, I.C.F.R. Bioactive compounds content and antimicrobial activities of wild edible Asteraceae species of the Mediterranean flora under commercial cultivation conditions. *Food Res. Int.* **2019**, *119*, 859–868. [CrossRef] [PubMed]
99. Petropoulos, S.A.; Fernandes, Â.; Dias, M.I.; Pereira, C.; Calhelha, R.; Di Gioia, F.; Tzortzakis, N.; Ivanov, M.; Sokovic, M.; Barros, L.; et al. Wild and cultivated *Centaurea raphanina* subsp. *mixta*: A valuable source of bioactive compounds. *Antioxidants* **2020**, *9*, 314.
100. Petropoulos, S.A.; Fernandes, Â.; Dias, M.I.; Pereira, C.; Calhelha, R.C.; Ivanov, M.; Sokovic, M.D.; Ferreira, I.C.F.R.; Barros, L. The Effect of Nitrogen Fertigation and Harvesting Time on Plant Growth and Chemical Composition of *Centaurea raphanina* subsp. *mixta* (DC.) Runemark. *Molecules* **2020**, *25*, 3175. [CrossRef]

101. Petropoulos, S.A.; Fernandes, Â.; Dias, M.I.; Pereira, C.; Calhelha, R.C.; Chrysargyris, A.; Tzortzakis, N.; Ivanov, M.; Sokovic, M.D.; Barros, L.; et al. Chemical composition and plant growth of *Centaurea raphanina* subsp. *mixta* plants cultivated under saline conditions. *Molecules* **2020**, *25*, 2204.
102. Petropoulos, S.; Levizou, E.; Ntatsi, G.; Fernandes, Â.; Petrotos, K.; Akoumianakis, K.; Barros, L.; Ferreira, I. Salinity effect on nutritional value, chemical composition and bioactive compounds content of *Cichorium spinosum* L. *Food Chem.* **2017**, *214*, 129–136. [CrossRef]
103. Petropoulos, S.; Karkanis, A.; Fernandes, Â.; Barros, L.; Ferreira, I.C.F.R.; Ntatsi, G.; Petrotos, K.; Lykas, C.; Khah, E. Chemical composition and yield of six genotypes of common purslane (*Portulaca oleracea* L.): An alternative source of omega-3 fatty acids. *Plant Foods Hum. Nutr.* **2015**, *70*, 420–426. [CrossRef]
104. Petropoulos, S.; Karkanis, A.; Martins, N.; Ferreira, I.C.F.R. Phytochemical composition and bioactive compounds of common purslane (*Portulaca oleracea* L.) as affected by crop management practices. *Trends Food Sci. Technol.* **2016**, *55*, 1–10. [CrossRef]
105. Tarsitano, E.; Calvano, G.; Cavalcanti, E. The Mediterranean Way a model to achieve the 2030 Agenda Sustainable Development Goals (SDGs). *J. Sustain. Dev.* **2019**, *12*, 108. [CrossRef]
106. Burlingame, B.; Dernini, S. *Sustainable Diets and Biodiversity. Directions and Solutions for Policy, Research and Action*; Burlingame, B., Dernini, S., Eds.; FAO: Rome, Italy, 2010; ISBN 9789251072882.
107. Trautwein, E.A.; McKay, S. The role of specific components of a plant-based diet in management of dyslipidemia and the impact on cardiovascular risk. *Nutrients* **2020**, *12*, 2671. [CrossRef]
108. Benvenuti, S.; Maggini, R.; Pardossi, A. Agronomic, Nutraceutical, and Organoleptic Performances of Wild Herbs of Ethnobotanical Tradition. *Int. J. Veg. Sci.* **2017**, *23*, 270–281. [CrossRef]
109. Johns, T.; Powell, B.; Maundu, P.; Eyzaguirre, P.B. Agricultural biodiversity as a link between traditional food systems and contemporary development, social integrity and ecological health. *J. Sci. Food Agric.* **2013**, *93*, 3433–3442. [CrossRef] [PubMed]
110. Zimmerer, K.S.; Vanek, S.J. Toward the integrated framework analysis of linkages among agrobiodiversity, livelihood diversification, ecological systems, and sustainability amid global change. *Land* **2016**, *5*, 10. [CrossRef]
111. Morales, P.; Ferreira, I.C.F.R.; Carvalho, A.M.; Sánchez-Mata, M.C.; Cámara, M.; Fernández-Ruiz, V.; Pardo-de-Santayana, M.; Tardío, J. Mediterranean non-cultivated vegetables as dietary sources of compounds with antioxidant and biological activity. *LWT Food Sci. Technol.* **2014**, *55*, 389–396. [CrossRef]
112. Guiomar, N.; Godinho, S.; Pinto-Correia, T.; Almeida, M.; Bartolini, F.; Bezák, P.; Biró, M.; Bjørkhaug, H.; Bojnec, S.; Brunori, G.; et al. Typology and distribution of small farms in Europe: Towards a better picture. *Land Use Policy* **2018**, *75*, 784–798. [CrossRef]
113. Petropoulos, S.; Fernandes, Â.; Vasileios, A.; Ntatsi, G.; Barros, L.; Ferreira, I. Chemical composition and antioxidant activity of *Cichorium spinosum* L. leaves in relation to developmental stage. *Food Chem.* **2018**, *239*, 946–952. [CrossRef]
114. Petropoulos, S.; Fernandes, Â.; Karkanis, A.; Ntatsi, G.; Barros, L.; Ferreira, I. Successive harvesting affects yield, chemical composition and antioxidant activity of *Cichorium spinosum* L. *Food Chem.* **2017**, *237*, 83–90. [CrossRef]
115. Petropoulos, S.A.; Fernandes, Â.; Dias, M.I.; Vasilakoglou, I.B.; Petrotos, K.; Barros, L.; Ferreira, I.C.F.R. Nutritional value, chemical composition and cytotoxic properties of common purslane (*Portulaca oleracea* L.) in relation to harvesting stage and plant part. *Antioxidants* **2019**, *8*, 293. [CrossRef]
116. Bonaccio, M.; Bes-Rastrollo, M.; de Gaetano, G.; Iacoviello, L. Challenges to the Mediterranean diet at a time of economic crisis. *Nutr. Metab. Cardiovasc. Dis.* **2016**, *26*, 1057–1063. [CrossRef]
117. Panagiotakos, D.B.; Pitsavos, C.; Chrysohoou, C.; Vlismas, K.; Skoumas, Y.; Palliou, K.; Stefanadis, C. The effect of clinical characteristics and dietary habits on the relationship between education status and 5-year incidence of cardiovascular disease: The ATTICA study. *Eur. J. Nutr.* **2008**, *47*, 258–265. [CrossRef]
118. Katsarou, A.; Tyrovolas, S.; Psaltopoulou, T.; Zeimbekis, A.; Tsakountakis, N.; Bountziouka, V.; Gotsis, E.; Metallinos, G.; Polychronopoulos, E.; Lionis, C.; et al. Socio-economic status, place of residence and dietary habits among the elderly: The Mediterranean islands study. *Public Health Nutr.* **2010**, *13*, 1614–1621. [CrossRef] [PubMed]
119. Panagiotakos, D.B.; Pitsavos, C.; Chrysohoou, C.; Vlismas, K.; Skoumas, Y.; Palliou, K.; Stefanadis, C. Dietary habits mediate the relationship between socio-economic status and CVD factors among healthy adults: The ATTICA study. *Public Health Nutr.* **2008**, *11*, 1342–1349. [CrossRef] [PubMed]

120. Bonasia, A.; Lazzizera, C.; Elia, A.; Conversa, G. Nutritional, biophysical and physiological characteristics of wild rocket genotypes as affected by soilless cultivation system, salinity level of nutrient solution and growing period. *Front. Plant Sci.* **2017**, *8*, 1–15. [CrossRef]
121. Ceccanti, C.; Landi, M.; Incrocci, L.; Pardossi, A.; Venturi, F.; Taglieri, I.; Ferroni, G.; Guidi, L. Comparison of three domestications and wild-harvested plants for nutraceutical properties and sensory profiles in five wild edible herbs: Is domestication possible? *Foods* **2020**, *9*, 1065. [CrossRef] [PubMed]
122. Ceccanti, C.; Brizzi, A.; Landi, M.; Incrocci, L.; Pardossi, A.; Guidi, L. Evaluation of Major Minerals and Trace Elements in Wild and Domesticated Edible Herbs Traditionally Used in the Mediterranean Area. *Biol. Trace Elem. Res.* **2020**. [CrossRef] [PubMed]
123. Cohen, S.; Koltai, H.; Selvaraj, G.; Mazuz, M.; Segoli, M.; Bustan, A.; Guy, O. Assessment of the Nutritional and Medicinal Potential of Tubers from Hairy Stork's-Bill. *Plants* **2020**, *9*, 1069. [CrossRef] [PubMed]
124. Karkanis, A.C.; Petropoulos, S.A. Physiological and growth responses of several genotypes of common purslane (*Portulaca oleracea* L.) under Mediterranean semi-arid conditions. *Not. Bot. Horti Agrobot. Cluj-Napoca* **2017**, *45*, 569–575. [CrossRef]
125. Karkanis, A.C.; Fernandes, A.; Vaz, J.; Petropoulos, S.; Georgiou, E.; Ciric, A.; Sokovic, M.; Oludemi, T.; Barros, L.; Ferreira, I. Chemical composition and bioactive properties of *Sanguisorba minor* Scop. under Mediterranean growing conditions. *Food Funct.* **2019**, *10*, 1340–1351. [CrossRef]
126. Finimundy, T.C.; Karkanis, A.; Fernandes, Â.; Petropoulos, S.A.; Calhelha, R.; Petrović, J.; Soković, M.; Rosa, E.; Barros, L.; Ferreira, I.C.F.R. Bioactive properties of *Sanguisorba minor* L. cultivated in central Greece under different fertilization regimes. *Food Chem.* **2020**, *327*, 127043. [CrossRef]
127. Ceccanti, C.; Landi, M.; Benvenuti, S.; Pardossi, A.; Guidi, L. Mediterranean wild edible plants: Weeds or "new functional crops"? *Molecules* **2018**, *23*, 2299. [CrossRef]
128. Petropoulos, S.; Fernandes, Â.; Karkanis, A.; Antoniadis, V.; Barros, L.; Ferreira, I. Nutrient solution composition and growing season affect yield and chemical composition of *Cichorium spinosum* plants. *Sci. Hortic.* **2018**, *231*, 97–107. [CrossRef]
129. Kyriacou, M.C.; Rouphael, Y.; Di Gioia, F.; Kyratzis, A.; Serio, F.; Renna, M.; De Pascale, S.; Santamaria, P. Micro-scale vegetable production and the rise of microgreens. *Trends Food Sci. Technol.* **2016**, *57*, 103–115. [CrossRef]
130. Di Gioia, F.; De Bellis, P.; Mininni, C.; Santamaria, P.; Serio, F. Physicochemical, agronomical and microbiological evaluation of alternative growing media for the production of rapini (*Brassica rapa* L.) microgreens. *J. Sci. Food Agric.* **2017**, *97*, 1212–1219. [CrossRef] [PubMed]
131. Di Gioia, F.; Renna, M.; Santamaria, P. Sprouts, Microgreens and "Baby Leaf" Vegetables. In *Minimally Processed Refrigerated Fruits and Vegetables*; Yildiz, F., Wiley, R., Eds.; Springer: New York, NY, USA, 2017; pp. 403–432, ISBN 9781493970162.
132. Di Gioia, F.; Petropoulos, S.A.; Ozores-Hampton, M.; Morgan, K.; Rosskopf, E.N. Zinc and Iron Agronomic Biofortification of Brassicaceae Microgreens. *Agronomy* **2019**, *9*, 677. [CrossRef]
133. Pires, T.C.S.P.; Dias, M.I.; Barros, L.; Calhelha, R.C.; Alves, M.J.; Oliveira, M.B.P.P.; Santos-Buelga, C.; Ferreira, I.C.F.R.; Barreira, J.C.M.; Santos-Buelga, C.; et al. Edible flowers as sources of phenolic compounds with bioactive potential. *Food Res. Int.* **2018**, *105*, 580–588. [CrossRef] [PubMed]
134. Renna, M.; Di Gioia, F.; Leoni, B.; Mininni, C.; Santamaria, P. Culinary assessment of self-produced microgreens as basic ingredients in sweet and savory dishes. *J. Culin. Sci. Technol.* **2017**, *15*, 126–142. [CrossRef]
135. Dinu, M.; Pagliai, G.; Casini, A.; Sofi, F. Mediterranean diet and multiple health outcomes: An umbrella review of meta-analyses of observational studies and randomised trials. *Eur. J. Clin. Nutr.* **2018**, *72*, 30–43. [CrossRef]
136. Becerra-Tomás, N.; Blanco Mejía, S.; Viguiliouk, E.; Khan, T.; Kendall, C.W.C.; Kahleova, H.; Rahelić, D.; Sievenpiper, J.L.; Salas-Salvadó, J. Mediterranean diet, cardiovascular disease and mortality in diabetes: A systematic review and meta-analysis of prospective cohort studies and randomized clinical trials. *Crit. Rev. Food Sci. Nutr.* **2020**, *60*, 1207–1227. [CrossRef]
137. Feldman, E. Mediterranean diet and frailty risk. *Integr. Med. Alert* **2018**, *21*, 37–40.
138. Ciancarelli, M.; Massimo, C.; Amicis, D.; Ciancarelli, I. Mediterranean Diet and Health Promotion: Evidence and current concerns. *Med. Res. Arch.* **2017**, *5*, 1–16. [CrossRef]

139. Di Daniele, N.D.; Noce, A.; Vidiri, M.F.; Moriconi, E.; Marrone, G.; Annicchiarico-Petruzzelli, M.; D'Urso, G.; Tesauro, M.; Rovella, V.; De Lorenzo, A.D. Impact of Mediterranean diet on metabolic syndrome, cancer and longevity. *Oncotarget* **2017**, *8*, 8947–8979. [CrossRef] [PubMed]
140. Van den Brandt, P.A. The impact of a Mediterranean diet and healthy lifestyle on premature. *Am. J. Clin. Nutr.* **2011**, *94*, 913–920. [CrossRef] [PubMed]
141. Pitsavos, C.; Panagiotakos, D.B.; Tzima, N.; Chrysohoou, C.; Economou, M. Adherence to the Mediterranean diet is associated with total antioxidant capacity in healthy adults: The ATTICA study. *Am. J. Clin. Nutr.* **2005**, *82*, 694–699. [CrossRef] [PubMed]
142. Esposito, K.; Maiorino, M.I.; Bellastella, G.; Panagiotakos, D.B.; Giugliano, D. Mediterranean diet for type 2 diabetes: Cardiometabolic benefits. *Endocrine* **2017**, *56*, 27–32. [CrossRef]
143. Buckland, G.; Travier, N.; Cottet, V.; González, C.A.; Luján-Barroso, L.; Agudo, A.; Trichopoulou, A.; Lagiou, P.; Trichopoulos, D.; Peeters, P.H.; et al. Adherence to the Mediterranean diet and risk of breast cancer in the European Prospective Investigation into Cancer and Nutrition cohort study. *Int. J. Cancer* **2013**, *2927*, 2918–2927. [CrossRef]
144. O'Connor, L.E.; Hu, E.A.; Steffen, L.M.; Selvin, E.; Rebholz, C.M. Adherence to a Mediterranean-style eating pattern and risk of diabetes in a U.S. prospective cohort study. *Nutr. Diabetes* **2020**, *10*. [CrossRef]
145. Billingsley, H.E.; Carbone, S. The antioxidant potential of the Mediterranean diet in patients at high cardiovascular risk: An in-depth review of the PREDIMED. *Nutr. Diabetes* **2018**, *8*, 1–8. [CrossRef]
146. Balabanos, D.; Savva, S.; Mitakou, S.; Mikropoulou, E.V.; Skaperda, Z.; Stagos, D.; Priftis, A.; Halabalaki, M.; Vougogiannopoulou, K.; Kouretas, D.; et al. Extracts from the Mediterranean food plants *Carthamus lanatus*, *Cichorium intybus*, and *Cichorium spinosum* enhanced GSH Levels and increased Nrf2 expression in human endothelial cells. *Oxid. Med. Cell. Longev.* **2018**, *2018*, 1–14.
147. Pugliese, G.; Barrea, L.; Laudisio, D.; Aprano, S.; Castellucci, B.; Framondi, L.; Di Matteo, R.; Savastano, S.; Colao, A.; Muscogiuri, G. Mediterranean diet as tool to manage obesity in menopause: A narrative review. *Nutrition* **2020**, *79–80*, 110991. [CrossRef]
148. Trevisan, M.; Krogh, V.; Grioni, S.; Farinaro, E. Mediterranean diet and all-cause mortality: A cohort of Italian men. *Nutr. Metab. Cardiovasc. Dis.* **2020**, *30*, 1673–1678. [CrossRef]
149. Schwingshackl, L.; Morze, J.; Hoffmann, G. Mediterranean diet and health status: Active ingredients and pharmacological mechanisms. *Br. J. Pharmacol.* **2020**, *177*, 1241–1257. [CrossRef] [PubMed]
150. Mitka, M. Mediterranean Diet May Reduce Stroke Risk in Individuals With Genetic Predisposition to Diabetes. *JAMA* **2013**, *310*, 1013. [CrossRef] [PubMed]
151. Vitale, M.; Vaccaro, O.; Masulli, M.; Bonora, E.; Del Prato, S.; Giorda, C.B.; Nicolucci, A.; Squatrito, S.; Auciello, S.; Babini, A.C.; et al. Polyphenol intake and cardiovascular risk factors in a population with type 2 diabetes: The TOSCA.IT study. *Clin. Nutr.* **2017**, *36*, 1686–1692. [CrossRef] [PubMed]
152. Alcalay, R.N.; Gu, Y.; Mejia-santana, H.; Cote, L.; Marder, K.S.; Scarmeas, N. The Association between Mediterranean Diet Adherence and Parkinson's Disease Participants and Methods. *Mov. Disord.* **2012**, *27*, 771–774. [CrossRef] [PubMed]
153. Singh, A.; Kukreti, R.; Saso, L.; Kukreti, S. Oxidative stress: A key modulator in neurodegenerative diseases. *Molecules* **2019**, *24*, 1583. [CrossRef] [PubMed]
154. Maraki, M.I.; Yannakoulia, M.; Stamelou, M.; Stefanis, L.; Xiromerisiou, G.; Kosmidis, M.H.; Dardiotis, E.; Hadjigeorgiou, G.M.; Sakka, P.; Anastasiou, C.A.; et al. Mediterranean diet adherence is related to reduced probability of prodromal Parkinson's disease. *Mov. Disord.* **2019**, *34*, 48–57. [CrossRef]
155. Paknahad, Z.; Sheklabadi, E.; Derakhshan, Y.; Bagherniya, M.; Chitsaz, A. The effect of the Mediterranean diet on cognitive function in patients with Parkinson's disease: A randomized clinical controlled trial. *Complement. Ther. Med.* **2020**, *50*, 102366. [CrossRef]
156. Scarmeas, N.; Stern, Y.; Tang, M.; Luchsinger, J.A. Mediterranean Diet and Risk for Alzheimer's Disease. *Ann. Neurol.* **2006**, *59*, 912–921. [CrossRef]
157. Panza, F.; Lozupone, M.; Solfrizzi, V.; Custodero, C.; Valiani, V.; D'Introno, A.; Stella, E.; Stallone, R.; Piccininni, M.; Bellomo, A.; et al. Contribution of mediterranean diet in the prevention of alzheimer's disease. *Role Mediterr. Diet Brain Neurodegener. Dis.* **2017**, 139–155. [CrossRef]
158. Scarmeas, N.; Luchsinger, J.A.; Mayeux, R.; Stern, Y. Mediterranean diet and Alzheimer disease mortality. *Neurology* **2007**, *69*, 1084–1093. [CrossRef]

159. Masana, M.F.; Koyanagi, A.; Haro, J.M.; Tyrovolas, S. n-3 Fatty acids, Mediterranean diet and cognitive function in normal aging: A systematic review. *Exp. Gerontol.* **2017**, *91*, 39–50. [CrossRef] [PubMed]
160. Limongi, F.; Siviero, P.; Bozanic, A.; Noale, M.; Veronese, N.; Maggi, S. The Effect of Adherence to the Mediterranean Diet on Late-Life Cognitive Disorders: A Systematic Review. *J. Am. Med. Dir. Assoc.* **2020**, *21*, 1402–1409. [CrossRef] [PubMed]
161. Nowbar, A.N.; Gitto, M.; Howard, J.P.; Francis, D.P.; Al-Lamee, R. Mortality from ischemic heart disease: Analysis of data from the world health organization and coronary artery disease risk factors from NCD risk factor collaboration. *Circ. Cardiovasc. Qual. Outcomes* **2019**, *12*, 1–11. [CrossRef] [PubMed]
162. Carbone, S.; Mauro, A.G.; Mezzaroma, E.; Kraskauskas, D.; Marchetti, C.; Buzzetti, R.; Van Tassell, B.W.; Abbate, A.; Toldo, S. A high-sugar and high-fat diet impairs cardiac systolic and diastolic function in mice. *Int. J. Cardiol.* **2015**, *198*, 66–69. [CrossRef] [PubMed]
163. Martínez-González, M.A.; Gea, A.; Ruiz-Canela, M. The Mediterranean Diet and Cardiovascular Health: A Critical Review. *Circ. Res.* **2019**, *124*, 779–798. [CrossRef] [PubMed]
164. Bédard, A.; Riverin, M.; Dodin, S.; Corneau, L.; Lemieux, S. Sex differences in the impact of the Mediterranean diet on cardiovascular risk profile. *Br. J. Nutr.* **2012**, *108*, 1428–1434. [CrossRef]
165. Estruch, R.; Martínez-González, M.Á.; Corella, D.; Salas-Salvadó, J.; Ruiz-Gutiérrez, V.; Covas, M.I.; Fiol, M.; Gómez-Gracia, E.; López-Sabater, M.C.; Vinyoles, E.; et al. Effects of a Mediterranean-Style Diet on Cardiovascular Risk Factors. *Ann. Intern. Med.* **2006**, *145*, 1–11. [CrossRef]
166. Lee, J.; Pase, M.; Pipingas, A.; Raubenheimer, J.; Thurgood, M.; Villalon, L.; Macpherson, H.; Gibbs, A.; Scholey, A. Switching to a 10-day Mediterranean-style diet improves mood and cardiovascular function in a controlled crossover study. *Nutrition* **2015**, *31*, 647–652. [CrossRef]
167. Anagnostis, P.; Sfikas, G.; Gotsis, E.; Karras, S.; Athyros, V.G. Is the Beneficial Effect of Mediterranean Diet on Cardiovascular Risk Partly Mediated through Better Blood Pressure Control? *Open Hypertens. J.* **2013**, *5*, 36–39. [CrossRef]
168. Buckland, G.; Travier, N.; Cotter, V.; González, C.A.; Luján-Barroso, L.; Agudo, A.; Trichopoulou, A.; Lagiou, P.; Trichopoulos, D.; Vilardell, M.; et al. Adherence to the mediterranean diet and risk of coronary heart disease in the spanish EPIC cohort study. *Am. J. Epidemiol.* **2009**, *170*, 1518–1529. [CrossRef]
169. Panagiotakos, D.B.; Chrysohoou, C.; Pitsavos, C.; Tzioumis, K.; Papaioannou, I.; Stefanadis, C.; Toutouzas, P. The association of Mediterranean diet with lower risk of acute coronary syndromes in hypertensive subjects. *Int. J. Cardiol.* **2002**, *82*, 141–147. [CrossRef]
170. Dontas, A.S.; Zerefos, N.S.; Panagiotakos, D.B.; Valis, D.A. Mediterranean diet and prevention of coronary heart disease in the elderly. *Clin. Interv. Aging* **2007**, *2*, 109–115. [CrossRef] [PubMed]
171. Zidan, Y.; Bouderbala, S.; Djellouli, F.; Lacaille-Dubois, M.A.; Bouchenak, M. *Portulaca oleracea* reduces triglyceridemia, cholesterolemia, and improves lecithin: Cholesterol acyltransferase activity in rats fed enriched-cholesterol diet. *Phytomedicine* **2014**, *21*, 1504–1508. [CrossRef] [PubMed]
172. Karimi, E.; Oskoueian, E.; Karimi, A.; Noura, R.; Ebrahimi, M. *Borago officinalis* L. flower: A comprehensive study on bioactive compounds and its health-promoting properties. *J. Food Meas. Charact.* **2018**, *12*, 826–838. [CrossRef]
173. Bisht, S.; Bhandari, S.; Bisht, N.S. *Urtica dioica* (L): An undervalued, economically important plant. *Agric. Sci. Res. J.* **2012**, *2*, 250–252.
174. Turati, F.; Carioli, G.; Bravi, F.; Ferraroni, M.; Serraino, D.; Montella, M.; Giacosa, A.; Toffolutti, F.; Negri, E.; Levi, F.; et al. Mediterranean diet and breast cancer risk. *Nutrients* **2018**, *10*, 326. [CrossRef] [PubMed]
175. Laudisio, D.; Barrea, L.; Muscogiuri, G.; Annunziata, G.; Colao, A.; Savastano, S. Breast cancer prevention in premenopausal women: Role of the Mediterranean diet and its components. *Nutr. Res. Rev.* **2020**, *33*, 19–32. [CrossRef]
176. Capurso, C.; Vendemiale, G. The Mediterranean Diet Reduces the Risk and Mortality of the Prostate Cancer: A Narrative Review. *Front. Nutr.* **2017**, *4*, 38. [CrossRef]
177. Coughlin, S.S.; Stewart, J.; Williams, L.B. A review of adherence to the Mediterranean diet and breast cancer risk according to estrogen- and progesterone-receptor status and HER2 oncogene expression. *Ann. Epidemiol. Public Heal.* **2018**, *1*, 1–13. [CrossRef]
178. Al Shaikh, A.; Braakhuis, A.J.; Bishop, K.S. The Mediterranean Diet and Breast Cancer: A Personalised Approach. *Healthcare* **2019**, *7*, 104. [CrossRef]

179. Di Gioia, F.; Petropoulos, S.A. Phytoestrogens, phytosteroids and saponins in vegetables: Biosynthesis, functions, health effects and practical applications. In *Advances in Food and Nutrition Research*; Elsevier: Amsterdam, The Netherlands, 2019.
180. Bosetti, C.; Turati, F.; Pont, A.D.; Ferraroni, M.; Polesel, J.; Negri, E.; Serraino, D.; Talamini, R.; La Vecchia, C.; Zeegers, M.P. The role of Mediterranean diet on the risk of pancreatic cancer. *Br. J. Cancer* **2013**, *109*, 1360–1366. [CrossRef] [PubMed]
181. Romanos-Nanclares, A.; Sánchez-Quesada, C.; Gardeazábal, I.; Martínez-González, M.Á.; Gea, A.; Toledo, E. Phenolic Acid Subclasses, Individual Compounds, and Breast Cancer Risk in a Mediterranean Cohort: The SUN Project. *J. Acad. Nutr. Diet.* **2020**, *120*, 1002–1015.e5. [CrossRef] [PubMed]
182. Jaramillo, S.; Muriana, F.J.G.; Guillen, R.; Jimenez-Araujo, A.; Rodriguez-Arcos, R.; Lopez, S. Saponins from edible spears of wild asparagus inhibit AKT, p70S6K, and ERK signalling, and induce apoptosis through G0/G1 cell cycle arrest in human colon cancer HCT-116 cells. *J. Funct. Foods* **2016**, *26*, 1–10. [CrossRef]
183. Mikropoulou, E.V.; Vougogiannopoulou, K.; Kalpoutzakis, E.; Sklirou, A.D.; Skaperda, Z.; lle Houriet, J.; Wolfender, J.L.; Trougakos, I.P.; Kouretas, D.; Halabalaki, M.; et al. Phytochemical composition of the decoctions of Greek edible greens (chórta) and evaluation of antioxidant and cytotoxic properties. *Molecules* **2018**, *23*, 1541. [CrossRef]
184. Bilušić, T.; Šola, I.; Rusak, G.; Poljuha, D.; Čikeš Čulić, V. Antiproliferative and pro-apoptotic activities of wild asparagus (*Asparagus acutifolius* L.), black bryony (*Tamus communis* L.) and butcher's broom (*Ruscus aculeatus* L.) aqueous extracts against T24 and A549 cancer cell lines. *J. Food Biochem.* **2019**, *43*, 1–9. [CrossRef]
185. Ryu, D.S.; Kim, S.H.; Lee, D.S. Anti-proliferative effect of polysaccharides from *Salicornia herbacea* on induction of G2/M arrest and apoptosis in human colon cancer cells. *J. Microbiol. Biotechnol.* **2009**, *19*, 1482–1489. [CrossRef]
186. Di Gioia, F.; Pinela, J.; de Haro Bailón, A.; Fereira, I.C.; Petropoulos, S.A. The dilemma of "good" and "bad" glucosinolates and the potential to regulate their content. In *Glucosinolates: Properties, Recovery, and Applications*; Galanakis, C.M., Ed.; Academic Press: London, UK, 2019; Volume 1, pp. 1–45, ISBN 9780128164938.
187. Mandrich, L.; Caputo, E. Brassicaceae-derived anticancer agents: Towards a green approach to beat cancer. *Nutrients* **2020**, *12*, 868. [CrossRef]
188. Petropoulos, S.A.; Di Gioia, F.; Ntatsi, G. Vegetable organosulfur compounds and their health promoting effects. *Curr. Pharm. Des.* **2017**, *23*, 1–26. [CrossRef]
189. Wade, A.T.; Elias, M.F.; Murphy, K.J. Adherence to a Mediterranean diet is associated with cognitive function in an older non-Mediterranean sample: Findings from the Maine-Syracuse Longitudinal Study. *Nutr. Neurosci.* **2019**, 1–12. [CrossRef]
190. Bourre, J.M. Diet, Brain Lipids, and Brain Functions: Polyunsaturated Fatty Acids, Mainly Omega-3 Fatty Acids. In *Handbook of Neurochemistry and Molecular Neurobiology*; Tettamanti, G., Goracci, G., Eds.; Springer: New York, NY, USA, 2009; pp. 410–441.
191. Bourre, J.M. Dietary omega-3 Fatty acids and psychiatry: Mood, behaviour, stress, depression, dementia and aging. *J. Nutr. Heal. Aging* **2005**, *9*, 31–38.
192. Bourre, J.M. Roles of unsaturated fatty acids (especially omega-3 fatty acids) in the brain at various ages and during ageing. *J. Nutr. Heal. Aging* **2004**, *8*, 163–174.
193. Mcmillan, L.; Owen, L.; Kras, M.; Scholey, A. Behavioural effects of a 10-day Mediterranean diet. Results from a pilot study evaluating mood and cognitive performance. *Appetite* **2011**, *56*, 143–147. [CrossRef] [PubMed]
194. Ferrer-Cascales, R.; Albaladejo-Blázquez, N.; Ruiz-Robledillo, N.; Rubio-Aparicio, M.; Laguna-Pérez, A.; Zaragoza-Martí, A. Low adherence to the mediterranean diet in isolated adolescents: The mediation effects of stress. *Nutrients* **2018**, *10*, 1894. [CrossRef] [PubMed]
195. Cano, A.; Marshall, S.; Zolfaroli, I.; Bitzer, J.; Ceausu, I.; Chedraui, P.; Durmusoglu, F.; Erkkola, R.; Goulis, D.G.; Hirschberg, A.L.; et al. The Mediterranean diet and menopausal health: An EMAS position statement. *Maturitas* **2020**, *139*, 90–97. [CrossRef] [PubMed]
196. Echeverría, G.; McGee, E.E.; Urquiaga, I.; Jiménez, P.; D'Acuña, S.; Villarroel, L.; Velasco, N.; Leighton, F.; Rigotti, A. Inverse associations between a locally validated Mediterranean diet index, overweight/obesity, and metabolic syndrome in Chilean adults. *Nutrients* **2017**, *9*, 862. [CrossRef]

197. Esposito, K.; Maiorino, M.I.; Bellastella, G.; Chiodini, P.; Panagiotakos, D.; Giugliano, D. A journey into a Mediterranean diet and type 2 diabetes: A systematic review with meta-analyses. *BMJ Open* **2015**, *5*. [CrossRef]
198. Núñez-Córdoba, J.M.; Valencia-Serrano, F.; Toledo, E.; Alonso, A.; Martínez-González, M.A. The Mediterranean diet and incidence of hypertension: The Seguimiento Universidad de Navarra (SUN) study. *Am. J. Epidemiol.* **2009**, *169*, 339–346. [CrossRef]
199. Bendinelli, B.; Masala, G.; Bruno, R.M.; Caini, S.; Saieva, C.; Boninsegni, A.; Ungar, A.; Ghiadoni, L.; Palli, D. A priori dietary patterns and blood pressure in the EPIC Florence cohort: A cross-sectional study. *Eur. J. Nutr.* **2019**, *58*, 455–466. [CrossRef]
200. Nissensohn, M.; Román-Viñas, B.; Sánchez-Villegas, A.; Piscopo, S.; Serra-Majem, L. The Effect of the Mediterranean Diet on Hypertension: A Systematic Review and Meta-Analysis. *J. Nutr. Educ. Behav.* **2016**, *48*, 42–53.e1. [CrossRef]
201. Hernáez, Á.; Castañer, O.; Elosua, R.; Pintó, X.; Estruch, R.; Salas-Salvadó, J.; Corella, D.; Arós, F.; Serra-Majem, L.; Fiol, M.; et al. Mediterranean Diet Improves High-Density Lipoprotein Function in High-Cardiovascular-Risk Individuals. *Circulation* **2017**, *135*, 633–643. [CrossRef]
202. Murie-Fernandez, M.; Irimia, P.; Toledo, E.; Martínez-Vila, E.; Buil-Cosiales, P.; Serrano-Martínez, M.; Ruiz-Gutiérrez, V.; Ros, E.; Estruch, R.; Martínez-González, M. ángel Carotid intima-media thickness changes with Mediterranean diet: A randomized trial (PREDIMED-Navarra). *Atherosclerosis* **2011**, *219*, 158–162. [CrossRef] [PubMed]
203. Romaguera, D.; Norat, T.; Vergnaud, A.C.; Mouw, T.; May, A.M.; Agudo, A.; Buckland, G.; Slimani, N.; Rinaldi, S.; Couto, E.; et al. Mediterranean dietary patterns and prospective weight change in participants of the EPIC-PANACEA project. *Am. J. Clin. Nutr.* **2010**, *92*, 912–921. [CrossRef] [PubMed]
204. Romaguera, D. Mediterranean diet and type 2 diabetes risk in the European prospective investigation into cancer and nutrition (EPIC) study: The interAct project. *Diabetes Care* **2011**, *34*, 1913–1918. [PubMed]
205. Schwingshackl, L.; Schwedhelm, C.; Galbete, C.; Hoffmann, G. Adherence to mediterranean diet and risk of cancer: An updated systematic review and meta-analysis. *Nutrients* **2017**, *9*, 1063. [CrossRef]
206. Kenfield, S.A.; Dupre, N.; Richman, E.L.; Stampfer, M.J.; Chan, J.M.; Giovannucci, E.L. Mediterranean diet and prostate cancer risk and mortality in the health professionals follow-up study. *Eur. Urol.* **2014**, *65*, 887–894. [CrossRef]
207. Trichopoulou, A.; Costacou, T.; Bamia, C.; Trichopoulos, D. Adherence to a Mediterranean Diet and Survival in a Greek Population. *N. Engl. J. Med.* **2003**, *348*, 2599–2608. [CrossRef]
208. Bhushan, A.; Fondell, E.; Ascherio, A.; Yuan, C.; Grodstein, F.; Willett, W. Adherence to Mediterranean diet and subjective cognitive function in men. *Eur. J. Epidemiol.* **2018**, *33*, 223–234. [CrossRef]
209. Fung, T.; Rexrode, K.; Mantzoros, C.; Manson, J.; Willet, W.; Hu, F. Mediterranean diet and incidence and mortality of coronary heart disease and stroke in women. *Circulation* **2009**, *119*, 1093–1100. [CrossRef]
210. Jennings, A.; Cashman, K.D.; Gillings, R.; Cassidy, A.; Tang, J.; Fraser, W.; Dowling, K.G.; Hull, G.L.J.; Berendsen, A.A.M.; De Groot, L.C.P.G.M.; et al. A Mediterranean-like dietary pattern with Vitamin D3 (10 μg/d) supplements reduced the rate of bone loss in older Europeans with osteoporosis at baseline: Results of a 1-y randomized controlled trial. *Am. J. Clin. Nutr.* **2018**, *108*, 633–640. [CrossRef]
211. Lassale, C.; Batty, G.D.; Baghdadli, A.; Jacka, F.; Sánchez-Villegas, A.; Kivimäki, M.; Akbaraly, T. Healthy dietary indices and risk of depressive outcomes: A systematic review and meta-analysis of observational studies. *Mol. Psychiatry* **2019**, *24*, 965–986. [CrossRef]
212. Carlos, S.; De La Fuente-Arrillaga, C.; Bes-Rastrollo, M.; Razquin, C.; Rico-Campà, A.; Martínez-González, M.A.; Ruiz-Canela, M. Mediterranean diet and health outcomes in the SUN cohort. *Nutrients* **2018**, *10*, 439. [CrossRef] [PubMed]
213. Anastasiou, C.A.; Yannakoulia, M.; Kosmidis, M.H.; Dardiotis, E.; Hadjigeorgiou, G.M.; Sakka, P.; Arampatzi, X.; Bougea, A.; Labropoulos, I.; Scarmeas, N. Mediterranean diet and cognitive health: Initial results from the Hellenic Longitudinal Investigation of Ageing and Diet. *PLoS ONE* **2017**, *12*, 1–18. [CrossRef] [PubMed]
214. Savanelli, M.C.; Barrea, L.; Macchia, P.E.; Savastano, S.; Falco, A.; Renzullo, A.; Scarano, E.; Nettore, I.C.; Colao, A.; Somma, C. Preliminary results demonstrating the impact of Mediterranean diet on bone health. *J. Transl. Med.* **2017**, *15*, 1–8. [CrossRef] [PubMed]

215. Estruch, R.; Ros, E.; Salas-Salvadó, J.; Covas, M.-I.; Corella, D.; Arós, F.; Gómez-Gracia, E.; Ruiz-Gutiérrez, V.; Fiol, M.; Lapetra, J.; et al. Primary Prevention of Cardiovascular Disease with a Mediterranean Diet Supplemented with Extra-Virgin Olive Oil or Nuts. *N. Engl. J. Med.* **2018**, *378*, e34. [CrossRef]
216. Sofi, F.; Abbate, R.; Gensini, G.F.; Casini, A. Accruing evidence on benefits of adherence to the Mediterranean diet on health: An updated systematic review and meta-analysis. *Am. J. Clin. Nutr.* **2010**, *92*, 1189–1196. [CrossRef] [PubMed]
217. Rodriguez, A.; Curhan, G.C.; Gambaro, G.; Taylor, E.N.; Ferraro, P.M. Mediterranean diet adherence and risk of incident kidney stones. *Am. J. Clin. Nutr.* **2020**. [CrossRef] [PubMed]
218. Kössler, T.; Weber, K.S.; Wölwer, W.; Hoyer, A.; Strassburger, K.; Burkart, V.; Szendroedi, J.; Roden, M.; Müssig, K.; Roden, M.; et al. Associations between cognitive performance and Mediterranean dietary pattern in patients with type 1 or type 2 diabetes mellitus. *Nutr. Diabetes* **2020**, *10*, 4–9. [CrossRef]
219. Corley, J.; Cox, S.R.; Taylor, A.M.; Hernandez, M.V.; Maniega, S.M.; Ballerini, L.; Wiseman, S.; Meijboom, R.; Backhouse, E.V.; Bastin, M.E.; et al. Dietary patterns, cognitive function, and structural neuroimaging measures of brain aging. *Exp. Gerontol.* **2020**, *142*, 111117. [CrossRef]
220. Kelly, M.E.; Duff, H.; Kelly, S.; McHugh Power, J.E.; Brennan, S.; Lawlor, B.A.; Loughrey, D.G. The impact ofsocial activities, social networks, social support and social relationships on the cognitive functioning of healthy older adults: A systematic review. *Syst. Rev.* **2017**, *6*. [CrossRef]
221. Wu, L.; Sun, D. Adherence to Mediterranean diet and risk of developing cognitive disorders: An updated systematic review and meta-analysis of prospective cohort studies. *Sci. Rep.* **2017**, *7*, 1–9. [CrossRef]
222. Godos, J.; Ferri, R.; Caraci, F.; Cosentino, F.I.I.; Castellano, S.; Galvano, F.; Grosso, G. Adherence to the mediterranean diet is associated with better sleep quality in Italian adults. *Nutrients* **2019**, *11*, 976. [CrossRef] [PubMed]
223. Gianfredi, V.; Nucci, D.; Tonzani, A.; Amodeo, R.; Benvenuti, A.L.; Villarini, M.; Moretti, M. Sleep disorder, Mediterranean Diet and learning performance among nursing students: inSOMNIA, a cross-sectional study. *Ann. Ig.* **2018**, *30*, 470–481. [PubMed]
224. Campanini, M.Z.; Guallar-Castillón, P.; Rodríguez-Artalejo, F.; Lopez-Garcia, E. Mediterranean diet and changes in sleep duration and indicators of sleep quality in older adults. *Sleep* **2017**, *40*, 1–9. [CrossRef]
225. Cavaliere, A.; De Marchi, E.; Banterle, A. Exploring the adherence to the mediterranean diet and its relationship with individual lifestyle: The role of healthy behaviors, pro-environmental behaviors, income, and education. *Nutrients* **2018**, *10*, 141. [CrossRef]
226. Kim, M.J.; Hall, C.M.; Kim, D.K. Predicting environmentally friendly eating out behavior by value-attitude-behavior theory: Does being vegetarian reduce food waste? *J. Sustain. Tour.* **2020**, *28*, 797–815. [CrossRef]
227. Kearney, J. Food consumption trends and drivers. *Philos. Trans. R. Soc. B Biol. Sci.* **2010**, *365*, 2793–2807. [CrossRef] [PubMed]
228. Coats, L.; Aboul-Enein, B.H.; Dodge, E.; Benajiba, N.; Kruk, J.; Khaled, M.B.; Diaf, M.; El Herrag, S.E. Perspectives of Environmental Health Promotion and the Mediterranean Diet: A Thematic Narrative Synthesis. *J. Hunger Environ. Nutr.* **2020**, 1–23. [CrossRef]
229. Dernini, S.; Berry, E.M. Mediterranean Diet: From a Healthy Diet to a Sustainable Dietary Pattern. *Front. Nutr.* **2015**, *2*, 1–7. [CrossRef]
230. Carlos, M.; Elena, B.; Teresa, I.M. Are Adherence to the Mediterranean Diet, Emotional Eating, Alcohol Intake, and Anxiety Related in University Students in Spain? *Nutrients* **2020**, *12*, 2224. [CrossRef]
231. Lanfranchi, M.; Calabrò, G.; De Pascale, A.; Fazio, A.; Giannetto, C. Household food waste and eating behavior: Empirical survey. *Br. Food J.* **2016**, *118*, 3059–3072. [CrossRef]
232. Sarti, S.; Terraneo, M.; Tognetti Bordogna, M. Poverty and private health expenditures in Italian households during the recent crisis. *Health Policy* **2017**, *121*, 307–314. [CrossRef] [PubMed]
233. Manios, Y.; Costarelli, V. Childhood Obesity in the WHO European Region. In *Epidemiology of Obesity in Children and Adolescents*; Moreno, L.A., Pigeot, I., Ahrens, W., Eds.; Springer International Publishing: New York, NY, USA, 2011; pp. 43–68, ISBN 9781441960399.
234. Blas, A.; Garrido, A.; Unver, O.; Willaarts, B. A comparison of the Mediterranean diet and current food consumption patterns in Spain from a nutritional and water perspective. *Sci. Total Environ.* **2019**, *664*, 1020–1029. [CrossRef] [PubMed]

235. Sánchez, P.H.; Ruano, C.; de Irala, J.; Ruiz-Canela, M.; Martínez-González, M.; Sánchez-Villegas, A. Adherence to the Mediterranean diet and quality of life in the SUN Project. *Eur. J. Clin. Nutr.* **2012**, *66*, 360–368. [CrossRef] [PubMed]
236. Veronese, N.; Notarnicola, M.; Cisternino, A.M.; Inguaggiato, R.; Guerra, V.; Reddavide, R.; Donghia, R.; Rotolo, O.; Zinzi, I.; Leandro, G.; et al. Trends in adherence to the Mediterranean diet in South Italy: A cross sectional study. *Nutr. Metab. Cardiovasc. Dis.* **2020**, *30*, 410–417. [CrossRef]
237. Renna, M.; Gonnella, M. Ethnobotany, Nutritional Traits, and Healthy Properties of Some Halophytes Used as Greens in the Mediterranean Basin. *Handb. Halophytes* **2020**, 1–19. [CrossRef]
238. Renna, M.; Montesano, F.; Signore, A.; Gonnella, M.; Santamaria, P. BiodiverSO: A Case Study of Integrated Project to Preserve the Biodiversity of Vegetable Crops in Puglia (Southern Italy). *Agriculture* **2018**, *8*, 128. [CrossRef]
239. Korre, M.; Tsoukas, M.A.; Frantzeskou, E.; Yang, J.; Kales, S.N. Mediterranean Diet and Workplace Health Promotion. *Curr. Cardiovasc. Risk Rep.* **2014**, *8*, 1–7. [CrossRef]
240. Avital, K.; Buch, A.; Hollander, I.; Brickner, T.; Goldbourt, U. Adherence to a Mediterranean diet by vegetarians and vegans as compared to omnivores. *Int. J. Food Sci. Nutr.* **2020**, *71*, 378–387. [CrossRef]
241. Chrysargyris, A.; Kloukina, C.; Vassiliou, R.; Tomou, E.M.; Skaltsa, H.; Tzortzakis, N. Cultivation strategy to improve chemical profile and anti-oxidant activity of *Sideritis perfoliata* L. subsp. *perfoliata*. *Ind. Crops Prod.* **2019**, *140*, 111694. [CrossRef]
242. Konuk, F.A. The influence of perceived food quality, price fairness, perceived value and satisfaction on customers' revisit and word-of-mouth intentions towards organic food restaurants. *J. Retail. Consum. Serv.* **2019**, *50*, 103–110. [CrossRef]
243. Lai, P.H.; Chuang, S.T.; Zhang, M.C.; Nepal, S.K. The non-profit sharing economy from a social exchange theory perspective: A case from World Wide Opportunities on Organic Farms in Taiwan. *J. Sustain. Tour.* **2020**, *28*, 1970–1987. [CrossRef]
244. Asian, S.; Hafezalkotob, A.; John, J.J. Sharing economy in organic food supply chains: A pathway to sustainable development. *Int. J. Prod. Econ.* **2019**, *218*, 322–338. [CrossRef]

Article

Effects of a Combination of Elderberry and Reishi Extracts on the Duration and Severity of Respiratory Tract Infections in Elderly Subjects: A Randomized Controlled Trial

Carlos Gracián-Alcaide [1], Jose A. Maldonado-Lobón [2], Elisabeth Ortiz-Tikkakoski [1], Alejandro Gómez-Vilchez [1], Juristo Fonollá [2,3], Jose L. López-Larramendi [4], Mónica Olivares [2] and Ruth Blanco-Rojo [2,*]

1 Nursing Home "Residencia de Mayores Claret", 18011 Granada, Spain; carlosgracian1@hotmail.es (C.G.-A.); elisabetzk@gmail.com (E.O.-T.); alejandrojgv14@gmail.com (A.G.-V.)
2 Research & Development Department, Biosearch Life, 18004 Granada, Spain; jamaldonado@biosearchlife.com (J.A.M.-L.); juristo@ugr.es (J.F.); molivares@biosearchlife.com (M.O.)
3 Faculty of Pharmacy, University of Granada, 18011 Granada, Spain
4 Commercial & Marketing Department, Biosearch Life, 18004 Granada, Spain; jllarramendi@biosearchlife.com
* Correspondence: rblanco@biosearchlife.com; Tel.: +34-913-802-973

Received: 29 October 2020; Accepted: 18 November 2020; Published: 21 November 2020

Abstract: Elderly people are particularly vulnerable to respiratory tract infections, so natural strategies to ameliorate the duration and severity of these infections are of great interest in this population. The objective of this study is to evaluate the efficacy of the consumption of a combination of elderberry and reishi extracts on the incidence, severity, and duration of respiratory tract infections in a group of healthy elderly volunteers. A randomized, double-blind, placebo-controlled pilot study was performed during the winter season. A group of 60 nursing home residents ≥65 years of age was randomly assigned to receive a combination of 1.5 g of elderberry +0.5 g of reishi or a placebo daily for 14 weeks. Data about the health conditions of the volunteers were evaluated and recorded by a medical doctor every 2 weeks. The incidence of respiratory infections was similar in both groups. However, volunteers in the extract group presented a significantly lower duration of common cold events (2.5 vs. 4.8 days, $p = 0.033$).and a significantly lower probability of having a high severity influenza-like illness event ($p = 0.039$). Moreover, the incidence of sleep disturbances was significantly lower in the extract group ($p = 0.049$). Therefore, the administration of a combination of elderberry and reishi extracts to the elderly population during the winter season might be used as a natural strategy to reduce the duration and severity of respiratory tract infections.

Keywords: elderberry; reishi; respiratory tract infections; common cold; influenza-like illness; respiratory infection symptoms; randomized controlled trial

1. Introduction

Respiratory tract infections, defined as any of a number of infectious diseases involving the respiratory tract, are the most common infections in humans of all ages, leading to a significant rate of mortality and morbidity worldwide [1]. According to the World Health Organization (WHO), approximately 290,000 to 650,000 deaths annually are caused by influenza virus infection alone [2]. Ominously, in the last 20 years, novel coronaviruses causing acute respiratory syndromes, such as SARS-CoV, MERS-CoV, and, currently, SARS-CoV-2, have emerged and triggered several global pandemics with high case fatality rates [3].

Among all populations, elderly people are at increased risk of acquiring respiratory infections and of experiencing a more severe disease course following infection due to coexisting chronic diseases, functional impairment, malnutrition, polypharmacy and immunosenescence [4,5]. The number of hospital admissions for respiratory infections is much higher among people aged ≥75 years, and it has been shown that the average length of hospital stay increases with age [1,6]. Moreover, these risks are even greater in older adults who reside in long-term care facilities [7] due to the close proximity of residents in combination with advanced age, multimorbidity and frailty [8]. Therefore, strategies that prevent respiratory infections, ameliorate their severity, or shorten the average duration of the disease will have the greatest benefit in the elderly population, especially older adults residing in long-term care facilities.

Natural substances and their derivatives have a long history of use for protection against infection and enhancement of the immune response [9]. Indeed, over the years, scientific evidence demonstrated the immunomodulatory and antiviral properties of several of these botanical and fungal extracts used in ancient medicine [9,10]. In this regard, elderberry (the whole fruit of *Sambucus nigra L.* containing multiple bioactive compounds) extract, which has been traditionally used to address cold and flu symptoms [11], has been shown in two recent meta-analyses to reduce overall symptom duration and severity when it is supplemented at the onset of upper respiratory infections in children and adults [12,13]. Moreover, preclinical studies have shown that the acidic polysaccharide-rich fraction contained in elderberry extract presented antimicrobial and antiviral effects, specifically against influenza viruses [14]. On the other hand, reishi (*Ganoderma lucidum*), the oldest known mushroom in ancient Chinese medicine [15], has been largely used to promote health and treat a large number of ailments [16]. Reishi has been reported to have a number of pharmacological effects, including immunomodulatory, anti-inflammatory, analgesic, antibacterial, and antiviral properties [17–20], which has been mainly attributed to its content in *G. lucidum* polysaccharide (GLPS) [17].

In addition, it has been suggested that the combination of different plant extracts may have a synergetic effect on the protection and treatment of respiratory infections [10]. On this subject, a drink containing *Echinacea purpurea* and *Sambucus nigra* extracts was demonstrated to be as effective as oseltamivir in the treatment of influenza virus infection in patients with early symptomatology [21]. Considering the individual properties of elderberry and reishi extracts, the combination of both compounds may have a promising effect on the protection against respiratory infection in the elderly population.

Therefore, the objective of the present study is to evaluate the efficacy of daily consumption of a combination of elderberry and reishi extracts on the incidence, severity and duration of respiratory tract infections in a group of healthy elderly volunteers living in a nursing home.

2. Materials and Methods

2.1. Study Design and Subjects

A randomized, double-blinded, placebo-controlled pilot trial was performed. The study was started in December 2019 and ended in March 2020. Volunteers were recruited from the nursing home "Residencia de Mayores Claret", which is located in Granada (Spain). The inclusion criteria were men and women older than 65 years of age who resided in a nursing home with medical service. The exclusion criteria were the presence of any disease or disorder that may affect the development and results of the study, namely, swallowing disorders, chronic pulmonary disease, need for enteral or parenteral nutrition, or allergy to some ingredient of the study product and use of any supplement or medication that could influence the outcomes of the trial. The study was conducted according to the Declaration of Helsinki, and the protocol was approved by the Regional Ethical Committee (Granada, Spain). Informed consent was obtained from all subjects. The trial was registered with the US Library of Medicine (http://www.clinicaltrials.gov) under the number NCT04386408.

The volunteers were randomly assigned to one of two groups: the individuals in the control group consumed a sachet containing a placebo (maltodextrin), whereas the individuals in the extract group consumed a sachet containing the combination of elderberry and reishi. The study products were provided as oral suspension sachets. The subjects were given treatments for 14 weeks, and they consumed one sachet per day (consuming the extract supplement or placebo) diluted in water at lunchtime. Participating subjects were instructed to not deviate from their regular habits during the 14 weeks of intervention. Neither the researchers nor the subjects knew which treatment sequence the subjects had been assigned to; the researchers were unblinded only at the end of the study. All volunteers were vaccinated against the flu during the same week, according to the vaccine campaign of 2019/2020.

2.2. Study Products

Each extract sachet (Elderpro ^{TM}, Biosearch S.A., Granada, Spain) contained 1.5 g of elderberry (*Sambucus nigra* L.) dried fruit juice standardized to 0.15% anthocyanosides and 0.5 g of reishi (*Ganoderma lucidum*) body aqueous extract standardized to 35% polysaccharides plus excipients (aspartame, sucralose and flavorings). Each placebo sachet contained 1.955 g of maltodextrin, plus the same excipients as extract sachet. Additionally, cochineal carmine and caramel dye were added to the placebo to ensure that the appearance of the placebo powder was identical to that of the extract powder. The powder mix of the extract and the placebo were provided in identical sachets with a code number that referred to the volunteer code according to the randomization. Elderberry extract was obtained through atomization of the concentrated elderberry juice obtained from the whole fruit by an aqueous method, whereas reishi extract was obtained from the fruity body of *G. lucidum* through the hydroalcoholic method. The content of anthocyanosides and polysaccharides in the extracts were analyzed by spectrophotometry. Elderberry and reishi extracts were provided by Biosearch Life (Granada, Spain), and sachets were prepared by HC Clover PS in Madrid (Spain).

2.3. Study Outcomes and Data Collection

The study's primary outcome was the incidence of respiratory infections, mainly influenza like illness (ILI) and common cold, during the intervention. The ILI diagnosis was based on the case definition used by the European Centre for Disease Prevention and Control [22] as follows: sudden onset of symptoms with one or more respiratory symptoms (cough, sore throat and/or nasal congestion) plus one or more systemic symptoms (fever, headache, myalgia and/or malaise). Common cold was defined as sudden onset of one or more respiratory symptoms (cough, sore throat and/or nasal congestion) with no systemic symptoms [23]. The total number of respiratory infections was the sum of common cold and ILI episodes.

Secondary outcomes included the determination of the severity and duration of the respiratory infections and incidence of symptoms related to them. The severity of the respiratory infections was determined according to the number of symptoms (previously defined) presenting simultaneously. Common cold was considered mild severity if subjects presented 1 symptom and high severity if they presented ≥2 symptoms simultaneously, whereas ILI was considered mild severity if subjects presented <4 symptoms simultaneously and high severity if subjects presented ≥4 symptoms simultaneously. The consumption of analgesics and antibiotics during the follow-up period was also recorded. Finally, some safety parameters, such as nausea, lack of appetite, sleep disturbances, and changes in weight and blood pressure, were recorded during the intervention.

All data about the health conditions of the volunteers and the consumption of medical treatments were evaluated and recorded by a medical doctor every 2 weeks in the case report form corresponding to each volunteer. Weight was determined using a Tanita BC-418 Body Composition Analyzer (Tanita., Tokyo, Japan) at baseline and at the end of the intervention. Systolic and diastolic blood pressures (BP) were measured with a validated digital automated blood pressure monitor (Medisana Healthcare, Barcelona, Spain). Compliance was assessed at the end of the intervention by comparing the number of sachets provided and the number returned. Adverse events, defined as any unfavorable, unintended effect, were recorded at the follow-up visits (at 2, 4, 6, 8, 10, 12 and 14 weeks).

2.4. Statistical Analysis

The normality of the distribution was tested for all measured variables by normal probability plots and the Shapiro–Wilk test. Data are presented as the mean (standard deviation) for continuous variables and as n (%) for categorical variables.

For comparisons between groups at the beginning of the study (extract vs. control), continuous variables were analyzed with Student's t-test or the nonparametric Kruskal–Wallis method, as appropriate, and categorical variables were analyzed with chi-square tests.

The occurrence of infections and symptoms were described using the incidence ratio (IR) and incidence rate ratio (IRR) with the 95% CI and p-value for the IRR. A Poisson regression model was applied to adjust the number of events by sex, age, and smoking habits, and additionally, by use of sleeping medication in the sleep disturbance parameter. Differences in the duration of respiratory infections between groups and changes in weight and blood pressure between baseline and the end of the intervention were determined by univariate model analysis adjusted by age, sex, and smoking habits and by BMI and hypertension medication in systolic and diastolic blood pressure variables.

A general alpha level of 0.05 was used as the cutoff point for statistical significance. Statistical analysis was carried out using SPSS software version 27.0 for Windows (SPSS, Chicago, IL, USA).

3. Results

3.1. Study Data, Compliance and Baseline Characteristics of the Subjects

A total of 60 older adults were recruited and randomly distributed into two groups: the control group (n = 30) and the extract group (n = 30). Before completion of the 14-week intervention period, 6 volunteers in the control group and 1 in the extract group discontinued the intervention and dropped out of the study for the reasons detailed in the study flow chart (Figure 1). No differences among the groups were detected between the number and the causes of withdrawal. The compliance rate was confirmed to be very high (≈100%). No adverse events resulting from the intake of either type of treatment were reported. Data were analyzed for all the subjects included in the study who had attended at least one of the follow-up visits (analysis per intention to treat, ITT).

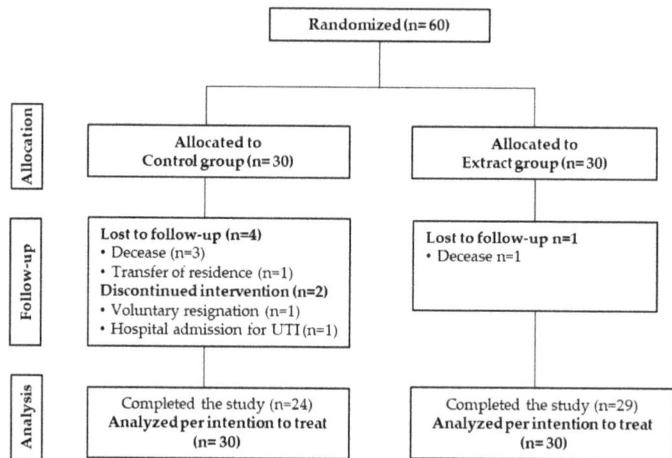

Figure 1. Diagram of the Consolidated Standards of Reporting Trials.

The baseline characteristics of the 60 older adults included in the statistical analyses are presented in Table 1. Most of the volunteers (71.7%) were older than 80 years, 20% were between 70 and 80 years old, and 8.3% were between 65 and 69 years old. No significant differences were detected between subjects in the study groups in the baseline characteristics.

Table 1. Baseline characteristics of the subjects participating in the study.

	Control Group (n = 30)	Extract Group (n = 30)	*p* between Groups
Age (years)	82.7 ± 9.2	85.9 ± 7.8	0.155
Sex			0.739
Men	5 (16.7%)	6 (20%)	
Women	25 (83.3%)	24 (60%)	
Weight (kg)	68.1 ± 13.9	63.6 ± 10.9	0.176
BMI (kg/m^2)	26.1 ± 4.6	25.4 ± 3.8	0.478
Smoking habit			0.109
Current smoker	4 (13.3%)	2 (6.7%)	
Former smoker	1 (3.3%)	6 (20%)	
No smokers	25 (83.3%)	22 (73.3%)	
Physical activity			0.519
Very low	23 (76.7%)	25 (83.3%)	
Low	7 (23.3%)	5 (16.7%)	
Systolic BP (mm Hg)	126.2 ± 14.7	130.3 ± 10.4	0.218
Diastolic BP (mm Hg)	70.7 ± 10.2	71.8 ± 9.4	0.667
Sleeping medication	10 (33.3%)	8 (26.7%)	0.389
Hypertension medication	19 (63.3%)	18 (60.0%)	0.518

Values are mean ± SD for continuous variables and n (%) for categorical variables. *p* indicates differences between groups.

3.2. Respiratory Tract Infections Incidence and Duration and Severity of Related Symptoms

Table 2 shows the incidence of respiratory infections during the intervention period (14 weeks). The incidence of total respiratory infections and common colds was 9.4% and 25.0% lower in the extract group than in the control group, but the differences were not significant ($p = 0.728$ and $p = 0.508$, respectively), whereas the incidence of ILI was very similar in both groups ($p = 0.978$).

Table 2. Incidence of respiratory tract infections.

	Control Group IR (SD)	Extract Group IR (SD)	IR Ratio (95% CI)	IR Decrease, %	*p* Value [c]
Common cold	0.321 (0.188)	0.240 (0.135)	0.750 (0.320–1.759)	25.0	0.508
ILI [a]	0.494 (0.188)	0.523 (0.174)	1.059 (0.514–2.182)	−0.06	0.978
Respiratory infections [b]	0.921 (0.273)	0.834 (0.224)	0.906 (0.523–1.570)	9.4	0.725

[a] Incidence of influenza-like illness; [b] Incidence of Common cold + ILI; [c] Poisson regression model adjusted by sex, age and smoking habits.

However, the duration and severity of the respiratory infections seemed to be lower in the group that received the extract than in the control group. Common cold events mean duration were significantly lower in the extract group than in the control group (2.5 ± 1.0 vs. 4.8 ± 1.4 days, $p = 0.033$), although we found no differences in ILI events (6.5 ± 3.1 vs. 5.7 ± 2.1 days, $p = 0.399$) (Figure 2a). When common cold and ILI events were classified as mild or high severity according to the number of simultaneous symptoms, we observed that subjects in the extract group presented a significantly lower probability of having a high severity ILI event [OR (95% CI) 0.227 (0.054–0.961), $p = 0.039$], although this was not significant in common cold events [OR (95% CI) 0.714 (0.086–5.959), $p = 0.755$] (Figure 2b).

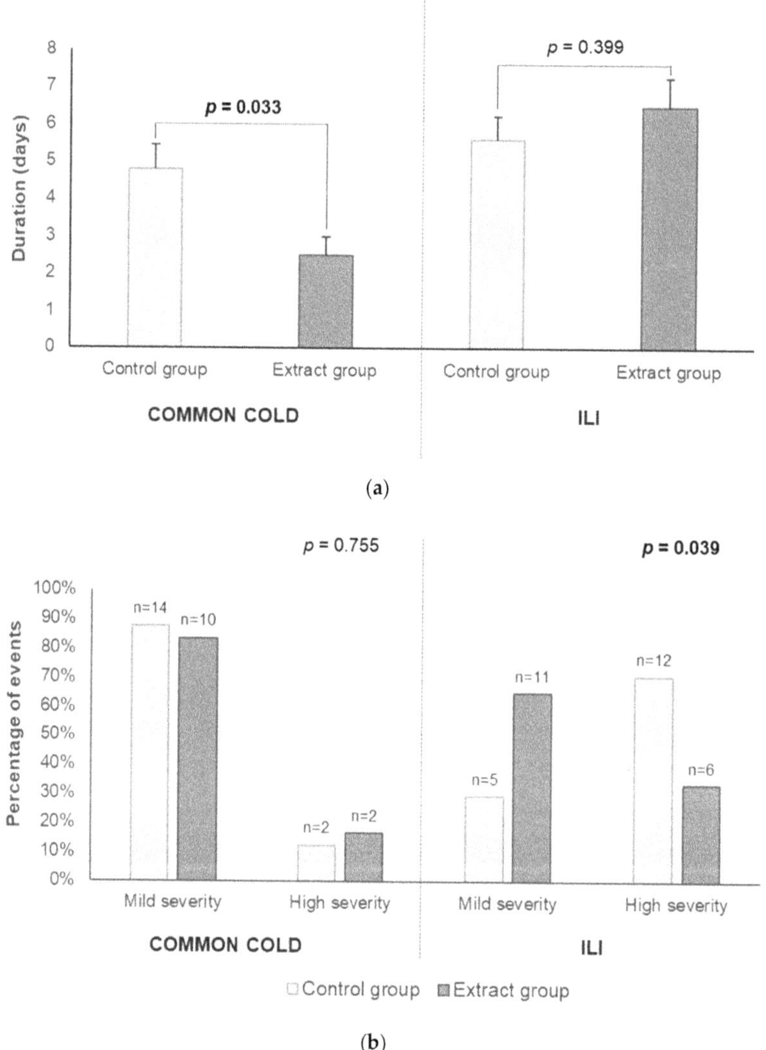

Figure 2. Duration and severity of respiratory tract infections. (**a**) Duration in days of common cold and influenza-like illness (ILI) events according to intervention groups. Values are represented as mean ± standard error of mean (SEM) (light grey bars for control group and dark grey bars for extract group). p-value indicates differences between groups for the duration of common cold or ILI events (univariate model adjusted by sex, age and smoking habits). (**b**) Percentage of events (common cold or ILI) that course with mild or high severity according to intervention groups (light grey bars for control group and dark grey bars for extract group). The number above each bar indicates the number of events in that group. p-value indicates differences in severity of common cold or ILI events between groups (chi-square test). Statistically significant values ($p < 0.05$) are given in bold.

Finally, the incidence of sore throat was significantly lower in the extract group, presenting the volunteers that consumed the extract an incidence ratio decreased by 69.8% compared to the control group., ($p = 0.043$, Table 3). With respect to other symptoms related to respiratory infections, their incidence was, in general, lower in the extract group than in the control group, except for the symptom tiredness and malaise, which had a higher incidence in the extract group, but differences were not statistically significant.

Table 3. Incidence of symptoms related to respiratory infections.

	Control Group IR (SD)	Extract Group IR (SD)	IR Ratio (95% CI)	IR Decrease, %	*p*-Value [d]
Cough	0.558 (0.202)	0.552 (0.177)	0.989 (0.495–1.976)	1.1	0.976
Nasal congestion	0.478 (0.200)	0.357 (0.141)	0.747 (0.334–1.6719	25.3	0.478
Sore throat	0.356 (0.211)	0.108 (0.073)	0.302 (0.095–0.963)	69.8	0.043
Fever	0.244 (0.112)	0.128 (0.717)	0.528 (0.52–1.834)	47.2	0.315
Headache	0.128 (0.937)	0.102 (0.070)	0.798 (0.266–2.396)	21.2	0.688
Muscle/Bone Pain	0.146 (0.92)	0.138 (0.794)	0.941 (0.327–2.713)	5.9	0.911
Tiredness/Malaise	0.180 (0.101)	0.342 (0.155)	1.897 (0.755–4.770)	−89.7	0.173
Local Respiratory Symptoms [a]	1.466 (0.352)	1.031 (0.239)	0.703 (0.440–1.124)	29.7	0.141
Systemic Symptoms [b]	0.754 (0.233)	0.656 (.189)	0.870 (0.500–1.511)	23	0.62
Total Symptoms [c]	2.185 (0.410)	1.808 (0.316)	0.828 (0.584–1.174)	27.2	0.289

[a] Sum of local respiratory symptoms (sore throat, cough and/or nasal congestion); [b] Sum of systemic symptoms (fever, headache, muscle/bone pain and/or tiredness/malaise); [c] Sum of all symptoms associated with respiratory infections (sore throat, cough, nasal congestion fever, headache, muscle/bone pain and/or tiredness/malaise); [d] Poisson regression model adjusted by sex, age, and smoking habits. Statistically significant values ($p < 0.05$) are given in bold.

The administration of antibiotics and analgesics to treat respiratory infections during the intervention was similar in both groups [IRR (95% CI): 0.993 (0.388–2.545) and 1.034 (0.411–2.980), respectively].

3.3. Safety Parameters

Regarding safety parameters (Table 4), we found no significant differences between groups in nausea or lack of appetite ($p = 0.484$ and $p = 0.873$, respectively) or in changes in body weight between baseline and the end of the intervention in both groups ($p = 0.842$ for the control group and $p = 0.981$ for the extract group). Systolic and diastolic BP tended to decrease in the extract group, although no significant differences were observed between baseline and the end of the intervention ($p = 0.079$ for systolic BP and $p = 0.092$ for diastolic BP). No changes were observed in the control group for these parameters ($p = 0.633$ for systolic BP $p = 0.494$ for diastolic BP). Interestingly, the incidence of sleep disturbances was significantly lower, 58.6%, in the extract group than in the control group [IR ratio (95% CI) 0.414 (0.172–0.983), $p = 0.049$].

Table 4. Incidence of symptoms related to respiratory infections.

	Control Group		Extract Group	
Nausea IR (SD) [a]	0.100 (0.057)		0.166 (0.074)	
Lack of appetite IR (SD) [a]	0.666 (0.149)		0.633 (0.145)	
Sleeping disturbance IR (SD) [a]	0.899 (0.325)		0.372 (0.152) *	
	Baseline	End of intervention	Baseline	End of intervention
Weight (kg) [b]	68.4 ± 14.2	67.3 ± 13.4	64.2 ± 10.6	63.8 ± 9.5
Systolic BP (mm Hg) [b]	126.8 ± 15	127.8 ± 10.7	130.6 ± 10.4	126.5 ± 9.1
Diastolic BP (mm Hg) [b]	70.5 ± 10.7	72 ± 8.1	72 ± 9.5	68.3 ± 8.7

[a] Poisson regression model adjusted by sex, age, and smoking habits; and additionally, by use of sleeping medication in the sleeping disturbance parameter. Asterisk (*) indicate $p < 0.05$ between control and extract groups. [b] Values are mean ± SD. Univariate models adjusted by sex, age and smoking in the weight variable; and additionally by BMI and hypertension medication in systolic and diastolic BP variables were used to evaluate differences between baseline and the end of the intervention at each group.

4. Discussion

The present study shows that daily consumption of a combination of elderberry and reishi extracts during the winter season is effective in reducing the duration and severity of respiratory infections in elderly people. Several clinical studies have shown the beneficial immunological effect of elderberry [12,13] and reishi [24], but to our knowledge, this is the first study that evaluated the combined effect of both extracts on the incidence, severity, and duration of respiratory infections in older adults, a population group at increased risk of this type of infection and a worse prognosis of the disease [5].

Elderberry (*Sambucus nigra* L fruit) has been indicated in several clinical trials to relieve the symptoms of the common cold and influenza when ingested at the onset of infection in children and adults [21,25–27]. Additionally, the effect of elderberry has been tested for prevention of respiratory symptoms in situations of high risk of infection, such as intercontinental flights, showing a reduction in the duration and severity of common cold events the travelers developed [28]. The therapeutic effects of elderberry have been attributed to its content of bioactive compounds, such as anthocyanins and other polyphenols, through several mechanisms [29,30]. First, elderberry extract has been reported to enhance the immune response by increasing both inflammatory and anti-inflammatory cytokines [31,32]. Second, it was observed that elderberry preparations exert antiviral activity, specifically against human influenza (H1N1) virus. It has been shown that several elderberry compounds directly bind the H1N1 virus particles, resulting in the inability of the particles to enter host cells and in a decreased viral load and spread [33,34]. Regarding the reishi (*Ganoderma lucidum*), it has been observed in vitro and in animal studies that the triterpenoids contained in the reishi exert anti-influenza activity by neuraminidase inhibition [35,36]. Moreover, a recent clinical study reported that a yogurt enriched with β-glucans from reishi increased the absolute count of peripheral blood total lymphocytes ($CD3^+$, $CD4^+$, and $CD8^+$ T cells) compared to a nonenriched yogurt in a group of healthy children [24]. Therefore, the results achieved in this randomized clinical trial could be related to the observed immunomodulatory and antiviral activities of elderberry and reishi. Moreover, this study adds scientific evidence to support the traditional use of these extracts to relieve respiratory infection symptoms.

In addition, a decline in immune function in older people has been attributed, at least in part, to an increase in oxidative stress [37]. Therefore, increases in dietary antioxidants, such as anthocyanosides of elderberry and other phenolic compounds present in elderberry and reishi extracts, may contribute to improving immune function in the elderly. Indeed, both elderberry and reishi extracts have been confirmed to exert antioxidant properties during antiradical activity assays in vitro [38,39]. Additionally, one study showed a significant increase in the total antioxidant capacity and total thiol concentration in the serum of a group of healthy subjects who consumed an elderberry infusion for 30 days with similar content of anthocyanins to our extract [40]. Although we did not analyze any antioxidant capacity biomarkers in our volunteers, we cannot discard a possible relationship between the reduction in the duration and severity of respiratory infections and the possible reduction in oxidative stress, as observed by other authors. In a crossover study performed in older adults, regular consumption of gold kiwifruit increased plasma antioxidants and reduced the duration and severity of some symptoms associated with respiratory infections [41]. However, further studies should be performed to test this hypothesis.

Interestingly, the older adults who took the combination of extracts reported fewer sleep problems than those who took the placebo. This effect could be attributed to reishi, which has been traditionally used for insomnia treatment [16]. Indeed, a randomized controlled trial described that the polysaccharide extract of *Ganoderma lucidum* improved the insomnia severity scores in patients with neurasthenia [42]. Moreover, *Ganoderma lucidum* extract has been shown to increase total sleep time and nonrapid eye movement sleep time in animal studies [43,44], with a probable mechanism linked to the modulation of cytokines such as tumor necrosis factor-α [44].

One strength of the present study was the design as a double-blind randomized controlled trial controlled by placebo. This methodology allowed us to avoid bias related to confounding factors (through a control group), selection bias (through randomization), and interpretation bias (through double blinding). However, a limitation of the study was the small sample size, limited by the fact that this is a pilot study. Additionally, the diagnosis of flu was based on a confluence of symptoms rather than viral detection. Moreover, further studies in other population groups should be performed.

5. Conclusions

In conclusion, the administration of a combination of elderberry and reishi extracts to the elderly population during the winter season reduces the duration and severity of respiratory infections. The study subjects did not suffer from any adverse effects and even presented a beneficial effect on night sleep. Therefore, the use of this combination of extracts from elderberry and reishi may be a natural, suitable, and safe strategy to short and ameliorate respiratory tract infections symptomatology in older adults, a particularly vulnerable population.

Author Contributions: Conceptualization, J.L.L.-L. and M.O.; methodology, M.O. and R.B.-R.; formal analysis, R.B.-R.; investigation, C.G.-A., E.O.-T. and A.G.-V.; resources, J.F. and J.A.M.-L.; data curation, J.A.M.-L. and R.B.-R.; writing—original draft preparation, R.B.-R.; writing—review and editing, M.O.; visualization, R.B.-R.; supervision, M.O.; project administration, M.O.; funding acquisition, M.O. All authors have read and agreed to the published version of the manuscript.

Funding: This research was funded by the Centre for the Development of Industrial Technology (CDTI) and co-financed by European Regional Development Fund (ERDF) through Operational Program for Spain 2014 to 2020 (Project number: IDI_20190733).

Acknowledgments: Authors wish to thank Ángela Inmaculada Gonzalez Medina, the Director of the nursing home "Residencia de Mayores Claret", for allowing the study to be carried out on the nursing home facilities.

Conflicts of Interest: J.A.M-L, J.L.L.-L., M.O. and R.B.-R. are workers of Biosearch Life. J.F. has been employee of Biosearch Life. The authors declare no other conflict of interest. The funders had no role in the design of the study; in the collection, analyses, or interpretation of data; in the writing of the manuscript, or in the decision to publish the results.

References

1. Nicholson, K.G.; Abrams, K.R.; Batham, S.; Medina, M.J.; Warren, F.C.; Barer, M.; Bermingham, A.; Clark, T.W.; Latimer, N.; Fraser, M.; et al. Randomised controlled trial and health economic evaluation of the impact of diagnostic testing for influenza, respiratory syncytial virus and Streptococcus pneumoniae infection on the management of acute admissions in the elderly and high-risk 18- to 64-year-olds. *Health Technol. Assess.* **2014**, *18*, 1–274. [PubMed]
2. Iuliano, A.D.; Roguski, K.M.; Chang, H.H.; Muscatello, D.J.; Palekar, R.; Tempia, S.; Cohen, C.; Gran, J.M.; Schanzer, D.; Cowling, B.J.; et al. Estimates of global seasonal influenza-associated respiratory mortality: A modelling study. *Lancet* **2018**, *391*, 1285–1300. [CrossRef]
3. Petersen, E.; Koopmans, M.; Go, U.; Hamer, D.H.; Petrosillo, N.; Castelli, F.; Storgaard, M.; Al Khalili, S.; Simonsen, L. Comparing SARS-CoV-2 with SARS-CoV and influenza pandemics. *Lancet Infect. Dis.* **2020**, *20*, e238–e244. [CrossRef]
4. Castle, S.C. Clinical relevance of age-related immune dysfunction. *Clin. Infect. Dis.* **2000**, *31*, 578–585. [CrossRef] [PubMed]
5. Dewan, S.K.; Zheng, S.; Xia, S.; Bill, K. Senescent remodeling of the immune system and its contribution to the predisposition of the elderly to infections. *Chin. Med. J.* **2012**, *125*, 3325–3331. [PubMed]
6. Masse, S.; Capai, L.; Falchi, A. Epidemiology of Respiratory Pathogens among Elderly Nursing Home Residents with Acute Respiratory Infections in Corsica, France, 2013–2017. *Biomed Res. Int.* **2017**, *2017*, 1423718. [CrossRef]
7. Childs, A.; Zullo, A.R.; Joyce, N.R.; McConeghy, K.W.; van Aalst, R.; Moyo, P.; Bosco, E.; Mor, V.; Gravenstein, S. The burden of respiratory infections among older adults in long-term care: A systematic review. *BMC Geriatr.* **2019**, *19*, 210. [CrossRef]
8. Rainwater-Lovett, K.; Chun, K.; Lessler, J. Influenza outbreak control practices and the effectiveness of interventions in long-term care facilities: A systematic review. *Influenza Other Respir. Viruses* **2014**, *8*, 74–82. [CrossRef]
9. Licciardi, P.V.; Underwood, J.R. Plant-derived medicines: A novel class of immunological adjuvants. *Int. Immunopharmacol.* **2011**, *11*, 390–398. [CrossRef]

10. Shahzad, F.; Anderson, D.; Najafzadeh, M. The Antiviral, Anti-Inflammatory Effects of Natural Medicinal Herbs and Mushrooms and SARS-CoV-2 Infection. *Nutrients* **2020**, *12*, 2573. [CrossRef]
11. Roxas, M.; Jurenka, J. Colds and influenza: A review of diagnosis and conventional, botanical, and nutritional considerations. *Altern. Med. Rev.* **2007**, *12*, 25–48. [PubMed]
12. Hawkins, J.; Baker, C.; Cherry, L.; Dunne, E. Black elderberry (*Sambucus nigra*) supplementation effectively treats upper respiratory symptoms: A meta-analysis of randomized, controlled clinical trials. *Complement. Ther. Med.* **2019**, *42*, 361–365. [CrossRef] [PubMed]
13. Harnett, J.; Oakes, K.; Carè, J.; Leach, M.; Brown, D.; Cramer, H.; Pinder, T.-A.; Steel, A.; Anheyer, D. The effects of *Sambucus nigra* berry on acute respiratory viral infections: A rapid review of clinical studies. *Adv. Integr. Med.* **2020**, *7*, 240–246. [CrossRef] [PubMed]
14. Kinoshita, E.; Hayashi, K.; Katayama, H.; Hayashi, T.; Obata, A. Anti-influenza virus effects of elderberry juice and its fractions. *Biosci. Biotechnol. Biochem.* **2012**, *76*, 1633–1638. [CrossRef] [PubMed]
15. Chang, S.-T.; Buswell, J.A. Ganoderma lucidum (Curt.: Fr.) P. karst.(Aphyllophoromycetideae)—A mushrooming medicinal mushroom. *Int. J. Med. Mushrooms* **1999**, *1*, 139–146. [CrossRef]
16. Wasser, S.P.; Coates, P.; Blackman, M.; Cragg, G.; Levine, M.; Moss, J.; White, J. Reishi or ling zhi (Ganoderma lucidum). In *Encyclopedia of Dietary Supplements*; Marcel Dekker: New York, NY, USA, 2005; pp. 680–690.
17. Xu, Z.; Chen, X.; Zhong, Z.; Chen, L.; Wang, Y. Ganoderma lucidum Polysaccharides: Immunomodulation and Potential Anti-Tumor Activities. *Am. J. Chin. Med.* **2011**, *39*, 15–27. [CrossRef]
18. Ko, H.-H.; Hung, C.-F.; Wang, J.-P.; Lin, C.-N. Antiinflammatory triterpenoids and steroids from Ganoderma lucidum and G. tsugae. *Phytochemistry* **2008**, *69*, 234–239. [CrossRef]
19. Mohammed, A.; Tanko, Y.; Mohammed, K.A.; Yaro, A.H. Studies of analgesic and anti-inflammatory effects of aqueous extract of Ganoderma lucidum in mice and wister rats. *IOSR J. Pharm. Biol. Sci.* **2012**, *4*, 54–57.
20. Yu, A.; Wong, C.-H.; Jan, J.T.; Cheng, Y.-S.E. Anti-Viral Effect of an Extract of Ganoderma Lucidum 2008. U.S. Patent Application, No. 11/859,729, 21 September 2006.
21. Rauš, K.; Pleschka, S.; Klein, P.; Schoop, R.; Fisher, P. Effect of an Echinacea-Based Hot Drink Versus Oseltamivir in Influenza Treatment: A Randomized, Double-Blind, Double-Dummy, Multicenter, Noninferiority Clinical Trial. *Curr. Ther. Res. Clin. Exp.* **2015**, *77*, 66–72. [CrossRef]
22. Influenza Case Definitions (2017) Stockholm: ECDC. Available online: http://ecdc.europa.eu/en/healthtopics/influenza/surveillance/Pages/influenza_case_definitions.aspx (accessed on 20 October 2020).
23. Eccles, R. Understanding the symptoms of the common cold and influenza. *Lancet Infect. Dis.* **2005**, *5*, 718–725. [CrossRef]
24. Henao, S.L.D.; Urrego, S.A.; Cano, A.M.; Higuita, E.A. Randomized Clinical Trial for the Evaluation of Immune Modulation by Yogurt Enriched with β-Glucans from Lingzhi or Reishi Medicinal Mushroom, *Ganoderma lucidum* (Agaricomycetes), in Children from Medellin, Colombia. *Int. J. Med. Mushrooms* **2018**, *20*, 705–716. [CrossRef] [PubMed]
25. Zakay-Rones, Z.; Varsano, N.; Zlotnik, M.; Manor, O.; Regev, L.; Schlesinger, M.; Mumcuoglu, M. Inhibition of several strains of influenza virus in vitro and reduction of symptoms by an elderberry extract (Sambucus nigra L.) during an outbreak of influenza B Panama. *J. Altern. Complement. Med.* **1995**, *1*, 361–369. [CrossRef] [PubMed]
26. Zakay-Rones, Z.; Thom, E.; Wollan, T.; Wadstein, J. Randomized study of the efficacy and safety of oral elderberry extract in the treatment of influenza A and B virus infections. *J. Int. Med. Res.* **2004**, *32*, 132–140. [CrossRef] [PubMed]
27. Kong, F. Pilot clinical study on a proprietary elderberry extract: Efficacy in addressing influenza symptoms. *Online J. Pharmacol. Pharmacokinet.* **2009**, *5*, 32–43.
28. Tiralongo, E.; Wee, S.S.; Lea, R.A. Elderberry supplementation reduces cold duration and symptoms in air-travellers: A randomized, double-blind placebo-controlled clinical trial. *Nutrients* **2016**, *8*, 182. [CrossRef]
29. Młynarczyk, K.; Walkowiak-Tomczak, D.; Łysiak, G.P. Bioactive properties of Sambucus nigra L. as a functional ingredient for food and pharmaceutical industry. *J. Funct. Foods* **2018**, *40*, 377–390. [CrossRef]
30. Vlachojannis, J.E.; Cameron, M.; Chrubasik, S. A systematic review on the sambuci fructus effect and efficacy profiles. *Phytother. Res.* **2010**, *24*, 1–8. [CrossRef]
31. Barak, V.; Halperin, T.; Kalickman, I. The effect of Sambucol, a black elderberry-based, natural product, on the production of human cytokines: I. Inflammatory cytokines. *Eur. Cytokine Netw.* **2001**, *12*, 290–296.

32. Barak, V.; Birkenfeld, S.; Halperin, T.; Kalickman, I. The effect of herbal remedies on the production of human inflammatory and anti-inflammatory cytokines. *Isr. Med Assoc. J. IMAJ* **2002**, *4*, 919–922.
33. Roschek, B.; Fink, R.C.; McMichael, M.D.; Li, D.; Alberte, R.S. Elderberry flavonoids bind to and prevent H1N1 infection in vitro. *Phytochemistry* **2009**, *70*, 1255–1261. [CrossRef]
34. Swaminathan, K.; Müller, P.; Downard, K.M. Substituent effects on the binding of natural product anthocyanidin inhibitors to influenza neuraminidase with mass spectrometry. *Anal. Chim. Acta* **2014**, *828*, 61–69. [CrossRef] [PubMed]
35. Zhu, Q.; Bang, T.H.; Ohnuki, K.; Sawai, T.; Sawai, K.; Shimizu, K. Inhibition of neuraminidase by Ganoderma triterpenoids and implications for neuraminidase inhibitor design. *Sci. Rep.* **2015**, *5*, 13195. [CrossRef] [PubMed]
36. Zhu, Q.; Amen, Y.M.; Ohnuki, K.; Shimizu, K. Anti-influenza effects of Ganoderma lingzhi: An animal study. *J. Funct. Foods* **2017**, *34*, 224–228. [CrossRef]
37. Hughes, D.A. Dietary antioxidants and human immune function. *Nutr. Bull.* **2000**, *25*, 35–41. [CrossRef]
38. Sidor, A.; Gramza-Michałowska, A. Advanced research on the antioxidant and health benefit of elderberry (Sambucus nigra) in food—A review. *J. Funct. Foods* **2015**, *18*, 941–958. [CrossRef]
39. Lin, Z.; Deng, A. Antioxidative and Free Radical Scavenging Activity of Ganoderma (Lingzhi). In *Ganoderma and Health*; Springer: Singapore, 2019; pp. 271–297.
40. Ivanova, D.; Tasinov, O.; Kiselova-Kaneva, Y. Improved lipid profile and increased serum antioxidant capacity in healthy volunteers after Sambucus ebulus L. fruit infusion consumption. *Int. J. Food Sci. Nutr.* **2014**, *65*, 740–744. [CrossRef]
41. Hunter, D.C.; Skinner, M.A.; Wolber, F.M.; Booth, C.L.; Loh, J.M.S.; Wohlers, M.; Stevenson, L.M.; Kruger, M.C. Consumption of gold kiwifruit reduces severity and duration of selected upper respiratory tract infection symptoms and increases plasma vitamin C concentration in healthy older adults. *Br. J. Nutr.* **2012**, *108*, 1235–1245. [CrossRef]
42. Tang, W.; Gao, Y.; Chen, G.; Gao, H.; Dai, X.; Ye, J.; Chan, E.; Huang, M.; Zhou, S. A Randomized, Double-Blind and Placebo-Controlled Study of a Ganoderma lucidum Polysaccharide Extract in Neurasthenia. *J. Med. Food* **2005**, *8*, 53–58. [CrossRef]
43. Chu, Q.-P.; Wang, L.-E.; Cui, X.-Y.; Fu, H.-Z.; Lin, Z.-B.; Lin, S.-Q.; Zhang, Y.-H. Extract of Ganoderma lucidum potentiates pentobarbital-induced sleep via a GABAergic mechanism. *Pharmacol. Biochem. Behav.* **2007**, *86*, 693–698. [CrossRef]
44. Cui, X.-Y.; Cui, S.-Y.; Zhang, J.; Wang, Z.-J.; Yu, B.; Sheng, Z.-F.; Zhang, X.-Q.; Zhang, Y.-H. Extract of Ganoderma lucidum prolongs sleep time in rats. *J. Ethnopharmacol.* **2012**, *139*, 796–800. [CrossRef]

Article

Temporal Changes and Correlations between Quality Loss Parameters, Antioxidant Properties and Enzyme Activities in Apricot Fruit Treated with Methyl Jasmonate and Salicylic Acid during Cold Storage and Shelf-Life

Ahmed Ezzat [1], Attila Hegedűs [2], Szilárd Szabó [3], Amin Ammar [4], Zoltán Szabó [5], József Nyéki [5], Bianka Molnár [5] and Imre J. Holb [5,6,*]

1 Department of Horticulture Kafr El-Shaikh, Faculty of Agriculture, Kafelelsehikh University, Kafr El-Sheikh 33516, Egypt; ahmed.kassem@agr.kfs.edu.eg
2 Department of Genetics and Plant Breeding, Faculty of Horticultural Science, Szent István University, 1118 Budapest, Hungary; hegedus.attila@uni-corvinus.hu
3 Department of Physical Geography and Geoinformatics, Debrecen, University of Debrecen, Egyetem tér 1., 4032 Debrecen, Hungary; szaboszilard.geo@gmail.com
4 Department of Food Science and Technology, Faculty of Agriculture, Kafelelsehikh University, Kafr El-Shaikh 33516, Egypt; amin.amar@agr.kfs.edu.eg
5 Institute of Horticulture, Faculty of Agronomy, University of Debrecen, P.O. Box. 36. H-4015 Debrecen, Hungary; zoltszab@agr.unideb.hu (Z.S.); nyekijo@agr.unideb.hu (J.N.); molnarbianka93@gmail.com (B.M.)
6 Hungarian Academy of Sciences, Centre for Agricultural Research, Plant Protection Institute Budapest, Herman Ottó út 15., 1022 Budapest, Hungary
* Correspondence: holbimre@gmail.com

Received: 21 October 2020; Accepted: 12 November 2020; Published: 14 November 2020

Abstract: The apricot storability is one of the largest challenges, which the apricot industry has to face all over the world; therefore, finding options for prolonging fruit quality during cold storage (CS) and shelf-life (SL) will help to decrease postharvest losses of apricot. The aim of this apricot fruit work was to study the temporal changes and correlations of 10 quality parameters (quality losses, antioxidant properties and enzyme activities) in the postharvest treatments of methyl jasmonate (MeJA) and salicylic acid (SA) under 1 °C CS (7, 14 and 21 days) and 25 °C SL (4 and 8 days after the 21-day CS) treatments. MeJA and SA significantly decreased the quality loss of chilling injury (CI) and fruit decay (FD) at all dates for both storage conditions. MeJA- and SA-treated fruits increased total antioxidant capacity (TAC), total soluble phenolic compounds (TSPC) and carotenoids contents (TCC) at all dates of both storage treatments. In contrast, the ascorbic acid content (AAC) increased only until days 14 and 4 in the CS and SL treatments, respectively. Among enzyme activity parameters, the activities of phenylalanine ammonia-lyase (PAL), peroxidase and superoxide dismutase (SOD) were significantly increased in the MeJA and SA treatments in all dates of both storage treatments. Catalase (CAT) activity increased in the SA and control treatments, while it decreased in the MeJA treatment in both storage conditions. In both the MeJA and the SA treatments, six pair-variables (FD vs. CI, PAL vs. CAT, PAL vs. SOD, TAC vs. SOD, TAC vs. FD, and AAC vs. CI) were significant in Pearson correlation and regression analyses among the 45 parameters pairs. Principal component analyses explained 89.3% of the total variance and PC1 accounted for 55.6% of the variance and correlated with the CI, FD, TAC, TSPC, TCC, PAL and SOD, indicating strong connections among most parameters. In conclusion, MeJA and SA are practically useful and inexpensive techniques to maintain quality attributes of CI, FD, TAC, TSPC, TCC, PAL, POD and SOD in apricot fruit during both CS and SL conditions.

Keywords: apricot; methyl jasmonate; salicylic acid; antioxidant property; enzyme activity; postharvest quality

1. Introduction

Minerals, vitamins, and antioxidant materials in fruit crops are essential sources for human health care [1,2]. The amount and longevity of these sources in stone fruits are highly dependent on the pre- and postharvest quality of the fruit e.g., [3–5]. Among stone fruit species, the post-harvest life of apricot (*Prunus armeniaca*) is short due to its climacteric nature [5]. Apricot fruit can keep its quality for 15–30 days after harvest if the fruit is stored at low temperature under cold storage (CS), e.g., [6–9], and this quality can be kept for up to a further 5 days at 25 °C under shelf-life (SL) conditions [10]. However, longer storage at low temperatures can cause symptoms of chilling injury and/or fruit decay incidence in the apricot fruit e.g., [5,7,11].

Jasmonic acid (JA) and salicylic acid (SA) have crucial roles in the growth of plants and in the regulation to various plant stresses e.g., [12,13], as well as in plant defense mechanisms, e.g., [14].

Methyl jasmonate (MeJA), as a natural plant growth regulator, e.g., [12,13,15–17], plays essential roles in many physiological mechanisms of plants, and it increases postharvest quality of horticultural crops including tropical fruits, such as avocado, papaya, grapefruit, mango, guava, pomegranates and loquat [16,18–24], and temperate fruits, such as apple, peach and cherry [25–28]. These studies demonstrated that MeJA reduced incidences of fruit decay and chilling injury but increased the amount of phenolic and antioxidant contents as well as defense-related enzymes in fruits during CS. One apricot study evaluated the effect of MeJA on the sensory quality and physico-chemical parameters of fruit during storage [8], but treatment effects on chilling injury, fruit decay, antioxidant capacity, phenolic contents and enzyme activities were not reported.

Salicylic acid (SA), together with its derivatives, is often applied to improve the quality of horticultural crops during storage [29–34]. Salicylic acid was shown to improve fruit quality and to decrease chilling injury and quality loss in some fruit crops, such as on peach [29,31,35], banana [30], loquat [36] and apricot [8,9,37]. The three apricot studies showed that SA improved some quality features of apricot fruit after harvest [8,9,37], but changes in enzyme activities under CS and the degree of quality loss and the variation of antioxidant properties under SL conditions were not investigated.

Only two studies carried out comparisons between MeJA and SA on evaluating quality features of cherry and apricot during storage [8,26] but such a comparative study on chilling injury, fruit decay, antioxidant capacity, phenolic contents and enzyme activities was not prepared for apricot under CS or SL storage conditions.

Several biological and physiological connections exist among the plant quality measurements, which can be expressed by presenting the correlation between the corresponding measurements such as between phenolic compounds, antioxidant properties and enzyme activity e.g., [38–41], which help to understand the background of physiological processes in plant organs. However, no attempt has been made to determine the potential inter-correlations between chilling injury, fruit decay, the parameters of antioxidant capacity, phenolic contents and enzyme activities for SA and MeJA in various storage conditions for apricot fruit.

The objective of this work was to evaluate and compare the effect of MeJA and SA on 10 fruit quality parameters of apricot under 1 °C CS (7, 14 and 21 days) and 25 °C SL (4 and 8 days after the 21 day CS) treatments. The 10 measured fruit quality parameters were classified into three groups: quality losses (chilling injury and fruit decay indices), antioxidant properties (total antioxidant capacity, total soluble phenol, total carotenoid and ascorbic acid content) and enzyme activities (phenylalanine ammonia-lyase (PAL), peroxidase (POD), superoxide dismutase (SOD), and catalase (CAT)). The inter-correlations among the 10 fruit quality parameters for treatments of SA and MeJA were also determined in the two storage conditions in order to clarify the physiological changes in apricot fruit during CS and SL storage.

2. Materials and Methods

2.1. Fruit Samples and Treatments

For all experiments, the cultivar 'Bergarouge' was used from a commercial apricot orchard (Nord-cot Ltd., Boldogkőváralja, Hungary). This cultivar has been planted in large areas in Hungary in the past few years. The mature fruit is firmer than most other cultivars, but the storage ability of this cultivar is similar to other commonly grown apricot cultivars in Hungary. The fruits were harvested 98 days from the anthesis (on 5 July) for the experiment. The fruits were packed in paper bags and transported to the laboratory on the same day of harvest. Healthy and uniform fruits (40 mm, equally colored, equally matured and equally shaped) were used for the experiment.

The concentrations of 0.2 mmol L^{-1} MeJA and 2 mmol L^{-1} SA were used for the experiments together with an untreated control (water treated). Three units of apricot fruits were used as MeJA, SA, and untreated control treatments. Each unit contained 150 fruits (3 treatments × 150 fruits = 450 fruits. Each unit was dipped into solutions of 0.2 mmol L^{-1} MeJA, 2 mmol L^{-1} SA, and distilled water (untreated control), respectively, for 15 min. Then, each treatment was divided into 2 further units (2 units × 75 fruits). The first unit, the CS treatment, was incubated at 1 °C and 95% RH; and then fruits were observed on days 7, 14 and 21. The second unit, the SL treatment, was incubated at 1°C and 95% RH for 21 days, then held at 25 °C for 4 and 8 days. All treatments were replicated three times (3 replicates × 450 fruits = 1350 fruits), and 2 repeats of the experiment were used (2 repeats × 1350 fruits).

2.2. Quality Loss Parameters

The two selected quality loss parameters were chilling injury (CI) and fruit decay (FD), which were evaluated for both CS and SL treatments at each assessment date.

2.2.1. Chilling Injury

Symptoms of CI occurred as flesh browning on 10 assessed fruits per treatments per assessment dates. Fruits were cut double parallel to the axial diameter and then the degree of CI symptoms was visually detected on the cut surface of fruits. The degree of flesh browning was classified into five categories as described by Wang et al. [31]: 0, flesh browning is not visible; 1, flesh browning is less than 25%; 2, the flesh browning is between ≥25% and <50%; 3, the flesh browning is between ≥50% and <75%; and 4, the flesh browning is more than ≥75%. Then an index of CI was calculated as CI index = Σ[(CI category) × (fruit number at the CI category)]/(4 × total fruit number).

2.2.2. Fruit Decay

A fruit surface with superficial browning was considered as a fruit decay symptom. The degree of FD symptoms was classified into five categories as described by Wang et al. [31]: 0, superficial browning is not visible; 1, the superficial browning is less than 25%; 2, the superficial browning is between ≥25% and <50%; 3, the superficial browning is between ≥50% and <75%; and 4, the superficial browning is more than ≥75%. Then an index of FD was calculated as FD index = Σ[(FD category) × (fruit number in the FD category)]/(4 × total fruit number).

2.3. Antioxidant Capacity and Related Parameters

Antioxidant capacity and related parameters were assessed at the CS and the SL treatments and at each assessment date. All reagents and materials for the analyses were purchased from Sigma-Aldrich (St. Louis, MO, USA).

2.3.1. Fruit Extract

Samples of apricot fruit were placed in the laboratory on the day of harvest, then the fruit were frozen at −18 °C. The extracts were prepared by homogenizing the fresh fruit for 5 min in 95% ethanol (1:3 *w*/*v*). Then, the homogenates were blended for 30 min at room temperature, and then centrifuged at 1500× *g* for 5 min. The supernatants were concentrated in an evaporator. The extract obtained from each fruit was adjusted to the same volume by reconstituting the samples in the water. The adjusted ratio was 0.20 mL extract/gram of fruit.

2.3.2. Total Antioxidant Capacity

The total antioxidant capacity (TAC) was measured spectrophotometrically by applying the ferric reducing antioxidant power (FRAP) method as described by Benzie and Strain [42]. Briefly, the FRAP reagent was prepared early before the measurements by adding 2.5 mL of a 10 mmol L^{-1} TPTZ (Sigma) solution into 40 mmol L^{-1} HCl plus 2.5 mL of 20 mmol L^{-1} $FeCl_3$ and 25 mL of 0.3 mol L^{-1} acetate buffer, pH 3.6, and then it was warmed up to 37 °C. The 0.04 mL sample supernatants were mixed with 0.2 mL of distilled water and 1.8 mL of FRAP reagent. The measurement used the reduction of the ferric-tripyridyltriazine complex [Fe^{3+}-TPTZ] to the ferrous form [Fe^{2+}-TPTZ] in acidic buffer (pH = 3.6). Fe^{2+}-TPTZ has an intensive color and this was monitored by measuring the absorption change at 593 nm. Then TAC was given as mg equivalents of ascorbic acid (AA) for 1 g fresh weight (FW) (mg AA g^{-1} FW).

2.3.3. Total Soluble Phenol Content

An Folin-Ciocalteu's (FC) reagent was applied to measure the amount of total soluble phenols (TSP) in the extracted fruits. The FC assay was used according to the study of Singleton and Rossi [43] in order to determine the TSP content as follows. The FC reagent was pre-diluted 10 times with distilled water, and 1.8 mL of the reagent was mixed with 40 µL of fruit extract solution and incubated for 5 min at 25 °C. Then, 1.2 mL of (7.5% *w*/*v*) sodium carbonate was mixed with the solutions. After 1 h at 25 °C, the absorbance of the samples was determined at 760 nm with a Hitachi UV2800 spectrophotometer (Tokyo, Japan). In order to express TSP in gallic acid equivalents (GAE), a calibration curve (R^2 = 0.995) was prepared with 20, 40, 60, 80 and 100 mg L^{-1} solutions of gallic acid. Then the content of TSP was given as mg GAEs for 100 g FW sample (GAE 100 g^{-1} FW).

2.3.4. Total Carotenoids Content

Total carotenoids content (TCC) was measured according to the study of Akin et al. [44] with a few modifications. In brief, 100 mL of methanol/petroleum ether (1:9, *v*/*v*) was homogenized with 5 g of fruit extract. Then this solution was poured into a separating funnel. The petroleum ether layer was filtrated through sodium sulphate and transferred to a flask. Then, TCC was measured with a Hitachi UV2800 spectrophotometer (Tokyo, Japan) at 450 nm. An extinction coefficient of 2500 was used to evaluate carotenoid content, and TCC was given as β-carotene equivalents (mg β-carotene 100 g^{-1} FW) according to Gross [45].

2.3.5. Ascorbic Acid Content

The ascorbic acid content (AAC) was determined spectrophotometrically using the dinitrophenylhydrazine (DNPH) method as described by Terada et al. [46]. Ten milligrams of each sample (2 mL extract) were added into a 100 mL flask contain 50 mL acetic acid solution, 5 drops of bromine water was added until the solution became colored, and then thiourea solution drops were added to it until a clear solution was obtained. Then 2, 4-dinitrophenyl hydrazine solution was added, and we completed the solution with acetic acid. The AAC was given as mg per 100 g FW.

2.4. Activity of Enzymes

Enzyme activities were measured using Sigma-Aldrich reagents (St. Louis, MO, USA) for all enzyme assessments.

2.4.1. PAL Activity

The activity of PAL was determined as described by Assis et al. [47]. The enzyme extract was prepared as 10 g of flesh from 10 fruits were homogenized in 25 mL of 50 mmol L^{-1} sodium borate buffer (5 mmol β-mercaptoethanol + 0.5 g polyvinyl pyrrolidone, PVPP; pH 8.8,). Then the enzyme extract (1 mL) together with L-phenylalanine (1 mL; 20 mmol L^{-1}) and sodium borate buffer (2 mL; 50 mmol L^{-1}) was incubated at 37 °C for 1 h. During the incubation, PAL catalyzes the reaction of the deamination of L-phenylalanine to trans-cinnamic acid. Then the reaction was stopped with HCl (1 mL; 1 mol L^{-1}), and the activity of PAL was given by measuring the amounts of trans-cinnamic acid spectrophotometrically at 290 nm. The raw enzyme preparation, together with L-phenylalanine, was not incubated and used as a blank sample. Each analysis was replicated three times and the activity of PAL was given as nmol cinnamic acid h^{-1} mg^{-1} protein.

2.4.2. POD Activity

The activity of POD samples was measured with the method described by Chance and Maehly [48]. Briefly, enzyme extract (0.5 mL) was mixed with 2 mL buffer containing 100 mmol L^{-1} sodium phosphate (pH 6.4) and 8 mmol L^{-1} guaiacol, and then the solution was placed into the incubator for 5 min at 30 °C. Then 1 mL of H_2O_2 (24 mmol L^{-1}) was added to the sample and the increasing absorbance was determined at 460 nm five times at 30, 60, 90, 120 and 150 s. Then, activity of POD was given as unit per gram FW per minute (U mg^{-1} FW min^{-1}).

2.4.3. SOD Activity

Supernatants were prepared for SOD activity measurements as frozen fruit tissues (1 g) were mixed with sodium phosphate buffer (5 mL; 50 mmol L^{-1}, pH 7.8) at 4 °C and then the supernatants was centrifuged (12,000× *g* for 20 min; 4 °C). The activity of SOD in these supernatants was measured as described by Rao et al. [49]. The 3 mL reaction mixture contained the following ingredients: sodium phosphate (50 mmol L^{-1}, pH 7.8), methionine (14 mmol L^{-1}), EDTA (3 µmol L^{-1}), nitro-blue-tetrazolium (NBT, 1 µmol L^{-1}), riboflavin (60 µmol L^{-1}) and crude enzyme extract (0.1 mL). Then the absorbance was detected at 560 nm for blue formazan. The quantity of enzyme that causes the 50% inhibition of NBT reduction was used as the SOD activity unit. The activity of SOD in the samples was given as U mg^{-1} protein.

2.4.4. CAT Activity

Catalase activity was measured as described by Abassi et al. [50] using the A and the B buffer solutions. In this process, 1 mL buffer A was added to a 50 µL enzyme extract in a cuvette, and 1 mL buffer B was added to a 50 µL enzyme extract in another cuvette. The change in the optical density in the cuvettes was measured spectrophotometrically at 240 nm for 45 s and 1 min when the extract was placed into the cuvettes. The differences between the values measured at 45 s and 1 min were defined as the activity of CAT. The activity of CAT in the samples was given as U mg-1 protein.

2.5. Statistical Analysis

2.5.1. ANOVA

A completely randomized design (CRD) was used to accomplish the experiments. Experimental data were analyzed by ANOVA using an SPSS program (SPSS Inc., Chicago, IL, USA). The effects of treatment (MeJA, SA, and untreated control), storage condition (CS and SL) and their interactions were evaluated on all parameters of quality losses, antioxidant properties and enzyme activities. Duncan's multiple range tests used separated means at $p < 0.05$ level.

2.5.2. Correlation and Regression Analysis among Parameters

In order to quantify the relationship between the fruit quality parameters, Pearson's correlation coefficients were determined for the relationships of the two quality loss parameters, the four antioxidant properties and the four enzyme activities in all combinations (Tables 2 and 3). Correlation analyses were performed separately for the two chemical (SA and MeJA) treatments. Statistical calculations using Pearson's correlation analysis were performed by Genstat 5 Release 4.1 (Lawes Agricultural Trust, IACR, Rothamsted, UK). Then, the best correlated variables were plotted and linear regression analyses were used in order to evaluate the strongest relationships among quality loss parameters, antioxidant properties and enzyme activities. The regression slopes were tested by a t-test in order to determine whether they are different between MeJA and SA treatments at $\alpha = 0.05$.

2.5.3. Principal Component Analysis

We conducted a standardized principal component analysis (PCA) based on the correlation matrix with the CI, FD, TAC, TSPC, TCC, AAC, PAL, POD, SOD, and CAT variables. All variables were standardized with transforming the values to z-scores. Model fit was tested with the root mean square residual (RMSR) [51]. Principal components (PCs) were visualized in biplot diagram. PCA was conducted with R 4.03 [52] with the psych [53], FactoMiner [54] and factoextra [55] packages (R Core Team, Vienna, Austria).

3. Results

3.1. Quality Loss Parameters

3.1.1. Chilling Injury

The degree of CI index was larger with the progression of assessment times in all chemical and storage treatments (Table 1). However, fruit pre-treated with MeJA and SA had lower CI indices at $p < 0.05$ compared to water-treated fruit for both storage treatments. Fruit treated with MeJA showed significantly lower CI index in the CS treatments at day 21 and in the SL treatments at day 8 compared to corresponding SA treatments.

Table 1. Indices of CI (chilling injury) and FD (fruit decay) values for the treatments of methyl jasmonate (MeJA, 0.2 mmol L^{-1}) and salicylic acid (SA, 2 mmol L^{-1}) on cultivar 'Bergarouge' apricot fruit at three cold storage (CS) dates (days 7, 14 and 21) and at two assessment dates (days 4 and 8) of shelf-life (SL) treatment.

Treatments	CS at 1 °C			SL at 25 °C	
	Day 7	Day 14	Day 21	Day 4	Day 8
CI index (%) [a]					
Control	3.12 a [b]	16.68 a	37.65 a	18.67 a	30.24 a
0.2 mmol MeJA	1.35 a	4.52 b	9.68 c	8.65 b	14.35 c
2 mmol SA	1.64 a	5.58 b	20.91 b	7.68 b	19.26 b
$LSD_{0.05}$ [c]	ns	3.56	6.83	4.29	3.98
FD index (%) [d]					
Control	16.05 a	56.24 a	96.25 a	66.36 a	100 a
0.2 mmol MeJA	2.11 b	6.57 b	16.98 b	16.35 b	36.65 b
2 mmol SA	2.33 b	7.58 b	15.61 b	20.36 b	35.36 b
$LSD_{0.05}$	1.74	5.45	4.72	8.31	3.34

[a] and [d] at day 0, both the CI and FD indices were zero. [b] Different letters represent significant differences at $p < 0.05$ for each column. [c] $LSD_{0.05}$: Least significant differences at $p = 0.05$ level.

3.1.2. Fruit Decay

Non-treated apricot fruit showed a great increase in FD index in both storage treatments reaching 96 and 100% after 21 and 8 days, respectively (Table 1). MeJA and SA treatments reduced superficial browning symptoms at $p < 0.05$ compared with control treatments. The FD indices were not significantly different between the treatments of SA and MeJA in either CS or SL treatments.

3.2. Antioxidant Capacity and Related Parameters

3.2.1. Total Antioxidant Capacity

The TAC of fruit increased in the MeJA and SA treatments until day 14 and day 4 in the CS and SL treatments, respectively, and then TAC is reduced (Figure 1A,B), while the TAC of control fruits decreased continuously under both CS and SL conditions (Figure 1A,B). Antioxidant capacity for control fruits was lower at $p < 0.05$ than the corresponding MeJA- and SA-treated fruits in both storage treatments. However, TAC values showed no significant differences between MeJA and SA treatments in either CS or SL treatments.

3.2.2. Total Soluble Phenol Content

The content of TSP was similar in the MeJA and SA treatments in both storage treatments, while the TSP content of control fruits decreased continuously in both treatments (Figure 1C,D). The TSP content was higher in the treatments of MeJA and SA at $p < 0.05$ compared to the control treatment in both storage treatments (Figure 1C,D).

3.2.3. Total Carotenoids Content

The contents of total carotenoids increased in both chemical (MeJA and SA) treatments until day 14 and day 4 in the CS and SL treatments, respectively, and then it started to decrease (Figure 1E,F), except for MeJA treatments for CS where the increase stopped at day 7. Salycilic acid treatments showed the highest total carotenoids content which was significantly higher at days 14 and 21 in the CS and at day 4 in the SL treatments compared to either the control or MeJA treatments (Figure 1E,F). In the control treatments, carotenoids slightly increased at days 7 and 14 in the CS and at day 4 in the SL treatments (Figure 1E,F), but these increases were negligible and non-significant compared to the increase in carotenoids in either MeJA or SA treatments.

3.2.4. Ascorbic Acid Content

AAC increased in the control fruits until days 14 and day 4 in the CS and SL treatments, respectively, then it sharply decreased (Figure 1G,H). AAC in the MeJA- and SA-treated fruits decreased continuously in both storage conditions (Figure 1G,H). However, by the last assessment dates (day 21 and day 8 for the CS and SL treatments, respectively), SA showed the highest ascorbic acid content, followed by the MeJA and then the control treatments, showing significant differences ($p < 0.05$) from each other (Figure 1G,H).

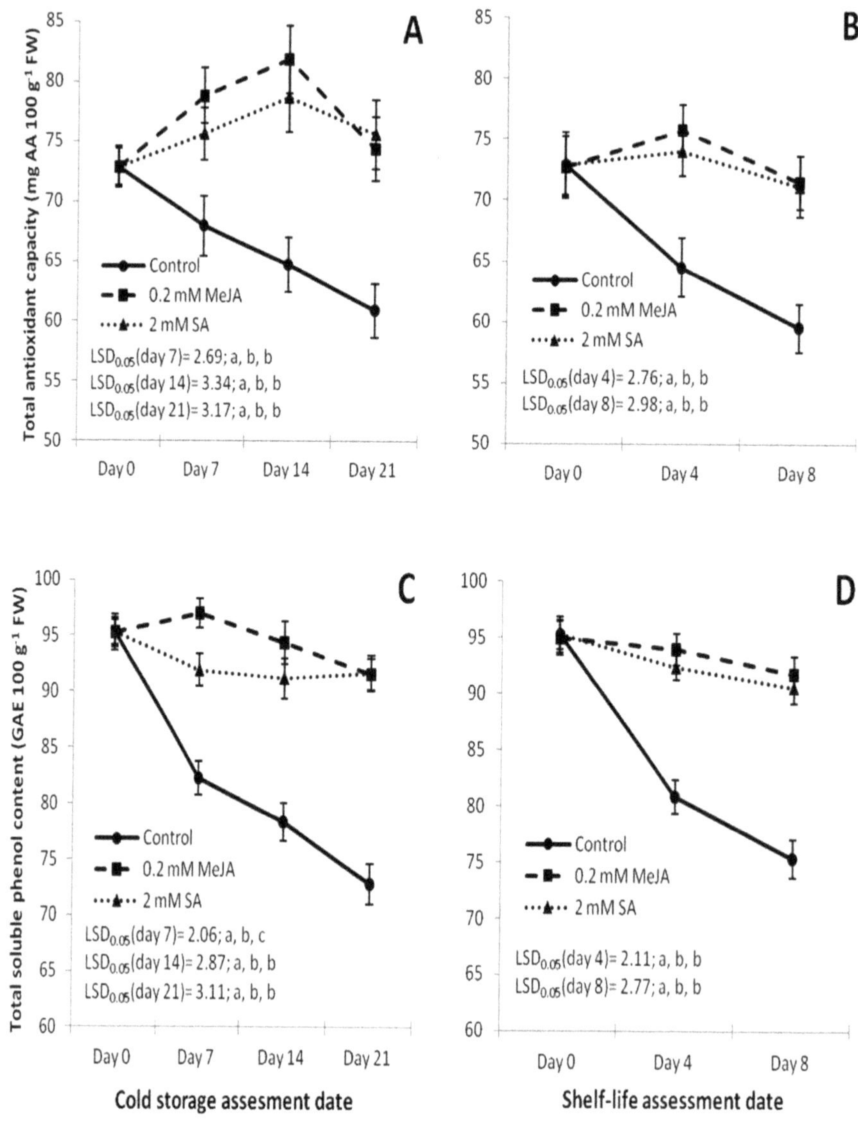

Figure 1. *Cont.*

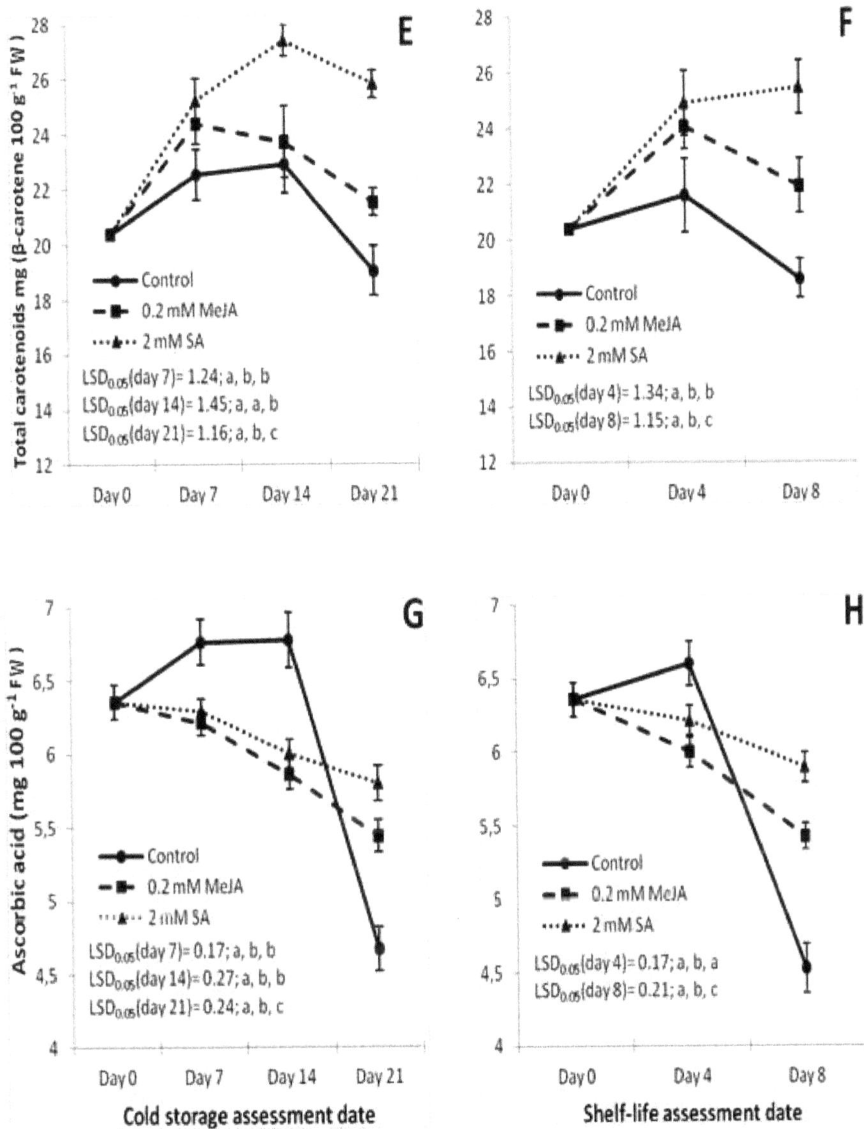

Figure 1. Effect of methyl jasmonate (MeJA, 0.2 mmol L^{-1}) and salicylic acid (SA, 2 mmol L^{-1}) on TSP (total soluble phenol) content, TAC (total antioxidant capacity) total carotenoids and AAC (ascorbic acid content) in cv. 'Bergarouge' in cold storage treatments (**A**, **C**,**E**,**G**) at 1 °C on the assessment days 7, 14 and 21, and in shelf-life (SL) treatments (**B**,**D**,**F**,**H**) at 25 °C on the assessment days 4 and 8. Standard deviation (SD) values are given for error bars. $LSD_{0.05}$ values were used to assess differences among the control, MeJA and SA treatments at $p < 0.05$. Control means water treated fruits. The first letter given after each $LSD_{0.05}$ value belongs to control treatments, second letter to MeJA; and third letter to SA treatments. Different letters represent significant differences among the treatments at $p < 0.05$.

3.3. Enzyme Activity

3.3.1. PAL Activity

The activity of PAL increased with assessment dates in the MeJA and SA treated fruits while it decreased in the water treated fruits in the two storage treatments (Figure 2A,B). The PAL activity was the highest in MeJA treatments followed by SA and control treatments in both storage treatments. The three treatments were different at $p < 0.05$ from each other at days 14 and 21 in the CS and at day 8 in the SL treatments.

3.3.2. POD Activity

POD activity increased in the MeJA and SA treatments until day 14 and until day 8 in the CS and SL treatments, respectively, then it decreased (Figure 2C,D). POD activity was inconsistent for the control treatment in both storage conditions. The activity of POD in the MeJA and SA treatments was similar, but it was higher at $p < 0.05$ in all assessment dates compared to the fluctuating levels of POD activity in the control treatments in both the CS and the SL treatments (Figure 2C,D).

3.3.3. SOD Activity

Activity of SOD increased in the chemical (MeJA and SA) treatments until day 14 and day 4 in the CS and SL treatments, respectively, and then enzyme activity started to decrease (Figure 2E,F), except for MeJA in the SL treatment, where the increase was continuous. A reduction in SOD activity was detected in the water-treated fruits in both CS and SL treatments (Figure 2E,F). Activity of SOD in the control treatments was lower at $p < 0.05$ than the corresponding MeJA and SA treatments in both CS and SL conditions. SOD activity was high in MeJA treatments, which was higher at $p < 0.05$ than the SA treatments at day 14 in the CS and at day 8 at the SL treatment.

3.3.4. CAT Activity

CAT activity increased in the SA and control treatments at all assessment dates in both storage treatments (Figure 2G,H). CAT activity in the MeJA-treated fruits decreased continuously in both CS and SL treatments, except for MeJA treatments between days 14 and 21 under CS conditions (Figure 2G,H). Activity of CAT in the MeJA treatments was lower at $p < 0.05$ compared to in the corresponding SA or control treatments in both CS and SL conditions. CAT activity was the highest in SA treatments, which was higher at $p < 0.05$ compared to the MeJA treatments at days 14 and 21 in the CS and at days 4 and 8 in the SL treatment (Figure 2G,H).

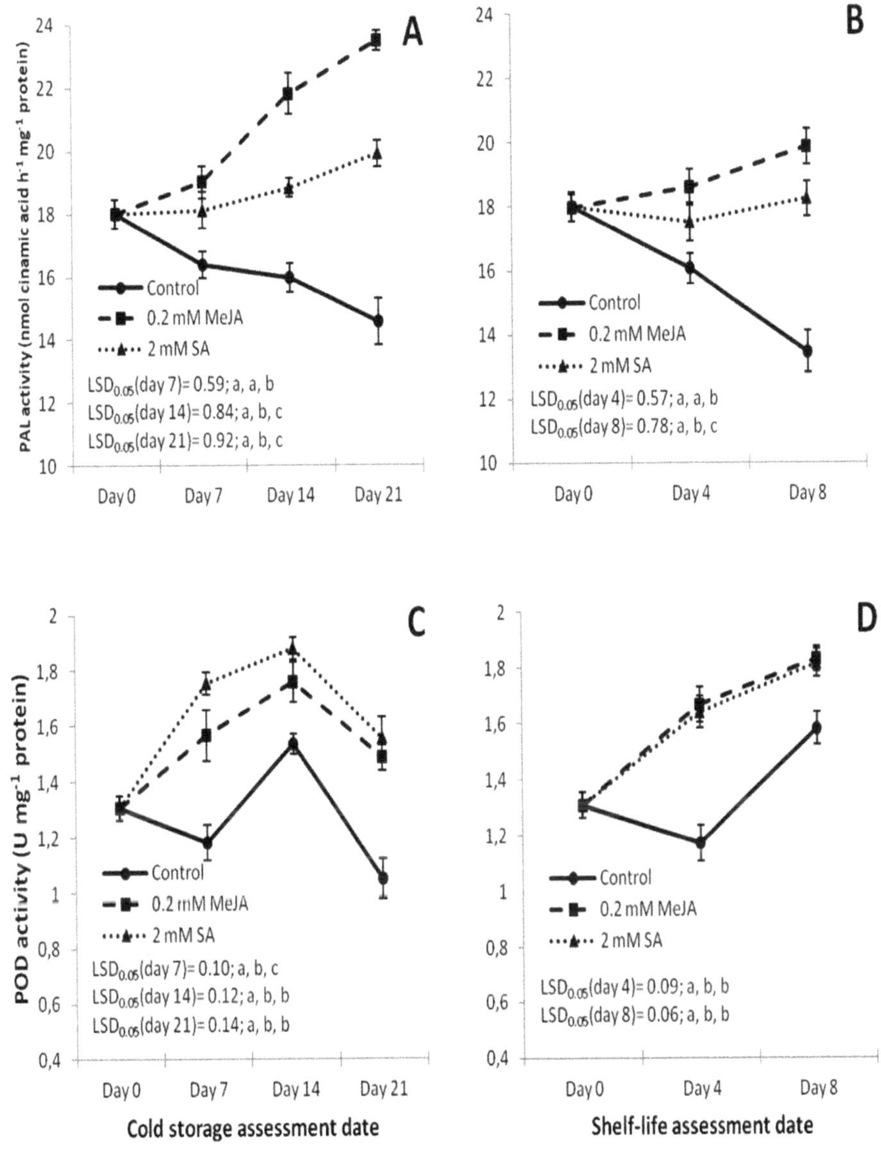

Figure 2. *Cont.*

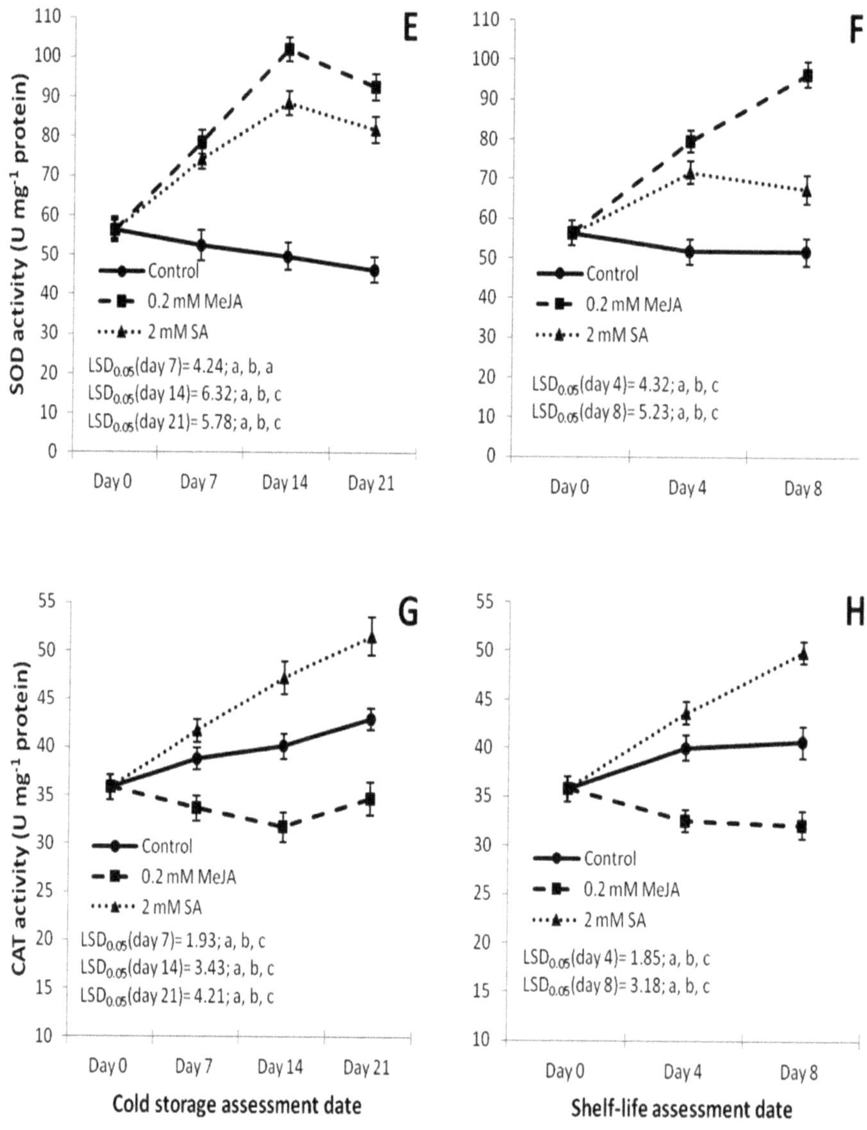

Figure 2. Effect of methyl jasmonate (MeJA; 0.2 mmol L^{-1}) and salicylic acid (SA, 2 mmol L^{-1}) on phenylalanine ammonia-lyase (PAL), peroxidase (POD), superoxide dismutase (SOD), and catalase activity (CAT) in cv. 'Bergarouge' in cold storage (CS) treatments (**A**,**C**,**E**,**G**) at 1 °C on the assessment days 7, 14 and 21, and in shelf-life (SL) treatments (**B**,**D**,**F**,**H**) at 25 °C on the assessment days 4 and 8. Other information is given in Figure 1 (symbols, error bars and $LSD_{0.05}$ values).

3.4. Relationship between Fruit Quality Parameters

Among the 45 pair-variables, Pearson's correlation coefficients of 23 and 9 pair-variables correlated significantly (at $p < 0.05$) in both chemical (MeJA and SA) treatments, respectively (Tables 2 and 3). Among these significantly pair-variables, six pair-variables were significant in both the MeJA and the SA. Among these six pair-variables, four were correlated positively (FD versus (vs.) CI, PAL vs. CAT,

PAL vs. SOD, and total antioxidant content vs. SOD) and two negatively (total antioxidant content vs. FD and ascorbic acid content vs. CI), indicating connections among fruit quality loss, antioxidant parameters and enzyme activities (Tables 2 and 3).

Table 2. Pearson correlation coefficients (r) and corresponding significance levels (p) amongst 10 fruit quality parameters in the methyl jasmonate (MeJA) treatments (0.2 mmol L^{-1}) on cv. 'Bergarouge' apricot. Data were combined for the assessment dates of days 7, 14 and 21 of cold storage (CS) and for the assessment dates of days 4 and 8 of shelf-life (SL) treatment.

	CI	FD	CAT	SOD	POD	PAL	AAC	TCC	TSPC
FD [a]	**0.967** [b,c]								
	<0.001								
CAT	−0.315	−0.423							
	>0.1	>0.1							
SOD	**0.818**	**0.813**	−0.005						
	0.013	**0.014**	>0.1						
POD	0.604	**0.732**	**−0.766**	0.331					
	0.079	**0.035**	**0.027**	>0.1					
PAL	0.273	0.124	**0.699**	**0.697**	−0.440				
	>0.1	>0.1	**0.047**	**0.048**	>0.1				
AAC	**−0.829**	**−0.776**	−0.129	**−0.921**	−0.241	−0.641			
	0.011	**0.024**	>0.1	**<0.001**	>0.1	0.068			
TCC	−0.668	−0.601	−0.194	**−0.825**	−0.108	−0.689	**0.772**		
	0.059	0.080	>0.1	**0.012**	>0.1	0.051	**0.025**		
TSPC	**−0.904**	**−0.799**	−0.012	**−0.850**	−0.294	−0.592	**0.900**	**0.747**	
	0.001	**0.016**	>0.1	**0.008**	>0.1	0.082	**0.001**	**0.032**	
TAC	**−0.963**	**−0.959**	0.323	**0.874**	−0.600	−0.290	**0.843**	**0.724**	**0.863**
	<0.001	**<0.001**	>0.1	**0.004**	0.080	>0.1	**0.009**	**0.040**	**0.005**

[a] Fruit quality parameters: CI = chilling injury index, FD = fruit decay index, CAT = catalase activity, SOD = superoxide dismutase activity, POD = peroxidase activity, PAL = phenylalanine ammonia-lyase activity, AAC = ascorbic acid content, TCC = total carotenoid content, TSPC = total soluble phenol content and TAC = total antioxidant capacity. [b] The significant ($p < 0.05$) correlation coefficient values are in bold. [c] $n = 20$

Table 3. Pearson correlation coefficients (r) and corresponding significance levels (p) amongst 10 fruit quality parameters in salicylic acid (SA) treatments (2 mmol L^{-1}) on cv. 'Bergarouge' apricot. Data were combined for the assessment dates of days 7, 14 and 21 of cold storage (CS) and for the assessment dates of days 4 and 8 of shelf-life (SL) treatment.

	CI	FD	CAT	SOD	POD	PAL	AAC	TCC	TSPC
FD [a]	**0.698** [b,c]								
	0.047								
CAT	**0.960**	0.576							
	<0.001	0.085							
SOD	0.167	−0.549	0.257						
	>0.1	0.1	>0.1						
POD	−0.171	0.260	−0.123	**−0.721**					
	>0.1	>0.1	>0.1	**0.039**					
PAL	0.630	−0.100	**0.696**	**0.711**	−0.381				
	0.071	>0.1	**0.048**	**0.042**	>0.1				
AAC	**−0.908**	−0.527	**−0.884**	−0.218	0.191	−0.646			
	<0.001	0.120	**0.003**	>0.1	>0.1	0.066			
TCC	0.305	0.036	0.307	0.253	−0.219	0.390	−0.298		
	>0.1	>0.1	>0.1	>0.1	>0.1	>0.1	>0.1		
TSPC	−0.412	−0.462	−0.374	0.238	−0.314	−0.071	0.432	−0.413	
	>0.1	>0.1	>0.1	>0.1	>0.1	>0.1	>0.1	>0.1	
TAC	−0.386	**−0.884**	−0.294	**0.738**	−0.525	0.393	0.246	0.008	0.577
	>0.1	**0.003**	>0.1	**0.033**	>0.1	>0.1	>0.1	>0.1	0.085

[a] Fruit quality parameters: CI = chilling injury index, FD = fruit decay index, CAT = catalase activity, SOD = superoxide dismutase activity, POD = peroxidase activity, PAL = phenylalanine ammonia-lyase activity, AAC = ascorbic acid content, TCC = total carotenoid content, TSPC = total soluble phenol content and TAC = total antioxidant capacity. [b] The significant ($p < 0.05$) correlation coefficient values are in bold. [c] $n = 20$

Thus, the relationships of the six pair-variables were further revealed by regression analysis (Figure 3). The linear regression analysis showed significant relationships for all six of the pair-variables with $r = 0.716–0.932$, $p = 0.03–0.001$ and with $r = 0.703–0.889$, $p = 0.03–0.001$ for the MeJA and SA treatments, respectively, but the slopes for all six of the pair-variables were not different between the MeJA and the SA treatments ($p = 0.654–0.116$ according to a t-test).

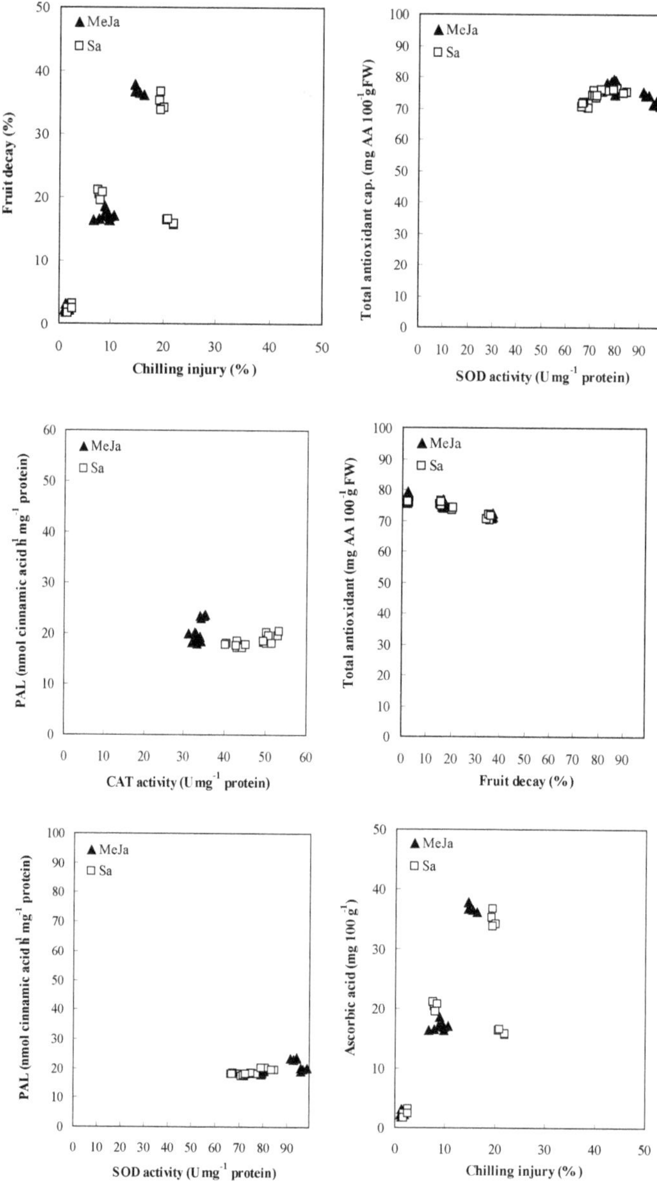

Figure 3. Relationships between fruit decay and chilling injury index, PAL (phenylalanine ammonia-lyase) and CAT (catalase activity), PAL and SOD (superoxide dismutase), total antioxidant content and SOD, total antioxidant content and fruit decay, and ascorbic acid content and chilling injury index in treatments of methyl jasmonate (MeJA, 0.2 mmol L^{-1}) and salicylic acid (SA, 2 mmol L^{-1}).

PCA explained the 89.3% of the total variance and the PCs had been justified. The RMSR was 0.05 indicating good fit. PC1 accounted for 55.6% of the variance and correlated with the CI, FD, TAC, TSPC, TCC (Carotenoids), PAL, and SOD. PC2 accounted for 19.7% of the variance and correlated with the POD and AAC. PC3 accounted for 14.0% of the variance and correlated with the CAT. The 95 ellipses of treatments indicated that the control group overlapped with the treatments (Figure 4); however, the difference was significant in the case of PC1 ($F = 13$, $df = 2$, $p > 0.001$) and non-significant for PC2 ($F = 1.518$, $df = 2$, $p = 0.251$). Accordingly, variables of PC1 were more efficient in discriminating the treatments than PC2, where the control and treated samples had similar values.

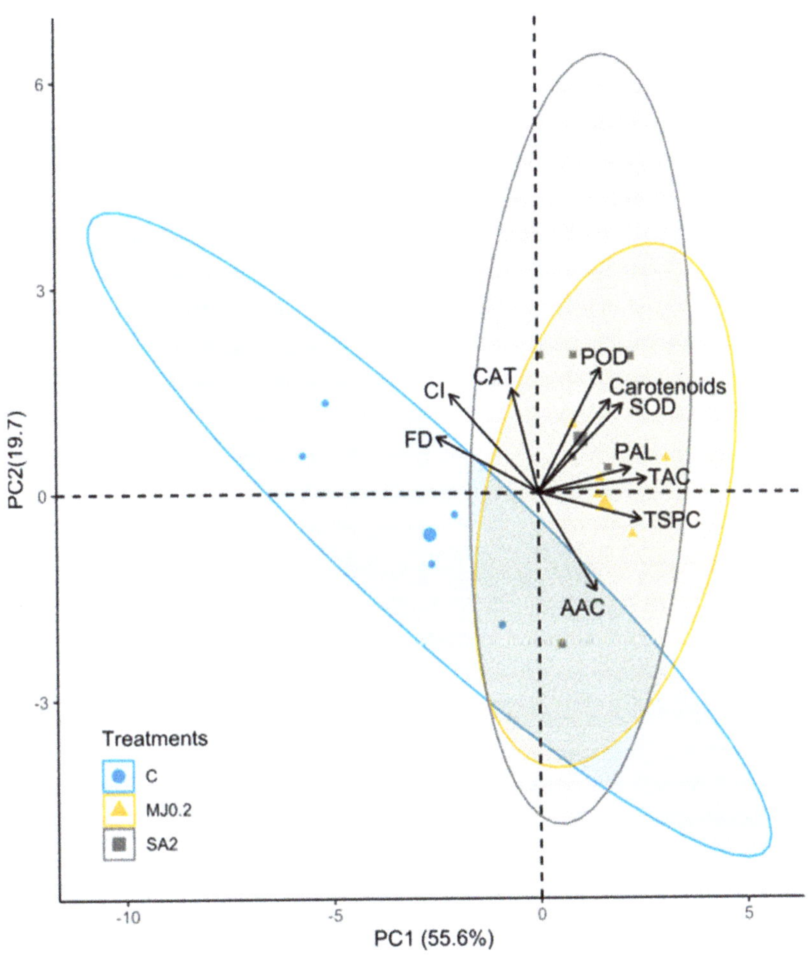

Figure 4. Biplot of principal component analyses (PCA) conducted on the measured fruit quality parameters: CI = chilling injury index, FD = fruit decay index, CAT = catalyse activity, SOD = superoxide dismutase activity, POD = peroxidase activity, PAL = phenylalanine ammonia-lyase activity, AAC = ascorbic acid content, Carotenoids = total carotenoid content, TSPC = total soluble phenol content and TAC = total antioxidant capacity. (C: control; MJ0.2: MeJA, 0.2 mmol L^{-1}; SA2: SA, 2 mmol L^{-1}); 95% ellipses).

4. Discussion

4.1. Quality Loss Parameters

In agreement with this study, MeJA treatment was demonstrated to decrease CI for fruits such as papaya, peach, pomegranates, and loquat, e.g., [20,22–24,27]. MeJA was shown to reduce CI throughout, promoting the expression of heat-shock proteins and regulating arginine metabolism [56]. Furthermore, CI reduction on the MeJA-treated fruits was also likely to be connected to the increase in polyamines (such as putrescine or spermidine) in the cells [15,34]. These polyamines maintain the ratio between the unsaturated and the saturated fatty acid, which have a key role in the membrane integrity and fluidity [57]. Acclimation to low temperatures is highly related to the increase in this fatty acid ratio in the membrane lipids [58], which could result in lower CI symptoms on the MeJA-treated fruits.

In SA treatments, previous studies on pomegranates and peach reported that SA and its derivates decreased CI in postharvest treatments [22,31,58], which is in agreement with our apricot study. The possible mechanism of the effect of SA on CI was studied by Ding et al. [59] and they indicated that MeSA probably had little direct effect on chilling damage, but it could provide indirectly a reduction in CI by inducing some defense-mechanism responses in the cells.

MeJA and MeSA treatments were also compared in the study of Sayyari et al. [22], but in contrast with this study, the authors concluded that differences among MeJA and MeSA treatments were non-significant for CI. The reasons for the different result are likely to be the differences among applied concentrations, fruit crops and SA derivates used in the two studies. Overall, our results indicate that both compounds (MeJA and SA) could be suitable in reducing CI in apricot fruit.

In accordance with our results, some studies also revealed that SA can decrease fruit decay on peach and cherry [26,60]. Salicylic acid is likely to induce the defense resistance system in fruit tissues, which result in lower FD compared to untreated fruits. In the case of MeJA, our results of the FD reduction by MeJA are also in agreement with the results of previous studies on grapefruit, peach, mango and strawberry. The study of Droby et al. [18] on grapefruit showed that fruit treated with different MeJA concentrations (1 to 50 mmol L^{-1}) had less FD compared to untreated fruit. They reported that the MeJA treatments had no direct toxicity effect on the fungus (*Penicillium digitatum*), which causes FD. They suggested that MeJA decreased FD indirectly by enhancing the natural resistance mechanisms of the fruit to fungus infection [18]. The effect of SA and MeJA in reducing FD was also reported by Yao and Tian [26]. The authors demonstrated that the sweet cherry fruit treated with SA (2 mM) or with MeJA (0.2 mM) had lower fruit damage caused by *Monilinia fructicola* than in the control treatment. The authors reported that SA and MeJA had direct inhibitory effects on the mycelial growth of *M. fructicola*. González-Aguilar et al. [19] and Ayala-Zavala et al. [61] showed that MeJA treatments reduced fruit decay in mango and strawberry, respectively, during both CS and SL periods. The possible mechanism of MeJA in reducing fruit decay may involve direct inhibition on the pathogen and/or indirect mechanism through induced resistance. The observed suppression of fruit decay of apricot in this study supported the hypothesis that MeJA can prolong the postharvest health of the fruit.

No previous studies compared the joint effect of MeJA and SA on fruit decay and we showed a similar efficacy for both compounds. We demonstrated significant correlation, significant regression and significant PCA (in PC1) relationships between FD and CI in both MeJA and SA treatments (Tables 2 and 3, Figures 3 and 4), which were not demonstrated by previous studies. The possible reason for the correlative connection between FD and CI may be that FD or CI can be reduced by similar mechanisms such as the activation of the defense mechanism in the fruit tissues by both MeJA and SA compounds. Overall, MeJA and SA treatments could be useful compounds for a joint reduction in chilling injury and fruit decay without adversely affecting fruit quality.

4.2. Antioxidant Capacity and Related Parameters

Similarly to our apricot study, antioxidant capacity including total phenol contents significantly increased by MeJA in CS and/or SL treatments in strawberry, peach and pomegranate fruits compared to untreated fruits [22,28,61].

In SA treatments, Sayyari et al. [22] demonstrated that SA significantly increased the TAC and TSP content of pomegranate fruits compared to control fruit. Previous studies showed that phenolic compounds, as an antioxidant substance, usually accumulated or at least maintained their levels under cold stress [62], which was the case in the SA and the MeJA treatments but not in the control treatments of this apricot study. The applications of MeJA or SA were able to keep the total phenol contents or were able to increase the antioxidant capacity during the storage periods (Figure 2). Our results indicate that postharvest applications of both MeJA and SA compounds have the potential to improve fruit health by reducing not only CI and FD but increasing the TAC of fruit tissues.

Ascorbic acid content of apricot fruit progressively increases during fruit maturity stages [63]. Our results demonstrate that SA- and/or MeJA-treated fruit had lower ascorbic acid contents at the beginning of storage. Reports on the effect of MeJA on AAC of fruits are not consistent as, for instance, González-Aguilar et al. [21] demonstrated that MeJA treatments did not change AAC, while Jin et al. [28] demonstrated that MeJA significantly influenced vitamin C levels of fruits after CS and SL treatments on peach compared to control treatments. In SA treatments, Sayyari et al. [58] demonstrated that AAC reduced significantly in the control and SA treatments at 0.7 and 1.4 mmol L^{-1}, while the AAC of the fruits was not changed in the dose of 2 mmol L^{-1}. Similarly, ascorbic acid levels remained the same by the end of storage in the SA treatments at 1 mmol L^{-1} in peaches and 2 mmol L^{-1} in oranges [31,64].

Previous fruit studies showed significant positive correlations among parameters of antioxidant capacity and phenol contents under storage conditions, e.g., [31,65]. One study [31] provided these correlations in postharvest application of SA and this study investigated these correlations for MeJA and SA jointly. For instance, the reduction in fruit decay was associated with increasing antioxidant capacity for both SA and MeJA treatments, which corroborated well with the significant correlation coefficients, the significant regression, and the significant PCA (in PC1) relationships between FD and TAC (Tables 2 and 3, Figures 3 and 4). In the case of CI vs. AAC pair-variable, the increasing ascorbic acid level in the fruit tissues was connected to a decreasing level of chilling injury, which was confirmed by the significant correlation coefficients and also by the significant regression relationship between CI and AAC for both MeJA and SA compounds (Tables 2 and 3, Figure 3).

4.3. Enzyme Activity

The key enzyme of phenyl alanine-lyase is included in the biosynthesis of phenolics such as flavonoids and phenols [66]. Previous studies showed that the high amounts of phenols and anthocyanins in apple and grape fruits resulted in an increased the level of PAL [67,68]. In this study, MeJA-treated fruits showed higher TAC and PAL activity compared to the water-treated fruits (Figures 1 and 2). These results indicate that antioxidant capacity increased by MeJA may also induce PAL activity, thus promoting phenolic metabolism. This result agrees with the studies of Yao and Tian [26]. Meng et al. [27] and Sayyari et al. [58] also demonstrated that MeJA increased the activity of PAL in fruits of sweet cherry, peach and pomegranate. In the case of SA, two studies [26,69] reported that SA had an essential role in systematic acquired resistance (SAR) induction by promoting defense and antioxidant enzymes such as PAL, which was also supported by our results in apricot fruit.

PAL and POD enzymes are involved in the biosynthesis of lignin; as a consequence, the activities of these two enzymes affect the fruit lignin content. Cao et al. [23] demonstrated that MeJA treatment promoted activities of PAL and POD as well as the lignin content of loquat fruit, which might also account for a lower decay index, which was also demonstrated in this apricot study (Table 1, Figure 2). Qin et al. [69] demonstrated that SA also increased the activities of POD and PAL, in agreement with this study (Figure 2). Yang et al. [35] showed that SA treatments prevent fruit from softening by increasing the activities of PAL and POD and the lignin content in fruit resulting in a higher firmness and a lower weight loss.

In addition, induction of systemic resistance to biotic or abiotic stresses can lead to direct activation of defense-related proteins. POD has an essential role in the structure components of cell walls and lignin formation. An increase in POD activity was reported for SA-treated (4 mM) mandarin fruit which was related to low FD [70]. Previous studies, e.g., [26,71,72], and this study also confirmed that either MeJA or SA treatment increased POD activity and decreased fruit decay (Figure 2; Table 1).

The SOD enzyme has several roles in plant cells, such as detoxifying ROS and lowering chilling injury [73,74]. SOD activity was high in chilling tolerant mandarin fruit during CS at 2.5 °C for 8 weeks [73]. Previous studies e.g., [23,24,75] showed that MeJA significantly increased the activity of SOD in loquat, peach and strawberry fruit in the storehouse compared to the control treatments, while Huang et al. [64] showed a similar effect of SA treatment for orange fruit. Our apricot study was in agreement with the above previous studies on SA or MeJA results on SOD activity. However, here we provided for the first time a comparison between MeJA and SA treatments on SOD activity which showed that SOD activity was generally higher in the MeJA treatments than in the SA treatments.

CAT is a key enzyme in the catalyzation of H_2O_2 into H_2O and O_2, and therefore it has a function to remove active oxygen species (AOS) from the cell during stress [64]. In agreement with our results, MeJA was reported to maintain high CAT activity in strawberry plants [75] and in loquat fruit during the CS period [23] compared to non-treated fruit. However, SA treatment's effects on CAT activity were not uniform in previous studies. Huang et al. [64] showed that SA pretreatment reduced CAT activities compared to the non-treated fruit. However, similarly to our apricot study, Mo et al. [76] demonstrated that SA treatments increased CAT activity in apple fruit compared to the non-treated control. This might be associated with the result of the study of Tian et al. [77] that SA can increase the CAT gene activity, i.e., the transcription and translation. In addition, not only CAT, but also ascorbate peroxidase is a key enzyme of fruits in the removal of H_2O_2. Although this peroxidase was not investigated in this study, future data on ascorbate peroxidase activity also contribute to the better understanding of the mechanism of the antioxidant capacity and of the defense systems involved.

No previous fruit studies investigated correlations among enzyme activity parameters for MeJA and SA jointly under postharvest treatments. However, this study confirmed that an increase in CAT was associated with increasing PAL activity for both SA and MeJA treatments, which corroborated well with the significant positive correlation coefficients and regression relationships between CAT and PAL (Tables 2 and 3, Figure 3). A similar relationship was detected for the pair-variable of SOD vs. PAL, where the increasing superoxide dismutase level in the fruit tissues was connected to an increasing level of phenylalanine ammonia-lyase, which was confirmed by significant correlation coefficients and also by significant regression as well as by significant PCA (in PC1) relationship between SOD and PAL for both MeJA and SA compounds (Tables 2 and 3, Figures 3 and 4).

Overall, our results on the activities of enzymes may imply that the four enzymes can promote various functions in the defense system and these may be jointly induced by MeJA and/or SA in apricot fruit. PAL, POD, SOD and CAT enzymes are substantially induced by the given MeJA and SA concentrations, and as a result, the storability of apricot fruit is increased.

5. Conclusions

This was the first study to investigate the effect of MeJA and SA jointly on quality losses, antioxidant properties and enzyme activities of apricot fruit in CS and SL conditions.

Our results show that treatments of both SA and MeJA decreased the CI and FD of apricot fruit in CS and SL periods. Both elicitors ensured high total antioxidant capacity, total polyphenolic and carotenoids contents, and enhanced the enzyme activity of PAL, POD, SOD and CAT in CS and SL conditions; as a result, the storability of apricot fruit is increased. Among the 45 pair-variables, Pearson's correlation coefficients of 23 and 9 pair-variables correlated significantly in the treatments of MeJA and SA, respectively. Among these pair-variables, six pair-variables (FD vs. CI, PAL vs. CAT,

PAL vs. SOD, TAC vs. SOD, TAC vs. FD, and AAC vs. CI) were significant in both the MeJA and the SA, indicating connections among fruit quality loss, antioxidant parameters and enzyme activities. The relationship between the six pair-variables was further confirmed by linear regression analysis with $r = 0.716–0.932$, $p = 0.03–0.001$ and with $r = 0.703–0.889$, $p = 0.03–0.001$ for the MeJA and SA treatments. PCA explained 89.3% of the total variance and PC1 accounted for 55.6% of the variance and correlated with the CI, FD, TAC, TSPC, TCC, PAL, and SOD variables. Variables of PC1 were efficient in discriminating the treatments. The treatments of MeJA and SA are practically useful and inexpensive techniques to maintain several quality attributes of apricot fruit during both CS and SL conditions.

Author Contributions: Conceptualization, A.E., I.J.H., A.A., Z.S. and J.N.; methodology, A.E.; A.H. and A.A.; software, S.S.; validation, A.E., I.J.H., A.H. and S.S.; formal analysis, I.J.H. and A.E.; investigation, A.E.; resources, I.J.H.; data curation, A.E.; writing—original draft preparation, A.E. and I.J.H.; writing—review and editing, S.S., J.N., Z.S. and B.M.; visualization, A.E. and I.J.H.; supervision, I.J.H.; project administration, B.M.; funding acquisition, I.J.H. All authors have read and agreed to the published version of the manuscript.

Funding: This research was funded by the Hungarian Scientific Research Funds (K 131478), by a János Bolyai Research Fellowship awarded to I. J. Holb, by the European Union and the State of Hungary, co-financed by the European Social Fund in the framework of TÁMOP-4.2.4.A/2-11/1-2012-0001 'National Excellence Program' under project number A2-SZJ-TOK-13-0061, and by The research was financed by the Thematic Excellence Programme of the Ministry for Innovation and Technology in Hungary, within the framework of the Space Sciences thematic programme of the University of Debrecen, and by the Thematic Excellence Programme of the Ministry for Innovation and Technology in Hungary (ED_18-1-2019-0028), within the framework of the climate change thematic programme of the University of Debrecen.

Acknowledgments: We also thank to Gyümölcsért Ltd.

Conflicts of Interest: The authors declare no conflict of interest.

References

1. Ercisli, S.; Akbulut, M.; Ozdemir, O.; Sengul, M.; Orhan, E. Phenolic and antioxidant diversity among persimmon (*Diospyrus kaki* L.) genotypes in Turkey. *Int. J. Food Sci. Nutr.* **2008**, *59*, 477–482. [CrossRef] [PubMed]
2. Ercisli, S.; Tosun, M.; Karlidag, H.; Dzubur, A.; Hadziabulic, S.; Aliman, Y. Color and antioxidant characteristics of some fresh fig (*Ficus carica* L.) genotypes from Northeastern Turkey. *Plant Foods Hum. Nutr.* **2012**, *67*, 271–276. [CrossRef] [PubMed]
3. Crisosto, C.H.; Mitchell, F.G.; Johnson, R.S. Factors in fresh market stone fruit quality. *Postharv. News Inform.* **1995**, *6*, 17N–21N.
4. Hacıseferoğulları, H.; Gezer, I.; Özcan, M.M.; Asma, B.M. Post-harvest chemical and physical–mechanical properties of some apricot varieties cultivated in Turkey. *J. Food Eng.* **2007**, *79*, 364–373. [CrossRef]
5. Muzzaffar, S.; Bhat, M.M.; Wani, T.A.; Wani, I.A.; Masoodi, F.A. Postharvest Biology and Technology of Apricot. In *Postharvest Biology and Technology of Temperate Fruits*; Mir, S., Shah, M., Mir, M., Eds.; Springer: Berlin, Germany, 2018.
6. Infante, R.; Meneses, C.; Defilippi, B.G. Effect of harvest maturity stage on the sensory quality of 'Palsteyn' apricot (*Prunus armeniaca* L.) after cold storage. *J. Hortic. Sci. Biotechnol.* **2008**, *83*, 828–832. [CrossRef]
7. Stanley, J.; Marshall, R.; Ogwaro, J.; Feng, R.; Wohlers, M.; Woolf, A. Postharvest storage temperatures impact significantly on apricot fruit quality. *Acta Hortic.* **2010**, *880*, 525–532. [CrossRef]
8. Ezzat, A.; Ammar, A.; Szabó, Z.; Holb, I. Salicylic acid treatment saves quality and enhances antioxidant properties of apricot fruit. *Hortic. Sci.* **2017**, *44*, 73–81.
9. Ezzat, A.; Ammar, A.; Szabó, Z.; Nyéki, J.; Holb, I.J. Postharvest treatments with methyl jasmonate and salicylic acid for maintaining physico-chemical characteristics and sensory quality properties of apricot fruit during cold storage and shelf-life. *Pol. J. Food Nutr. Sci.* **2017**, *67*, 159–166. [CrossRef]
10. Egea, M.I.; Matrinez-Madrid, P.; Sanchez-Bel, M.A.; Romojaro, F. The influence of electron-beam ionization on ethylene metabolism and quality parameter in apricot (*Prunus armeniaca* L., cv 'Builda'). *Swiss Soc. Food Sci. Technol.* **2007**, *40*, 1027–1035.
11. Crisosto, C.H.; Mitchell, F.G.; Zhiguo, J. Susceptibility to chilling injury of peach, nectarine, and plum cultivars grown in California. *HortScience* **1999**, *34*, 1116–1118. [CrossRef]

12. Turner, J.G.; Ellis, C.; Devoto, A. The jasmonate signal pathway. *Plant Cell Suppl.* **2002**, *14*, 153–164. [CrossRef] [PubMed]
13. Delker, C.; Stenzel, I.; Hause, B.; Miersch, O.; Feussner, I.; Wasternack, C. Jasmonate biosynthesis in *Arabidopsis thaliana*—Enzymes, products, regulation. *Plant Biol.* **2006**, *8*, 297–306. [CrossRef] [PubMed]
14. Pieterse, C.M.J.; Van der Does, D.; Zamioudis, C.; Leon-Reyes, A.; Van Wees, S.C.M. Hormonal modulation of plant immunity. *Annu. Rev. Cell Dev. Biol.* **2012**, *28*, 489–521. [CrossRef]
15. Wang, S.Y.; Buta, J.G. Methyl jasmonate reduces chilling injury in *Cucurbita pepo* through its regulation of abscisic acid and polyamine levels. *Environ. Exp. Bot.* **1994**, *34*, 427–432. [CrossRef]
16. Meir, S.; Philosoph-Hadas, S.; Lurie, S.; Droby, S.; Akerman, M.; Zauberman, G.; Shapiro, B.; Cohen, E.; Fuchs, Y. Reduction of chilling injury in stored avocado, grapefruit, and bell pepper by methyl jasmonate. *Can. J. Bot.* **1996**, *74*, 870–874. [CrossRef]
17. Jong-Joo, C.; Do, C.Y. Methyl jasmonate as a vital substance in plants. *Trends Gen.* **2003**, *19*, 409–413.
18. Droby, S.; Porat, R.; Cohen, L.; Weiss, B.; Shapira, B.; Philosoph-Hadas, S.; Meir, S. Suppressing green mold decay in grape fruit with postharvest jasmonates application. *J. Am. Soc. Hortic. Sci.* **1999**, *124*, 184–188. [CrossRef]
19. González-Aguilar, G.A.; Buta, J.G.; Wang, C.Y. Methyl jasmonate reduces chilling injury symptoms and enhances colour development of 'Kent' mangoes. *J. Sci. Food Agric.* **2001**, *81*, 1244–1249. [CrossRef]
20. González Aguilar, G.A.; Buta, J.G.; Wang, C.Y. Methyl jasmonate and modified atmosphere packaging (MAP) reduce decay and maintain post harvest quality of papaya 'Sunrise'. *Postharv. Biol. Technol.* **2003**, *28*, 361–370. [CrossRef]
21. González-Aguilar, G.A.; Tizuado-Hernándoz, M.E.; Zavaleeta-Gatica, R.; Martínez-Téllez, M.A. Methyl jasmonate treatments reduce chilling injury and activate the defense response of guava fruits. *Biochem. Biophysiol. Res. Commun.* **2004**, *313*, 694–701. [CrossRef]
22. Sayyari, M.; Babalar, M.; Kalantari, S.; Martínez-Romero, D.; Guillén, F.; Serrano, M.; Valero, M. Vapour treatments with methyl salicylate or methyl jasmonate alleviated chilling injury and enhanced antioxidant potential during postharvest storage of pomegranates. *Food Chem.* **2011**, *124*, 964–970. [CrossRef]
23. Cao, S.; Zheng, Y.; Yang, Z.; Wang, K.; Rui, H. Effect of methyl jasmonate on quality and antioxidant activity of postharvest loquat fruit. *J. Sci. Food Agric.* **2009**, *89*, 2064–2070. [CrossRef]
24. Jin, P.; Duan, Y.; Wang, L.; Wang, J.; Zheng, Y. Reducing chilling injury of loquat fruit by combined treatment with hot air and methyl jasmonate. *Food Bioprocess Technol.* **2014**, *7*, 2259–2266. [CrossRef]
25. Fan, X.; Mattheis, J.P.; Fellman, J.K. Responses of apples to postharvest jasmonate treatments. *J. Am. Soc. Hortic. Sci.* **1998**, *123*, 421–425. [CrossRef]
26. Yao, H.J.; Tian, S.P. Effects of pre- and postharvest application of SA or MeJA on inducing disease resistance of sweet cherry fruit in storage. *Postharv. Biol. Technol.* **2005**, *35*, 253–262. [CrossRef]
27. Meng, X.; Han, J.; Wang, Q.; Tian, S.P. Changes in physiology and quality of peach fruits treated by methyl jasmonate under low temperature stress. *Food Chem.* **2009**, *114*, 1028–1035. [CrossRef]
28. Jin, P.; Wang, K.; Shang, H.; Tong, J.; Zheng, Y. Low temperature conditioning combined with methyl jasmonate treatment reduces chilling injury of peach fruit. *J. Sci. Food Agric.* **2009**, *89*, 1690–1696. [CrossRef]
29. Li, L.P.; Han, T. The effects of salicylic acid in the storage of peach. *Food Sci.* **1999**, *7*, 61–63.
30. Srivastava, M.K.; Dwivedi, U.N. Delayed ripening of banana fruit by salicylic acid. *Plant Sci.* **2000**, *158*, 87–96. [CrossRef]
31. Wang, L.; Chena, S.; Kong, W.; Li, S.; Archbold, D. Salicylic acid pretreatment alleviates chilling injury and affects the antioxidant system and heat shock proteins of peaches during cold storage. *Postharv. Biol. Technol.* **2006**, *41*, 244–251. [CrossRef]
32. Shafiee, M.; Taghavi, T.S.; Babalar, M. Addition of salicylic acid to nutrient solution combined with postharvest treatments (hot water, salicylic acid, and calcium dipping) improved postharvest fruit quality of strawberry. *Sci. Hortic.* **2010**, *124*, 40–45. [CrossRef]
33. Valero, D.; Diaz-Mula, H.M.; Zapata, P.J.; Castillo, S.; Guillen, F.; Martinez-Romero, D.; Serrano, M. Postharvest treatments with salicylic acid, acetylsalicylic acid or oxalic acid delayed ripening and enhanced bioactive compounds and antioxidant capacity in sweet cherry. *J. Agric. Food Chem.* **2011**, *59*, 5483–5489. [CrossRef]
34. Wei, Y.; Liu, Z.; Su, Y.; Liu, D.; Ye, X. Effect of salicylic acid treatment on postharvest quality, antioxidant activities, and free polyamines of asparagus. *J. Food Sci.* **2011**, *76*, S126–S132. [CrossRef] [PubMed]

35. Yang, Z.; Cao, S.; Zheng, Y.; Jiang, Y. Combined salicyclic acid and ultrasound treatments for reducing the chilling injury on peach fruit. *J. Agric. Food Chem.* **2012**, *60*, 1209–1212. [CrossRef] [PubMed]
36. Cai, C.X.; Li, L.P.; Chen, K.S. Acetyl salicylic acid alleviates chilling injury of postharvest loquat (*Eriobotrya japonica* Lindl.) fruit. *Eur. Food Res. Technol.* **2005**, *223*, 533–539. [CrossRef]
37. Satraj, A.; Masud, T.; Abassi, K.S.; Mahmood, T.; Ali, A. Effect of different concentrations of salicylic acid on keeping quality of apricot cv. 'Habi' at ambient storage. *J. Biol. Food Sci. Res.* **2013**, *2*, 66–78.
38. Maisuthisakul, P.; Pasuk, S.; Ritthiruangdejc, P. Relationship between antioxidant properties and chemical composition of some Thai plants. *J. Food Compos. Anal.* **2008**, *21*, 229–240. [CrossRef]
39. Du, G.; Li, M.; Ma, F.; Lian, D. Antioxidant capacity and the relationship with polyphenol and vitamin C in Actinidia fruits. *Food Chem.* **2009**, *113*, 557–562. [CrossRef]
40. Sulaiman, S.F.; Yusoff, N.A.; Eldeen, I.M.; Seow, E.M.; Sajak, A.A.B.; Supriatno Ooi, K.L. Correlation between total phenolic and mineral contents with antioxidant activity of eight Malaysian bananas (*Musa* sp.). *J. Food Compos. Anal.* **2011**, *24*, 1–10. [CrossRef]
41. Ulewicz-Magulska, B.; Wesolowski, M. Total phenolic contents and antioxidant potential of herbs used for medical and culinary purposes. *Plant Foods Hum. Nutr.* **2019**, *74*, 61–67. [CrossRef]
42. Benzie, I.F.; Strain, J. The ferric reducing ability of plasma (FRAP) as a measure of 'antioxidant power': The FRAP assay. *Anal. Biochem.* **1996**, *239*, 70–76. [CrossRef] [PubMed]
43. Singleton, V.L.; Rossi, J.A. Colorimetry of total phenolics with phosphomolybdic phosphotungstic acid 'reagents'. *Am. J. Enol. Vitic.* **1965**, *16*, 144–158.
44. Akin, E.B.; Karabulut, I.; Topcu, A. Some compositional properties of main Malatya apricot (*Prunus armeniaca,* L.) varieties. *Food Chem.* **2008**, *107*, 939–948. [CrossRef]
45. Gross, J. *Carotenoids: Pigments in Fruits*; Academic: London, UK, 1987.
46. Terada, M.; Watanabe, Y.; Kunitoma, M.; Hayashi, E. Differential rapid analysis of ascorbic acid and ascorbic acid 2-sulfate by dinitrophenilhydrazine method. *Ann. Clinic. Biochem.* **1978**, *84*, 604–608. [CrossRef]
47. Assis, J.S.; Maldonado, R.; Mnoz, T.; Escribano, M.I.; Merodio, C. Effect of high carbon dioxid concentration on PAL activity and phenol content in ripening cherimoya fruit. *Postharv. Biol. Technol.* **2001**, *23*, 33–39. [CrossRef]
48. Chance, B.; Maehly, A.C. Assay of catalases and peroxidases. *Method Enzymol.* **1995**, *2*, 764–817.
49. Rao, M.V.; Paliyath, G.; Ormord, D.P. Ultraviolet B and ozon induced biochemical changes in antioxidant enzymes of *Arabidopsis thaliana*. *Plant Physiol.* **1996**, *110*, 125–136. [CrossRef]
50. Abassi, N.A.; Kushad, M.M.; Endress, S. Active oxygen-scavenging enzymes activities in developing apple flowers and fruits. *Sci. Hortic.* **1998**, *74*, 183–194. [CrossRef]
51. Basto, M.; Pereira, J.M. An SPSS R-menu for ordinal factor analysis. *J. Stat. Softw.* **2012**, *46*, 1–29. [CrossRef]
52. R Core Team. *R: A Language and Environment for Statistical Computing*; R Foundation for Statistical Computing: Vienna, Austria, 2020; Available online: https://www.R-project.org/ (accessed on 10 May 2020).
53. Revelle, W. *Psych: Procedures for Personality and Psychological Research*; Northwestern University: Evanston, IL, USA, 2015; Available online: http://CRAN.R-project.org/package=psych (accessed on 12 June 2015).
54. Lê, S.; Josse, J.; Husson, F. FactoMineR: An R package for multivariate analysis. *J. Stat. Softw.* **2008**, *25*, 30294. Available online: https://www.jstatsoft.org/article/view/v025i01 (accessed on 25 October 2008).
55. Kassambara, A.; Mundt, A. Factoextra: Extract and Visualize the Results of Multivariate Data Analyses. R Package Version 1.0.6. 2019. Available online: https://CRAN.R-project.org/package=factoextra (accessed on 1 June 2019).
56. Zhang, X.; Sheng, J.; Li, F.; Meng, D.; Shen, L. Methyl jasmonate alters arginine catabolism and improves postharvest chilling tolerance in cherry tomato fruit. *Postharv. Biol. Technol.* **2012**, *64*, 160–167. [CrossRef]
57. Mirdehghan, S.H.; Rahemi, M.; Castillo, S.; Martínez-Romero, D.; Serrano, M.; Valero, D. Pre-storage application of polyamines by pressure or immersion improves shelf-life of pomegranate stored at chilling temperature by increasing endogenous polyamine levels. *Postharv. Biol. Technol.* **2007**, *44*, 26–33. [CrossRef]
58. Sayyari, M.; Babalar, M.; Kalantari, S.; Serrano, M.; Valero, D. Effect of salicylic acid treatment on reducing chilling injury in stored pomegranates. *Postharv. Biol. Technol.* **2009**, *53*, 152–154. [CrossRef]
59. Ding, C.K.; Wang, C.Y.; Gross, K.C.; Smith, D.L. Jasmonate and salicylate induce the expression of pathogenesis related-proteingenes and increase resistance to chilling injury in tomato fruit. *Planta* **2002**, *214*, 895–900. [CrossRef]

60. Chan, Z.; Tian, S. Induction of H_2O_2-metabolizing enzymes and total protein synthesis by antagonistic yeast and salicylic acid in harvested sweet cherry fruit. *Postharv. Biol. Technol.* **2006**, *39*, 314–320. [CrossRef]
61. Ayala-Zavala, J.F.; Wang, S.Y.; Wang, C.Y.; González-Aguilar, G.A. Methyl jasmonate in conjunction with ethanol treat- ment increases antioxidant capacity, volatile compounds and post-harvest life of strawberry fruit. *Eur. Food Res. Technol.* **2005**, *221*, 731–738. [CrossRef]
62. Kondo, S.; Kittikorn, M.; Kanlayanarat, S. Preharvest antioxidant activities of tropical fruit and the effect of low temperature storage on antioxidants and jasmonates. *Postharv. Biol. Technol.* **2005**, *36*, 309–318. [CrossRef]
63. Hegedűs, A.; Pfeiffer, N.; Abrankó, L.; Blázovics, A.; Pedryc, A.; Stefanovits-Bányai, E. Accumulation of antioxidants in apricot fruit through ripening: Characterization of a genotype with enhanced functional properties. *Biol. Res.* **2011**, *44*, 339–344. [CrossRef]
64. Huang, R.-H.; Liu, J.-H.; Lu, Y.-M.; Xia, R.-X. Effect of salicylic acid on the antioxidant system in the pulp of 'Cara cara' navel orange (*Citrus sinensis* (L.) Osbeck) at different storage temperatures. *Postharv. Biol. Technol.* **2008**, *47*, 168–175. [CrossRef]
65. Ezzat, A.; El-Sherif, A.R.; Doaa Elgear, D.; Szabó, S.Z.; Holb, I.J. A comparison of fruit and leaf parameters of apple in three orchard training systems. *Zemdirb. Agric.* **2020**, *107*, 373–382. [CrossRef]
66. Dixon, R.A.; Paiva, N.L. Stress-induced phenylpropanoid metabolism. *Plant Cell* **1995**, *7*, 1085–1097. [CrossRef] [PubMed]
67. Ju, Z.G.; Yuan, Y.B.; Lieu, C.L.; Xin, S.H. Relationships among phenylalanine ammonia-lyase activity, simple phenol concentrations and anthocyanin accumulation in apple. *Sci. Hortic.* **1995**, *61*, 215–226. [CrossRef]
68. Hiratsuka, S.; Onodera, H.; Kawai, Y.; Kubo, T.; Itoh, H.; Wada, R. Enzyme activity changes during anthocyanin synthesis in 'Olympia' grape berries. *Sci. Hortic.* **2001**, *90*, 255–264. [CrossRef]
69. Qin, G.Z.; Tian, S.P.; Xu, Y.; Wan, Y.K. Enhancement of biocontrol efficacy of antagonistic yeasts by salicylic acid in sweet cherry fruit. *Physiol. Mol. Plant. Pathol.* **2003**, *62*, 147–154. [CrossRef]
70. Haider, S.T.A.; Ahmad, S.; Khan, S.; Anjum, M.A.; Nasir, M.; Naz, S. Effects of salicylic acid on postharvest fruit quality of "Kinnow" mandarin under cold storage. *Sci. Hortic.* **2020**, *259*, 108843. [CrossRef]
71. Wills, R.; McGlasson, B.; Graham, D.; Joyce, D. *Postharvest, An Introduction to the Physiology and Handling of Fruit and Vegetables and Ornamentals*, 4th ed.; University of New South Wales Press Ltd.: Sydney, Australia, 1998; pp. 125–156.
72. Peng, L.; Jiang, Y. Exogenous salicylic acid inhibits browning of fresh-cut Chinese water chestnut. *Food Chem.* **2006**, *94*, 535–540. [CrossRef]
73. Sala, J.M. Involvement of oxidative stress in chilling injury in cold-stored mandarin fruits. *Postharv. Biol. Technol.* **1998**, *13*, 255–261. [CrossRef]
74. Wang, S.Y.; Bowman, L.; Ding, M. Methyl jasmonate enhances antioxidant activity and flavonoid content in blackberries (*Rubus* sp.) and promotes antiproliferation of human cancer cells. *Food Chem.* **2008**, *107*, 1261–2619. [CrossRef]
75. Wang, S.Y. Methyl jasmonate reduces water stress in strawberry. *J. Plant Growth Regul.* **1999**, *18*, 127–134. [CrossRef]
76. Mo, Y.; Gong, D.; Liang, L.; Han, R.; Xie, J.; Li, W. Enhanced preservation effects of sugar apple fruits by salicylic acid treatment during post-harvest storage. *J. Sci. Food Agric.* **2008**, *88*, 2693–2699. [CrossRef]
77. Tian, S.; Qin, G.; Li, B.; Wang, Q.; Meng, X. Effects of salicylic acid on disease resistance and postharvest decay control of fruits. *Stewart Postharv. Rev.* **2007**, *6*, 1–7.

Publisher's Note: MDPI stays neutral with regard to jurisdictional claims in published maps and institutional affiliations.

Article

Phytochemical Characterization of Blue Honeysuckle in Relation to the Genotypic Diversity of *Lonicera* sp.

Jacek Gawroński [1], Jadwiga Żebrowska [1], Marzena Pabich [2], Izabella Jackowska [2], Krzysztof Kowalczyk [1] and Magdalena Dyduch-Siemińska [1,*]

1 Department of Genetics and Horticultural Plant Breeding, Institute of Plant Genetics, Breeding and Biotechnology, Faculty of Agrobioengineering, University of Life Sciences in Lublin, Akademicka 15 Street, 20-950 Lublin, Poland; jacek.gawronski@up.lublin.pl (J.G.); jadwiga.zebrowska@up.lublin.pl (J.Ż.); krzysztof.kowalczyk@up.lublin.pl (K.K.)

2 Department of Chemistry, Faculty of Food Science and Biotechnology, University of Life Sciences in Lublin, Akademicka 15 Street, 20-950 Lublin, Poland; marzena.pabich@up.lublin.pl (M.P.); izabella.jackowska@up.lublin.pl (I.J.)

* Correspondence: magdalena.dyduch@up.lublin.pl

Received: 20 August 2020; Accepted: 16 September 2020; Published: 18 September 2020

Abstract: The phytochemical characteristic analysis of a group of 30 haskap berry genotypes was carried out bearing in mind the concern for the consumption of food with high nutraceutical value that helps maintain good health. Phytochemical fruit composition and antioxidant activity were assessed by the Folin–Ciocalteau, spectrophotometric, DPPH (1,1-diphenyl-2-picrylhydrazyl) as well as ABTS (2,2′-azino-bis(3-ethylbenzothiazoline-6-sulfonic acid) method. Evaluation of antioxidant activity was referred to as the Trolox equivalent. The observed differences in the content of phenolics, flavonoids, vitamin C and antioxidant activity allowed us to select genotypes which, due to the high level of the analyzed compounds, are particularly recommended in everyone's diet. In addition, the analysis of the prospects of increasing the analyzed phytochemical properties, estimated by parameters such as heritability and genetic progress, indicates the effectiveness of breeding in relation to each of the analyzed traits. The results of the presented research can be used in the implementation of future breeding programs for this valuable species.

Keywords: ABTS; bioactive compounds; DPPH; flavonoids; genetic distance; genotypic and phenotypic correlation; nutraceutical value; phenolics; UPGMA

1. Introduction

The constantly growing awareness of consumers regarding the need for healthy eating and consuming food with health-promoting properties leads to the search and characterization of new species with this direction of utilization. Species belonging to the genus *Lonicera* such as *Lonicera caerulea* var. *edulis*, *L. caerulea* var. *kamtschatica*, *L. caerulea* var. *altaica*, *L. caerulea* var. *byarnikovae* and *L. caerulea* var. *emphyllocalyx*, as well as their hybrids, collectively known as *Lonicera caerulea* L., also known as haskap, blue honeysuckle, honeyberry or sweet berry honeysuckle, are representative of such plants. The fruits of this species, like strawberry, blueberry, blackberry or blackcurrant, belong to the so-called superfruits—fruits, which due to the presence of bioactive compounds, are very desirable in the human diet [1,2]. Health-promoting properties of the haskap berries include protective effects against cardiovascular and neurodegenerative diseases, osteoporosis, type 2 diabetes as well as antimicrobial, anticarcinogenic and anti-inflammatory activity [3]. Combining biological traits of individual species allowed us to obtain cultivars with large fruits without a bitter taste, not falling off the bush after ripening, fertile, more resistant to diseases, suitable for mechanical harvesting and storage, as well as significantly different in terms of phytochemical content. The assessment of the content of health-promoting phytochemical compounds, such as phenolic acids, flavonoids and anthocyanins has so far been performed usually for a small group of cultivars. Cehula et al. [4] studied 14 cultivars, Wang et al. [5] 7 cultivars, Senica et al. [6] 4 cultivars, Auzanneau et al. [7] 7 cultivars, Khattab et al. [8] 3 cultivars, Sochor et al. [9] 19 cultivars, Wojdyło et al. [10] 8 cultivars, Rupasinghe et al. [11] 3 cultivars and Kusznierewicz et al. [12] 6 cultivars. So far, the largest group of 30 cultivars has been analyzed by Kucharska et al. [13]. The level of the analyzed phytochemical properties, as indicated by numerous authors [12,14–16] depends, on one hand, on the properties of the cultivar (genotype), while on the other hand, it is the result of many environmental factors (soil, fertilization, etc.). The analysis of the influence of each of these factors allows us to determine the extent to which a given phytochemical property results from the impact of genetic factors, and to which it is a result of the effect of the environment, which can be expressed by the heritability coefficient, and which has not been presented so far for this species.

The attractive biological value of *Lonicera* fruits, which consists of the values of individual cultivars, causes a rapid increase in the cultivation area of this species. According to the available data, the global cultivation area in 2017 amounted approximately to 5500 ha. The largest planted area was in China and North Korea, in total, 2000 ha, Poland—1800 ha, Canada—1000 ha, Russia—400 ha and Japan—160 ha [17]. The cultivation area of this species is constantly growing and, for example, in Russia, this area increased to 735 ha in 2019 and its further increase to about 2000 ha is planned in 2022 [18]. Poland, with a cultivation area of approximately 4000 ha in 2019, is a world leader in the production of haskap berries [19]. The Commission Implementing Regulation (EU) 2018/1991 of 13 December 2018 authorizing the introduction of *Lonicera caerulea* L. berries to the market as a traditional food from a third country in accordance with the Regulation of the European Parliament and of the Council (EU) 2015/2283 [20] is likely to support a further increase in the cultivation area of this species. The growing area of cultivation, as well as the valuable properties of the fruits in terms of human health, require, as many authors postulate, the analysis of the biological potential of new genotypes of this species, their suitability for breeding, and thus an indication of the possibility of increasing the content of biologically active compounds in the future [11,13,21]. Therefore, the aim of this study was, firstly, to evaluate the content of polyphenols, flavonoids, vitamin C and antioxidant activity of berry in a large group (30) of haskap genotypes of various origin, including 10 that were analyzed for the first time. Secondly, the estimation of variability, as well as the not-yet presented genotypic correlation, heritability and the genetic advance of characteristic in a group of genotypes studied was evaluated. For this purpose, a large group of genotypes was necessary because it increased the precision of inference. Thirdly, this study can be a guide in the implementation of breeding work aimed at increasing the biological potential of this species and helpful in selecting a cultivar for those who want to consume fruits with a high nutraceutical value.

2. Materials and Methods

The Institute of Plant Genetics and Biotechnology has held a collection of cultivars and breeding clones of the genus *Lonicera* since 2007. The experiment used the bushes in the fourth year of fruiting planted in 2015 at the Experimental Station, University of Life Sciences in Lublin (51°13′59″ φN, 22°34′0″ λE, elevation: 225.48 m). The research covered a diverse group of genotypes including cultivars of Polish, Slovak, Russian and Canadian origin, as well as breeding clones, the list of which is presented in Table 1. Thirty genotypes were selected for analyses from which the fruits were harvested in June 2019. Three fruit sub-samples were collected from each genotype (at the start, in the middle and the end of the harvest period selecting fully pigmented fruits). Fruit sub-samples were frozen and stored at −20 °C until analysis (10 days). Directly before the analysis, the samples were mixed and the resulting average sample, weighting 150 g, was the material for testing the content of phytochemical compounds such as phenolics, flavonoids, vitamin C and antioxidant activity.

Table 1. List of genotypes tested.

No.	Genotype	Country of Origin	No.	Genotype	Country of Origin
1	1-17-59	RUS	16	Jugana	RUS
2	Amphora	RUS	17	K100	POL
3	Amur	SVK	18	Karina	POL
4	Aurora	CAN	19	LeningradskijVelikan	RUS
5	BakczarskijVelikan	RUS	20	Nimfa	RUS
6	BerryBlue	CZE	21	Polar Jevel	CAN
7	Blue Velvet	RUS	22	Siniczka	RUS
8	Borealis	CAN	23	Sinoglaska	RUS
9	Brązowa	POL	24	T3	RUS
10	Czarna	POL	25	T5	RUS
11	DoczVelikana	RUS	26	Uspiech	RUS
12	HoneyBee	CAN	27	Valhova	RUS
13	Indigo Gem	CAN	28	Vostorg	RUS
14	IndigoTreat	CAN	29	Warta	POL
15	Jolanta	POL	30	Zielona	POL

2.1. Fruit Extract Preparation for Polyphenols and Antioxidant Activity Determination

Before the experiment, whole frozen fruit samples (150 g) were homogenized with a blender (PHILIPS). The appropriate fruit material (2.5 g) was successively extracted three times with 8 mL acidified (0.1% (*v*/*v*) formic acid) 80% (*v*/*v*) methanol for 30 min (Multi-Rotator RS-60 (bioSan)) at room temperature. The supernatant was filtered using a vacuum pump and was poured together. The final volume was 25 mL. The obtained extract was used for the determination of total polyphenols (phenolics, flavonoids) as well as DPPH (1,1-diphenyl-2-picrylhydrazyl) and ABTS (2,2′-azino-bis(3-ethylbenzothiazoline-6-sulfonic acid) assay.

2.2. Total Phenolics Content (TPC)

The total phenolics content of fruit extracts was determined according to the method of as Khattab et al. [8]. The sample (0.1 mL each) was mixed with 0.4 mL of distilled water, and 2.5 mL 10% Folin–Ciocalteau reagent (*v*/*v*) and incubated at room temperature for 5 min before being neutralized by 2.0 mL of 7.5% (*w*/*v*) sodium carbonate solution. The reaction mixture was incubated at room temperature for 2 h and the absorbance was measured at 760 nm using a UV–VIS spectrophotometer (Shimadzu A-160, Shimadzu Corp., Kioto, Japan). The absorbance of each extract was reduced by the absorbance of the blank sample that contained the distilled water instead of the Folin–Ciocalteau reagent. Gallic acid in 50% (*v*/*v*) methanol solution at a range of concentrations (10–100 mg·L^{-1} was used as a standard and a calibration curve was drawn. The content of total phenolics was expressed as mg gallic acid equivalent (GAE)·100 g^{-1} of fresh weight. All samples were analyzed in five replications.

2.3. Flavonoids Content

Total flavonoids content of fruit extracts was determined using a spectrophotometric method [11]. The extract or standard (quercetin) (0.5 mL) was diluted to 3.7 mL with distilled water and mixed with 0.15 mL of 5% (*w*/*v*) sodium nitrite ($NaNO_2$) solution at time zero. After incubation at room temperature for 5 min, 0.15 mL of 10% (*w*/*v*) aluminum chloride ($AlCl_3$) was added and the mixtures were vigorously shaken. After 1 min, 1 mL of 1 M sodium hydroxide (NaOH) was added, mixed, and the absorbance at 510 nm was measured immediately versus a blank. The blank consisted of 50% (*v*/*v*) methanol instead of sample. Quercetin in 50% (*v*/*v*) methanol solution at a range of concentrations (10–100 mg·L^{-1}) was used as a standard and a calibration curve was drawn. Total flavonoid content in the extract was expressed as mg quercetin equivalent (QE)·100 g^{-1} fresh sample. All samples were analyzed in five replicates.

2.4. Antioxidant Activity

As suggested by Schlesier et al. [22] that the assessment of antioxidant activity should be carried out using at least two methods, the DPPH and ABTS methods were used in this work.

2.4.1. DPPH Assay

The antiradical activity of extracts was determined with the DPPH• radical (1,1-diphenyl-2-picrylhydrazyl), according to the method of Brand-Williams et al. [23] slightly modified by Khattab et al. [8]. Fruit extracts or standard (Trolox) (0.1 mL) were mixed with 2.9 mL of 0.1 mM DPPH solution and incubated in the shade at room temperature for 10 min. The absorbance was measured at 515 nm. Solution with 100% methanol, instead of samples, was used as a control. The absorbance of each extract was reduced by the absorbance of the blank sample that contained the extract and 100% methanol instead of the DPPH solution. Antioxidant activity was calculated based on Trolox standards at a range of concentrations 0.4–1.0 mM and expressed as a mmol Trolox equivalent (TE)·100 g^{-1} fresh sample. All samples were analyzed in five replications.

2.4.2. ABTS (2,2′-azino-bis(3-ethylbenzothiazoline-6-sulfonic acid) Assay

The antioxidant activity was also determined with the cation radical $ABTS^{+\bullet}$ (2.2′-azino-bis (3-ethylbenzothiazoline-6-sulfonic acid)), in line with the methodology described by Re et al. [24]. $ABTS^{+\bullet}$ was generated by reacting ABTS aqueous solution (7 mM) with potassium persulfate (2.45 mM) in the dark for 12–16 h. The initial solution was diluted until absorbance reached 0.7 (±0.02) at 734 nm. The reaction mixture contained a 3 mL $ABTS^{+\bullet}$ solution and a 0.03 mL extract or standard (Trolox). Measurements were made after 10 min at room temperature. Antioxidant activity was calculated based on Trolox standards at a range of concentrations 0.1–1.0 mM and expressed as a mmol Trolox equivalent (TE)·100 g^{-1} fresh sample. All samples were performed in five replicates.

2.5. Fruit Extract Preparation for Vitamin C Content Determination

From the homogenized fruits (150 g) was prepared a 5 g sample which was mixed with 40 mL of 3% metaphosphoric acid. After stirred for 30 min (Multi-Rotator RS-60 (bioSan)) at room temperature the solution was filtered using a vacuum pump. The final volume was made up to 50 mL with the 3% metaphosphoric acid and the obtained extract was used for the determination of vitamin C content.

Vitamin C was determined with the spectrofluorimetric method [25] with necessary modifications. The oxidizing solution (100 mL) was prepared by mixing of I_2 (1.3 g), KI (40%, 10 mL) solution, HCl (7 M, 0.1 mL) and distilled water. The derivatization reagent was prepared by dissolving OPDA (o-phenyloenediamine, 10 mg) in 10 mL of 0.005 M H_2SO_4. Sample (2 mL) was mixed with 0.3 mL of a 0.005 M solution of iodine in potassium iodide. After being vortexed for 1 min, 0.3 mL of 0.01 M $Na_2S_2O_3$ was added. After adjusting the pH of the samples to approximately 6.0, the derivatization was carried out by adding 0.3 mL of OPDA solution and stirring for 30 min at room temperature. The final volume was made up to 100 mL with the distilled water. The analysis was performed on Cary Eclipse (Varian, Palo Alto, CA, USA) spectrofluorimeter at an excitation wavelength $\lambda = 365$ nm and an emission wavelength $\lambda = 425$ nm. The vitamin C content was calculated based on a standard curve obtained using an aqueous solution of L-ascorbic acid standard in concentrations of 10, 20, 50 and 100 mg·L^{-1}. The vitamin C content in the extract was expressed as mg L-ascorbic acid·100 g^{-1} fresh sample. All samples were analyzed in five replications.

2.6. Statistical Analysis of the Data

All data obtained in the experiment were statistically analyzed by ANOVA (analysis of variance) and the significance of differences between means was established by the Fishers LSD test at $p \leq 0.05$ with Statistica 13.1 [26] statistical software. The cluster analysis based on the content of phenolics, flavonoids, vitamin C and antioxidant activity was conducted using the Unweighted Pair-Group Method with

Arithmetic Mean (UPGMA) method available in the same software [26]. Fruit phytochemical content and antioxidant activity are quantitative features, so the genetic estimation of these traits may be assessed with the use of genetic parameters like: heritability (H), genetic advance (GA), phenotypic and genotypic coefficient of variability (PCV and GCV respectively). These parameters are commonly used in creative breeding which is focused on obtaining new genetically improved cultivars (genotypes). Estimation of these coefficients will help to know the gene action affecting the concerned traits. Heritability is a genetic parameter used for the evaluation of genetic determination of quantitatively inherited traits. Heritability values are useful to indicate the reliability of the phenotypic value of characteristic to predict its genetic value, high values indicate a greater possibility of genetic advance (GA) obtained with the selection. Generally, heritability indicates the effectiveness with which the selection of genotypes could be based on phenotypic performance. Genetic advance (GA) expected from the selection is a precise indicator of the improvement of features in genotypic value for the new breeding population compared with the base population under one cycle of selection at a given selection intensity. Genetic advance (GA) can be estimated in population by multiplication of the values of heritability (H), phenotypical standard deviation of traits (∂_p) in population and selection intensity coefficient (k), which value depends on the percent of selection intensity (in our research 20%, k = 1.4), so that GA = $k \cdot \partial_p \cdot H$. In our study, all analyzed traits were assessed by their phenotypic and genotypic value. The phenotypic value of an individual is the values of the traits observed and measured in the test sample. For each quantitative trait, a part resulting from the action of genes can be separated from its phenotypic value and is called the genotypic value. The genotypic and phenotypic coefficient of variation (GCV, PCV) was calculated using the formula of Burton and DeVane [27], heritability (H) and genetic advance (GA) were calculated according to Allard [28] in an Excel spreadsheet. Relationships between fruit characteristics were estimated using Pearson's correlation coefficient (r). Correlation coefficients measure the mutual relationship between a pair of variables, independently of other variables being considered. Correlation coefficients between characteristics were estimated both on the basis of phenotypic values (phenotypic correlation) and genotypic values (genotypic correlation). Genotypic and phenotypic correlation coefficients were computed using META-R software [29].

3. Results and Discussion

The results obtained in this study (Table 2) clearly indicate that the content of bioactive compounds and antioxidant activity was significantly different in the studied group of genotypes.

Among the analyzed properties, the highest value of the coefficient of variation (CV, 32.8%) was observed for vitamin C, while flavonoid and phenolic compound values were much lower and very similar (19.2 and 19.8, respectively). Antioxidant activity determined by the ABTS method was characterized by greater variability when compared to the DPPH method. Vitamin C content ranged from 8.5 to 29.7 mg L·100 g^{-1} fw which meant over a three times higher content of this compound in cultivars such as Zielona, Karina or clone T5 compared to the cultivar Brązowa. In the study of Molina et al. [30], the average ascorbic acid level was 24.8 mg·100 g^{-1} fw; Caprioli et al. [31] recorded an average level of ascorbic acid of 22.5 mg·100 g^{-1}, while Wojdyło et al. [10] obtained 17 mg·100 g^{-1} fw. Thus, these values differed only slightly from the mean for all genotypes tested in this experiment. Apart from cultivar, works of other authors indicated a possible significant impact on the level of this property of factors such as locality and years of research, resulting in different content of this compound in the range: 17–25 mg·100 g^{-1} [32]; 31.9–44.5 mg·100 g^{-1} [33]; 3.19–32.12 mg·100 g^{-1} [10]. However, the cited values are significantly lower compared to the previously recorded high content in the range from 28.56 to 86.9 mg·100 g^{-1} by Pokorná-Juríková and Matuškovič [34] for *Lonicera kamtschatica*, cultivar Gerda 25 and exceptionally high content in the range of 67.66–186.61 mg·100 g^{-1} fw obtained in the study on cultivars and clones by Jurikowa et al. [35]. Nevertheless, it can be indicated (in principle) that this genetic background in most cultivars has a decisive influence on the level of synthesis of this compound, while other genotypes are more variable in this respect. For example, vitamin C content tested in Switzerland was the highest in the cultivar Indigo Gem, lower in IndigoTreat and the lowest in Berry Blue [7] and the same order of cultivars in terms of this compound content was observed in our research. This confirmed the essential role of the genotype of the cultivar in shaping this property, as pointed out by Jurikova et al. [36]. Increasing the level of vitamin C in fruits seems to be the desired direction of changes in new cultivars, and this is related to reports that the consumption of high vitamin C doses results in enhanced antisenescence and anti-atherosclerotic effects [37].

Table 2. Total phenolics, flavonoids, vitamin C content and antioxidant activity determined by the DPPH (1,1-diphenyl-2-picrylhydrazyl) and ABTS (2,2′-azino-bis(3-ethylbenzothiazoline-6-sulfonic acid) assays in fruits of 30 genotypes of blue honeysuckle. The results are given in mg gallic acid equivalent (GAE)·100 g^{-1} fw (phenolics), mg quercetin equivalent (QE)·100 g^{-1} fw (flavonoids), mg L-ascorbic acid·100 g^{-1} fw (Vitamin C), mmol Trolox equivalent (TE)·100 g^{-1} fw (DPPH and ABTS).

Genotypes	Phenolics	Flavonoids	Vitamin C	Antioxidant Activity	
				DPPH	ABTS
1-17-59	934.1 ± 21.8	1137.5 ± 19.9	9.8 ± 3.6	2.2 ± 0.01	5.5 ± 0.17
Amphora	727.9 ± 43.4	1030.5 ± 9.2	9.6 ± 0.7	2.0 ± 0.03	4.5 ± 0.28
Amur	509.8 ± 24.5	722.1 ± 18.0	24.1 ± 0.9	1.6 ± 0.02	2.6 ± 0.10
Aurora	422.3 ± 27.8	609.8 ± 13.5	15.6 ± 1.7	1.2 ± 0.05	2.6 ± 0.70
Bakczarskij Velikan	642.1 ± 24.6	1054.0 ± 17.2	16.5 ± 0.3	1.9 ± 0.04	4.1 ± 0.16
Berry Blue	501.0 ± 18.7	684.1 ± 4.3	11.1 ± 0.7	1.4 ± 0.03	2.5 ± 0.02
Blue Velvet	518.5 ± 28.4	679.0 ± 20.3	22.9 ± 0.6	1.4 ± 0.06	3.1 ± 0.05
Borealis	480.6 ± 30.6	679.2 ± 32.1	17.3 ± 0.6	1.5 ± 0.08	3.4 ± 0.05
Brązowa	470.2 ± 19.2	436.9 ± 6.6	8.8 ± 0.9	1.4 ± 0.01	2.9 ± 0.06
Czarna	746.5 ± 10.5	1002.5 ± 17.1	21.1 ± 2.5	1.9 ± 0.06	3.3 ± 0.18
Docz Velikana	642.2 ± 9.4	938.7 ± 8.0	17.6 ± 1.1	1.8 ± 0.04	3.3 ± 0.20
Honey Bee	497.1 ± 18.9	740.9 ± 9.1	21.2 ± 1.2	1.6 ± 0.05	3.2 ± 0.11
Indigo Gem	477.4 ± 18.7	680.7 ± 6.6	24.3 ± 0.2	1.4 ± 0.08	3.2 ± 0.06
Indigo Treat	756.6 ± 32.0	969.8 ± 15.3	15.4 ± 3.7	1.9 ± 0.05	4.4 ± 0.43
Jolanta	512.5 ± 6.9	626.1 ± 14.6	24.7 ± 2.6	1.6 ± 0.06	2.2 ± 0.84
Jugana	659.4 ± 17.3	863.8 ± 12.2	12.6 ± 4.3	1.9 ± 0.03	3.4 ± 0.24
K100	491.0 ± 14.5	727.8 ± 6.0	19.0 ± 0.2	1.4 ± 0.04	3.0 ± 0.05
Karina	606.7 ± 33.9	899.5 ± 12.8	27.7 ± 4.0	1.6 ± 0.05	3.6 ± 0.08
Leningradskij Velikan	616.6 ± 9.9	966.5 ± 12.3	23.8 ± 1.3	1.6 ± 0.05	3.7 ± 0.02
Nimfa	591.6 ± 13.6	902.0 ± 13.4	24.6 ± 1.7	1.7 ± 0.04	3.5 ± 0.10
Polar Jevel	461.3 ± 34.8	707.0 ± 7.9	8.5 ± 1.0	1.4 ± 0.06	2.5 ± 0.08
Siniczka	610.1 ± 24.1	906.8 ± 12.7	17.5 ± 0.9	1.8 ± 0.06	3.3 ± 0.14
Sinoglaska	732.9 ± 20.6	960.1 ± 6.1	13.9 ± 1.4	1.9 ± 0.03	3.3 ± 0.27
T3	658.8 ± 36.1	924.7 ± 7.5	21.9 ± 0.4	1.8 ± 0.06	3.5 ± 0.35
T5	506.6 ± 14.2	798.5 ± 12.4	26.8 ± 2.3	1.4 ± 0.05	2.9 ± 0.12
Uspiech	616.1 ± 20.4	949.8 ± 25.4	18.2 ± 0.8	1.8 ± 0.07	2.9 ± 0.52
Valhova	700.3 ± 21.7	1023.7 ± 18.6	14.1 ± 0.4	1.9 ± 0.05	3.6 ± 0.18
Vostorg	537.5 ± 0.9	766.4 ± 4.3	22.2 ± 0.3	1.6 ± 0.02	2.5 ± 0.06
Warta	462.7 ± 4.5	708.6 ± 15.5	17.6 ± 0.7	1.3 ± 0.04	2.8 ± 0.08
Zielona	688.6 ± 24.5	929.9 ± 10.3	29.7 ± 0.1	1.8 ± 0.03	3.5 ± 0.21
LSD	29.0	18.1	2.6	0.06	0.31
Mean	592.6	834.2	18.6	1.7	3.3
Standard deviation	118.0	160.9	6.1	0.3	0.7
Range	422.3–934.1	436.9–1137.5	8.5–29.7	1.2–2.2	2.2–5.5
CV (%)	19.8	19.2	32.8	15.1	21.7

The largest and most diverse group of polyphenols are flavonoids, which occur in the form of free molecules or bound to sugars. Flavonoids are characterized by a wide spectrum of health-promoting activities and are used in the pharmaceutical, medical and cosmetic industries. This is due to their antioxidant, anti-inflammatory, antimutagenic and anticarcinogenic properties combined with their ability to modulate key functions of cellular enzymes. They are used to combat diseases such as cancer, Alzheimer's disease and atherosclerosis [38]. The following genotypes were characterized by a high content of flavonoids, exceeding 1000 mg·100 g^{-1} fw: clone 1-17-59, cvs Bakczarskij Velikan, Amphora, Valhova and Czarna, and this value was more than twice as high as that observed in the weakest cultivar in this respect, Brązowa. Different contents of this compound in the analyzed group of 12 genotypes (of Russian origin) were observed by Rop et al. [14], who emphasized the pronounced cultivar variability in this regard. The cultivars Borealis and Indigo Gem tested by Rupasinghe et al. [11] in Canada were characterized by total flavonoid content at the level of 699.29 and 638.55 mg QE·100 g^{-1} fw, respectively. The same cultivars in Poland showed a similar content of this group of compounds, while flavonoid content in the cultivars Leningradskij Velikan and Nimfa (3.27 and 3.11 g·kg $^{-1}$ fw, respectively), tested in the Czech Republic [14], was much lower compared to the results obtained in the present work. The opposite was found in Berry Blue, Borealis and Indigo Gem, for which Rupasinghe et al. [16] obtained significantly higher values (1156.6, 1582.8 and 1128.5–1327.0 mg QE·100 g^{-1} fw respectively) compared to the results in this study. Therefore, it should be noted that a large variation in the content of flavonoids may result from cultivar differences and growing locations and conditions, as also indicated by the authors cited above.

The role of plant phenolics in the prevention against chronic diseases such as cardiovascular, diabetes and neurodegenerative diseases has been suggested in many studies. Honeysuckle berry fruit contains triterpenoic acids, β-carotene, catechol, flavonols, chlorogenic acid and many other acids [39]. In the current study, the group of genotypes with a high total phenolic content included: clone 1-17-59, cvs. IndigoTreat, Czarna, Sinoglaska, Amphora, and Valhova, while low values of this trait were recorded in cultivars Aurora, Polar Jevel and Warta. The list of 19 cultivars in terms of total polyphenol content presented by Gołba et al. [39], indicating that cultivars of Canadian origin such as Aurora, Borealis and Honey Bee were characterized by a lower content of this group of compounds compared to cultivars of Russian origin. This general tendency was also confirmed in our research, with the exception of the cultivar IndigoTreat, where polyphenol content was the highest among the cultivars studied. In the study by Senica et al. [40], the content of this group of compounds in Canadian cultivars ranged from 362.2 mg GAE·100 g^{-1} fw to 471 mg GAE·100 g^{-1} fw, while the values observed by us were close to or slightly above the upper range. Kusznierewicz et al. [12] showed that some genotypes analyzed in different locations were characterized by a similar level of total anthocyanin content, total phenolic content and antioxidant activity, while others significantly varied in this respect. For example, the cultivar Berry Blue, tested in Canada, showed a lower level of total phenolic content compared to the cultivar Indigo Gem [8], whereas in our research, it was Berry Blue that was more effective in this regard. In contrast, earlier studies by Orincak et al. [41] suggested that different cultivation conditions did not seem to significantly influence the content of this compound. Therefore, while the variability of chemical compositions in the cultivars tested under the same conditions results from their genetic diversity [42], additionally, changes in environmental conditions may also result in an increased or decreased level of their synthesis. Additional factors affecting the observed level of properties include different years (differences in climatic conditions) [34] and maturity stage of the fruit, as late-harvested fruits had significantly higher polyphenolic content then early harvested berries [43,44]. Moreover, Kithma et al. [15] found that the impact of the harvesting date on polyphenol composition was very distinct. Considering that the total phenolic content for the studied group of genotypes was on average 592 mg 100 g^{-1} fw, the daily intake of phenolic extract calculated by Jurgoński et al. [45] at the level of 0.8 g can be successfully met by eating fruits of this species.

The consumption of products containing high levels of antioxidants shows a positive effect against cancers and inflammatory diseases. The honeysuckle berry serves as a rich source of free radical scavengers [39]. The genotypes 1-17-59 and Amphora were characterized by the highest antioxidant activity, determined both by the DPPH and ABTS methods. In turn, the lowest antioxidant activity was found for the cultivar Aurora using the DPPH method and the cultivar Jolanta using the ABTS method. In the group of genotypes studied by Rop et al. [14], the cultivars Leningradskij Velikan and Nimfa had the lowest antioxidant activity. In our research, on the other hand, Nimfa belonged to the group of intermediate activity, and Leningradskij Velikan to a slightly lower antioxidant activity.

In turn, Auzanneau et al. [7] considered IndigoTreat and Uspiech as the best cultivars in terms of antioxidant activity, as determined by DPPH and ABTS, and Berry Blue as the weakest in this respect, which was also demonstrated in our research. Moreover, the latter authors pointed out that different antioxidant activities in individual years of research could be the result of weather conditions. Research by Rupasinghe et al. [11] and Bakowska-Barczak et al. [21] indicated that the antioxidant potential of haskap berries was higher compared to fruits of other berry plants. According to Khattab et al. [8], it resulted from the high phenolic and anthocyanin content, as the highest radical scavenging activity was found in cultivars with a high content of these compounds. In relation to phenolic compounds, this was confirmed by the results of this study listed in Table 3, where high and statistically significant correlations, both at the phenotype and genotype levels, were found between polyphenol content and antioxidant activity analyzed by the DPPH and ABTS methods (phenotypic 0.94, genotypic 0.95; phenotypic 0.72, genotypic 0.82, respectively).

Table 3. Phenotypic (right side) and genotypic (left side) correlation matrix between bioactive compounds studied in 30 blue honeysuckle genotypes.

	Flavonoids	Phenolics	Vitamin C	DPPH	ABTS
Flavonoids	1.0000	0.8797 *	0.0838 ^{ns}	0.8302 *	0.7085 *
Phenolics	0.8850 *	1.0000	−0.0343 ^{ns}	0.9417 *	0.7183 *
Vitamin C	−0.0246 ^{ns}	−0.1843 ^{ns}	1.0000	−0.0751 ^{ns}	−0.0615 ^{ns}
DPPH	0.8551 *	0.9520 *	−0.1911 ^{ns}	1.0000	0.6584 *
ABTS	0.7495 *	0.8222 *	−0.2235 ^{ns}	0.7435 *	1.0000

* significant at $p \leq 0.05$; ^{ns} not significant.

Such correlation also concerned flavonoid content and both methods measuring antioxidant activity. In addition, similarly to Rupasinghe et al. [11], a highly positive correlation between total phenolic content and total flavonoid content was observed. Therefore, according to many authors, the high antioxidant activity of the Kamchatka berry, previously determined on the basis of a smaller number of genotypes compared to this study, was caused by both the high level of total flavonoid and total polyphenol contents [9,14,46,47]. On the other hand, correlations between vitamin C and polyphenol content were negative, however, this direction of interactions was confirmed by the study of Senica et al. [40]. Although correlations at the phenotypic level are helpful in determining the relationship between the studied traits, it should be noted that apart from the genetic component, they also contain an environmental determinant, thus the desired direction of trait-level change should not be fully expected during the selection of breeding materials. Therefore, it is important to analyze the relationship directly at the genotypic level. In our research, in most cases, genotypic correlation coefficients were higher than phenotypic ones, which indicated, as reported in the example of other species of berry plants by Mishra et al. [48], that the effects of environment suppressed the phenotypic relationship between these characters. Moreover, knowledge of the degree of genetic correlations between important traits had a great impact on the selection of improved genotypes in breeding programs [49,50] and enabled indirect selection. According to Connor et al. [51], the total phenolic content was suitable for indirect selection for antioxidant activity in blueberry, indicating the possibility of also obtaining new blue honeysuckle genotypes with higher levels of phytochemical compounds.

The content of total phenolics, flavonoids, vitamin C as well as antioxidant activity analyzed by means of cluster analysis (Figure 1) revealed three main cluster groups.

The first one comprised only the breeding clone 1-17-59, which showed the remarkably high content of phenolics, flavonoids as well as antioxidant activity in both testing methods. The second one was composed of thee genotypes—Amfora, IndigoTreat and Bakczarskij Velikan—for which the values of the analyzed parameters were slightly lower. Two sub-clusters could be distinguished within the third cluster, the first one composed of 11 genotypes and the second one composed of 15 genotypes. It should be noted that within the analyzed clusters, there was no genotype clustering with a common country of origin, similar to what was observed by Sochor et al. [9].

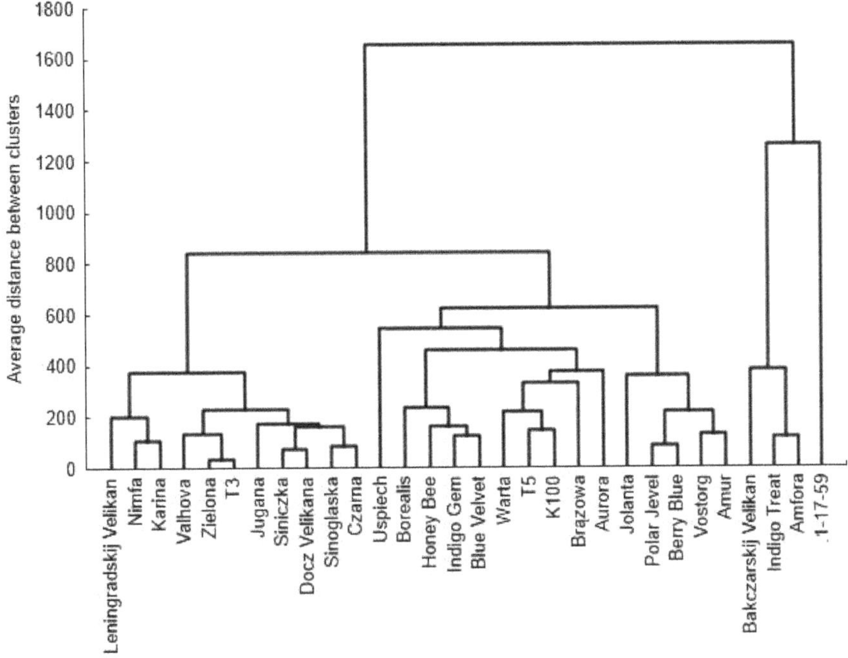

Figure 1. Dendrogram estimating distance among 30 blue honeysuckle genotypes based on phenolic acid, flavonoid, vitamin C content and antioxidant activity.

The estimation of genetic components of variance allows for the understanding of the function of genes affecting the quantitative traits. Among them, the most important parameters are genotypic and phenotypic variances, heritability [52], and heritability coupled with high genetic advance as a percentage of the mean provides better information than single parameters [53]; the values of these parameters are presented in Table 4.

Table 4. Estimates of variance, heritability (H) and genetic advance of fruit phytochemical content and antioxidant activity in 30 blue honeysuckle genotypes.

Traits	Phenotypic Variance (σ^2_P)	Phenotypic Coefficent of Variation (PCV %)	Genotypic Variance (σ^2_G)	Genotypic Coefficient of Variation (GCV %)	Heritability (H)	Genetic Advance (GA)	Genetic Advance as Percentage of Mean (GAM %)
Flavonoids	26,615	19.56	26,406	19.48	0.99	226.61	27.16
Phenolics	14,302	20.18	13,756	19.79	0.96	161.14	27.19
Vitamin C	110.21	56.34	105.72	55.19	0.95	14.09	75.67
DPPH	64,266	15.32	62,133	15.06	0.96	343.13	20.74
ABTS	528,371	22.08	468,188	20.78	0.88	901.74	27.39

Phenotypic variance values expressed by the phenotypic coefficient of variation (PCV%) for all traits were only slightly higher than genotypic variance values expressed by the genotypic coefficient of variation (GCV%), which indicated a low environmental impact on the level of these properties. The values of the heritability coefficient ranged from 0.88 for the assessment of antioxidant activity using the ABTS method to 0.99 for flavonoid content. These values should be considered high because Falconer [54] considered heritability above 0.5 already as high. Such high heritability values of the analyzed parameters are, according to Vieira et al. [50], useful to indicate the reliability of the phenotypic value in predicting the genetic value, since high values suggest a greater possibility of gain with the selection. Moreover, due to the insignificant influence of environmental conditions, they may constitute the basis for direct selection [55]. High heritability coupled with high genetic advance as a percentage of the mean was observed for vitamin C content, and thus selection would be very effective for this trait. For phenolic and flavonoid content and antioxidant activity, satisfactory selection effects can also be obtained, because heritability and genetic gain are still high. Since this is the first report regarding heritability and genetic advance for the content of phenols, flavonoids, vitamin C and antioxidant activity in blue honeysuckle, the results can be applied only to other species of berry plants. For example, in physico-chemical studies of various characteristics of strawberry fruits, Vieira et al. [50] observed heritability in the range 0.79–0.31, while Mishra et al. [48] estimated ascorbic acid heritability in this species at the level of 0.76. In the red raspberry, heritability (narrow-sense) estimates of Connor et al. [56] were H = 0.54 for antioxidant activity and 0.48 for total phenolic content; these estimates implied that a rapid response to selection was possible. Connor et al. [56] estimated the heritability of blueberries to be 0.43 for antioxidant activity and 0.46 for total phenolics. Antioxidant capacity and total phenolic content (TPC) were at the same level of heritability (0.55) in blackcurrants [57].

4. Conclusions

Honeysuckle fruits are a valuable source of health-promoting compounds that can be used as nutraceuticals. The use of the genus *Lonicera* fruits in the diet as a potential source of bioactive compounds with health-promoting properties can be extremely beneficial for consumers. The research presented in this article extends and updates the existing reports on honeysuckle berries. Our research revealed that some of the genotypes analyzed for the first time (1-17-59, DoczVelikana, Jugana, Polar Jevel, Valhova) were characterized by a higher level of phenolic compounds, flavonoids and antioxidant activity compared to the already analyzed genotypes. This shows that breeding studies allow for a significant increase in the content of biologically active compounds with health-promoting properties in the fruits of this species. It can be concluded, based on the analysis of the group of genotypes studied, that there is a potential for obtaining new genotypes with increased content of all analyzed properties, including particularly high vitamin C content.

Author Contributions: Conceptualization, J.G. and K.K.; methodology, J.Ż.; software, J.G.; validation, M.P., I.J. formal analysis, J.G.; investigation, M.P.; resources, J.G.; data curation, J.G.; writing—original draft preparation, M.D.-S.; writing—review and editing, M.D.-S.; project administration, J.Ż.; funding acquisition, K.K. All authors have read and agreed to the published version of the manuscript.

Funding: The work was funded from the statutory activity of the University of Life Sciences in Lublin.

Conflicts of Interest: The authors declare no conflict of interest.

References

1. Chmiel, T.; Abogado, D.; Wardencki, W. Optimization of capillary isotachophoretic method for determination of major macroelements in blue honeysuckle berries (*Lonicera caerulea* L.) and related products. *Anal. Bioanal. Chem.* **2014**, *406*, 4965–4986. [CrossRef]
2. Chang, S.K.; Alasalvar, C.; Shahidi, F. Superfruits: Phytochemicals, antioxidant efficacies, and health effects—A comprehensive review. *Crit. Rev. Food Sci. Nutr.* **2018**, *59*, 1–25. [CrossRef]
3. Cory, H.; Passarelli, S.; Szeto, J.; Tamez, M.; Mattei, J. The Role of Polyphenols in Human Health and Food Systems: A Mini-Review. *Front. Nutr.* **2018**, *5*, 87. [CrossRef] [PubMed]
4. Cehula, M.; Juríková, T.; Žiarovská, J.; Mlček, J.; Kysel, M. Evaluation of genetic diversity of edible honeysuckle monitored by RAPD in relation to bioactive substances. *Potravin. Slovak J. Food Sci.* **2019**, *13*, 490–496. [CrossRef]
5. Wang, Y.; Xie, X.; Ran, X.; Chou, S.; Jiao, X.; Li, E.; Zhang, Q.; Meng, X.; Li, B. Comparative analysis of the polyphenols profiles and the antioxidant and cytotoxicity properties of various blue honeysuckle varieties. *Open Chem. J.* **2018**, *16*, 637–646. [CrossRef]

6. Senica, M.; Stampar, F.; Mikulic-Petkovsek, M. Blue honeysuckle (*Lonicera cearulea* L. subs. *edulis*) berry; A rich source of some nutrients and their differences among four different cultivars. *Sci. Hortic.* **2018**, *238*, 215–221. [CrossRef]
7. Auzanneau, N.; Weber, P.; Kosińska-Cagnazzo, A.; Andlauer, W. Bioactive compounds and antioxidant capacity of *Lonicera caerulea* berries: Comparison of seven cultivars over three harvesting years. *J. Food Compost. Anal.* **2017**, *66*, 81–89. [CrossRef]
8. Khattab, R.; Su-Ling Brooks, M.; Ghanem, A. Phenolic Analyses of Haskap Berries (*Lonicera caerulea* L.): Spectrophotometry Versus High Performance Liquid Chromatography. *Int. J. Food Prop.* **2016**, *19*, 1708–1725. [CrossRef]
9. Sochor, J.; Jurikova, T.; Pohanka, M.; Skutkova, H.; Baron, M.; Tomaskova, L.; Balla, S.; Klejdus, B.; Pokluda, R.; Mlcek, J.; et al. Evaluation of antioxidant activity, polyphenolic compounds, amino acids and mineral elements of representative genotypes of *Lonicera edulis*. *Molecules* **2014**, *19*, 6504–6523. [CrossRef]
10. Wojdyło, A.; Nallely, P.; Jáuregui, N.; Carbonell-Barrachina, A.; Oszmiański, J.; Golis, T. Variability of Phytochemical Properties and Content of Bioactive Compounds in *Lonicera caerulea* L. var. kamtschatica Berries. *J. Agric. Food. Chem.* **2013**, *49*, 12072–12084. [CrossRef]
11. Rupasinghe, H.P.V.; Yu, L.J.; Bhullar, K.S.; Bors, B. Haskap (*Lonicera caerulea*): A new berry crop with high antioxidant capacity. *Can. J. Plant Sci.* **2012**, *92*, 1311–1317. [CrossRef]
12. Kusznierewicz, B.; Piekarska, A.; Mrugalska, B.; Konieczka, P.; Namieśnik, J.; Bartoszek, A. Phenolic Composition and Antioxidant Properties of Polish Blue-Berried Honeysuckle Genotypes by HPLC-DAD-MS, HPLC Postcolumn Derivatization with ABTS or FC, and TLC with DPPH Visualization. *J. Agric. Food Chem.* **2012**, *60*, 1755–1763. [CrossRef] [PubMed]
13. Kucharska, A.Z.; Sokół-Łętowska, A.; Oszmiański, J.; Piórecki, N.; Fecka, I. Iridoids, Phenolic Compounds and Antioxidant Activity of Edible Honeysuckle Berries (*Lonicera caerulea* var. *kamtschatica* Sevast.). *Molecules* **2017**, *22*, 405. [CrossRef] [PubMed]
14. Rop, O.; Řezníček, V.; Mlček, J.; Juríková, T.; Balík, J.; Sochor, J.; Kramářová, D.T. Antioxidant and radical oxygen species scavenging activities of 12 cultivars of blue honeysuckle fruit. *Hortic. Sci.* **2011**, *38*, 63–70. [CrossRef]
15. Kithma, A.B.; Rupasinghe, H.P. Polyphenols composition and anti-diabetic properties in vitro of haskap (*Lonicera caerulea* L.) berries in relation to cultivar and harvesting date. *Food Compost. Anal.* **2020**, *88*, 103402. [CrossRef]
16. Rupasinghe, H.P.; Boehm, M.M.A.; Sekhon-Loodu, S.; Parmar, I.; Bors, B.; Jamieson, A.R. Anti-Inflammatory Activity of Haskap Cultivars is Polyphenols-Dependent. *Biomolecules* **2015**, *5*, 1079–1098. [CrossRef]
17. Cassells, L.J. Experiences and conclusions from the last seven years of North American haskap cultivation: Varieties, fertilization and market trends. In Proceedings of the Haskap Conference 2017, Ozarow Mazowiecki, Poland, 9 November 2017.
18. Czernienko, A. Trends in the development of industrial horticulture of haskap in Russia, and the assessment of varieties in terms of market needs. In Proceedings of the 3rd International Haskap Conference, Jachranka, Poland, 7 November 2019.
19. Od Pięciu Lat Polska Jest Największym Producentem Jagody Kamczackiej na Świecie. Available online: http://www.sadyogrody.pl/owoce/101/ (accessed on 7 July 2020).
20. Official Journal of the European Union 17.12.2018. Available online: www.efsa.europa.eu (accessed on 7 July 2020).
21. Bakowska-Barczak, A.M.; Marianchuk, M.; Kolodziejczyk, P. Survey of bioactive components in Western Canadianberries. *Can. J. Physiol. Pharm.* **2007**, *85*, 1139–1152. [CrossRef]
22. Schlesier, K.; Harwat, M.; Böhm, V.; Bitsch, R. Assessment of antioxidant activity by using different in vitro methods. *Free Radic. Res.* **2002**, *36*, 177–187. [CrossRef]
23. Brand-Williams, W.; Cuvelier, M.; Berset, C. Use of a free radical method to evaluate antioxidant activity. *LWT Food Sci. Technol.* **1995**, *28*, 25–30. [CrossRef]
24. Re, R.; Pellegrini, N.; Proteggente, A.; Pannala, A.; Yang, M.; Rice-Evans, C.A. Antioxidant activity applying an improved ABTS radical cation decolorization assay. *Free Radical Biol. Med.* **1999**, *26*, 1231–1237. [CrossRef]
25. Wu, X.; Diao, Y.; Sun, C.; Yang, J.; Wang, Y.; Sun, S. Fluorimetricdetermination of ascorbicacid with *o*-phenylenediamine. *Talanta* **2003**, *59*, 95–99. [CrossRef]
26. Statistica 13.1. 2017. StatSoftPolska. Available online: www.statsoft.pl (accessed on 15 May 2020).

27. Burton, G.W.; Devane, E.M. Estimating heritability in tall fescue (Festucaarundinacea) from replicated clonal material. *J. Agron.* **1953**, *45*, 478–481. [CrossRef]
28. Allard, R.W. *Principles of Plant Breeding*, 2nd ed.; Wiley and Sons: New York, NY, USA, 1999.
29. Alvarado, G.; López, M.; Vargas, M.; Pacheco, Á.; Rodríguez, F.; Burgueño, J.; Crossa, J. META-R (Multi Environment Trail Analysis with R for Windows) Version 6.04, hdl:11529/10201, CIMMYT Research Data & Software Repository Network, V23. 2015. Available online: https://data.cimmyt.org/dataset.xhtml?persistentId=hdl:11529/10201 (accessed on 1 May 2020).
30. Molina, A.K.; Vega, E.N.; Pereira, C.; Dias, M.I.; Heleno, S.A.; Rodrigues, P.; Fernandes, I.P.; Barreiro, M.F.; Kostić, M.; Soković, M.; et al. Promising Antioxidant and Antimicrobial Food Colourants from *Lonicera caerulea* L. var. *Kamtschatica*. *Antioxidants* **2019**, *8*, 394. [CrossRef] [PubMed]
31. Caprioli, G.; Iannarelli, R.; Innocenti, M.; Bellumori, M.; Fiorini, D.; Sagratini, G.; Vittori, S.; Buccioni, M.; Santinelli, C.; Bramucci, M.; et al. Blue honeysuckle fruit (*Lonicera caerulea* L.) from eastern Russia: Phenolic composition, nutritional value and biological activities of its polar extracts. *Food Funct.* **2016**, *7*, 1892–1903. [CrossRef]
32. Bors, B.; Thomson, J.; Sawchuk, E.; Reimer, P.; Sawatzky, R.; Sander, T. *Haskap Breeding and Production—Final Report*; Saskatchewan Agriculture Reginan Canada: Moose Jaw, SK, Canada, 2012; pp. 1–142.
33. Małodobry, M.; Bieniasz, M.; Dziedzic, E. Evaluation of the yield and some components in the fruit of blue honeysuckle (*Lonicera caerulea* var. edulis Turcz. Freyn.). *Folia Hortic. Ann.* **2010**, *22*, 45–50. [CrossRef]
34. Pokorna, T.; Matuskovic, J. Assesment of nutritional value of *Lonicera kamtschatica* and *Lonicera edulis* fruits using fuzzy clustering method I. *Acta Hortic. Reg.* **2007**, *10*, 1–4.
35. Jurikova, T.; Sochor, J.; Rop, O.; Mlček, J.; Balla, S.; Szekeres, L.; Žitný, R.; Zitka, O.; Adam, W.; Kizek, R. Evaluation of Polyphenolic Profile and Nutritional Value of Non-Traditional Fruit Species in the Czech Republic—A Comparative Study. *Molecules* **2012**, *17*, 8968–8981. [CrossRef]
36. Jurikova, T.; Matuskovic, J.; Gazdik, Z. Effect of irrigation on intensirty of respiration and study of sugar and organic content in different development stages of *Lonicera kamtschatica* and *Lonicera edulis* berries. *Hortic. Sci.* **2009**, *36*, 14–20. [CrossRef]
37. Kim, S.M.; Lim, S.M.; Yoo, J.A.; Woo, M.J.; Cho, K.H. Consumption of high-dose vitamin C (1250 mg per day) enhances functional and structural properties of serum lipoprotein to improve anti-oxidant, anti-atherosclerotic, and anti-aging effects via regulation of anti-inflammatory microRNA. *Food Funct.* **2015**, *6*, 3604–3612. [CrossRef]
38. Panche, A.N.; Diwan, A.D.; Chandra, S.R. Flavonoids: An overview. *J. Nutr. Sci.* **2016**, *5*, 47. [CrossRef]
39. Gołba, M.; Sokół-Łętowska, A.; Kucharska, A. Health Properties and Composition of Honeysuckle Berry *Lonicera caerulea* L. An Update on Recent Studies. *Molecules* **2020**, *25*, 749. [CrossRef] [PubMed]
40. Senica, M.; Bavec, M.; Stampar, F.; Mikulic-Petkovsek, M. Blue honeysuckle (*Lonicera caerulea* subsp. *Edulis* (Turcz. ex Herder) Hultén.) berries and changes in their ingredients across different locations. *J. Sci. Food Agric.* **2018**, *98*, 3333–3342. [CrossRef] [PubMed]
41. Orincak, J.; Matuskovic, J.; Jurcak, S. *Possibilities of Species Lonicera caerulea in Utilization of the Secondary Metabolism in Food and Pharmaceutical Processing*, 1st ed.; SPU: Nitra, Slovakia, 2003; pp. 210–219.
42. Paulovicsová, B.; Turianica, I.; Juríková, T.; Baloghová, M.; Matuškovič, J. Antioxidant properties of selected less common fruit species. *Sci. Pap. Anim. Sci. Biotechnol.* **2009**, *42*, 608–614.
43. Ochmian, I.; Skupień, K.; Grajkowski, J.; Smolik, M.; Ostrowska, K. Chemical composition and physical characteristics of fruits of two cultivars of blue honeysuckle (*Lonicera caerulea* L.) in relation to their degree of maturity and harvest date. *Not. Bot. Horti Agrobot.* **2012**, *40*, 155–162. [CrossRef]
44. Ochmian, I.; Smolik, M.; Dobrowolska, A.; Rozwarski, R.; Kozos, K.; Chełpiński, P.; Ostrowska, K. The Influence of Harvest Date on Fruit Quality of Several Cultivars of Blue Honeysuckle Berries. *EJPAU* **2013**, *16*, 1.
45. Jurgoński, A.; Juśkiewicz, J.; Zduńczyk, Z. An anthocyanin-rich extract from Kamchatka honeysuckle increases enzymatic activity within the gut and ameliorates abnormal lipid and glucose metabolism in rats. *Nutrition* **2013**, *29*, 898–902. [CrossRef]
46. Jurikova, T.; Rop, O.; Mlcek, J.; Sochor, J.; Balla, S.; Szekeres, L.; Hegedusova, A.; Hubalek, J.; Adam, V.; Kizek, R. Phenolic profile of edible honeysuckle berries (genus *Lonicera*) and their biological effects. *Molecules* **2012**, *17*, 61–79. [CrossRef]

47. Gazdik, Z.; Krska, B.; Adam, V.; Saloun, J.; Pokorna, T.; Reznicek, V.; Horna, A.; Kizek, R. Electrochemical Determination of the Antioxidant Potential of Some Less Common Fruit Species. *Sensors* **2008**, *8*, 7564–7570. [CrossRef]
48. Mishra, P.; Ram, R.; Kumar, N. Genetic variability, heritability, and genetic advance in strawberry (*Fragaria* × *ananassa* Duch.). *Turk. J. Agric. For.* **2015**, *39*, 451–458. [CrossRef]
49. Masny, A.; Pruski, K.; Żurawicz, E.; Mądry, W. Breeding value of selected dessert strawberry (*Fragaria* × *ananassa* Duch.) cultivars for ripening time, fruit yield and quality. *Euphytica* **2016**, *207*, 225–243. [CrossRef]
50. Vieira, S.D.; Araujo, A.L.R.; Souza, D.C.; Resende, L.V.; Leite, M.E.; Resende, J.T.V. Heritability and Combining Ability Studies in Strawberry Population. *J. Agric. Sci.* **2019**, *11*, 457–469. [CrossRef]
51. Connor, A.M.; Luby, J.J.; Tong, C.B.S.; Finn, C.E.; Hancock, J.F. Variation and heritability estimates for antioxidant activity, total phenolic content and anthocyanin content in blueberry progenies. *J. Am. Soc. Hortic. Sci.* **2002**, *1*, 82–88. [CrossRef]
52. Cruz, C.D.; Regazzi, A.J.; Carneiro, P.C. *Biometric Models Applied to Genetic Improvement*; UFV: Viçosa, Brazil, 2012; p. 514.
53. Baye, B.; Ravishankar, R.; Singh, H. Variability and association of tuber yield and related traits in potato (*Solanumtubersum* L.). *J. Agric. Sci.* **2005**, *18*, 103–121.
54. Falconer, D.S. *Introduction to Quantitative Genetics*; Pearson Education Limited: London, UK, 1996.
55. Kumar, R.; Kumar, S.; Singh, A.K. Genetic variability anddiversity studies in snapdragon (*Antirrhinum majus*) under tarai conditions of Uttarakhand. *Indian. J. Agric. Sci.* **2012**, *82*, 535–537.
56. Connor, A.M.; Stephens, M.J.; Hall, H.K.; Alspach, P.A. Variation and Heritabilities of Antioxidant Activity and Total Phenolic Content Estimated from a Red Raspberry Factorial Experiment. *J. Am. Soc. Hortic. Sci.* **2005**, *130*, 403–411. [CrossRef]
57. Currie, A.; Langford, G.; Mcghie, T.; Apiolaza, L.A.; Snelling, C.; Braithewaite, B.; Vather, R. Inheritance of Antioxidants in a New Zealand Blackcurrant (*Ribesnigrum* L.) population. In Proceedings of the 13th Australasian Plant Breeding Conference, Christchurch, New Zealand, 18–21 April 2006; pp. 218–225.

MDPI
St. Alban-Anlage 66
4052 Basel
Switzerland
Tel. +41 61 683 77 34
Fax +41 61 302 89 18
www.mdpi.com

Applied Sciences Editorial Office
E-mail: applsci@mdpi.com
www.mdpi.com/journal/applsci